RETHINKING AIDS

THE TRAGIC COST OF PREMATURE CONSENSUS

Robert S. Root-Bernstein

THE FREE PRESS
A Division of Macmillan, Inc.
New York

Maxwell Macmillan Canada
Toronto

Maxwell Macmillan International
New York   Oxford   Singapore   Sydney

For Michèle, for reasons too multitudinous to mention

In experimental science it is always a mistake not to doubt when facts do not compel you to affirm.

—Louis Pasteur

The fatal tendency of mankind to leave off thinking about a thing which is no longer doubtful is the cause of half their errors.

—John Stuart Mill

People who claim to be *absolutely* convinced that their stand is the only right one are dangerous. Such conviction is the essence not only of dogmatism, but of its more destructive cousin, fanaticism. It blocks off the user from learning new truth, and it is a dead giveaway of unconscious doubt.

—Rollo May, M.D.

When we find ourselves entertaining an opinion about which there is a feeling that even to enquire into it would be absurd, unnecessary, undesirable, or wicked—we may know that the opinion is a non-rational one.

—Wilfred Trotter, M.D.

The reason for inventing a new theory is to drive us out of the hypotheses in which we hitherto have taken refuge into the state of thoroughly conscious ignorance which is the prelude to every real advance in science.

CONTENTS

FIGURES AND TABLES

TABLES

PREFACE

We do not understand AIDS. By *understand* I mean something very specific: having sufficient knowledge to intervene in or control the disease process. Epidemiologists must be able to predict accurately when outbreaks will occur and who is at highest risk. Microbiologists must be able to prove the underlying causes of the disease. Immunologists must be able to explain how the immune system fails and what may be done to protect it. Anthropologists and sociologists must be able to pinpoint behavioral patterns and cultural environments that put people at risk. And public health officials and physicians must be able to implement effective preventative measures and cures. By these criteria, we do not understand AIDS; in fact we are profoundly ignorant. I hope that by identifying the extent and nature of our ignorance we will be able to do something about it. In science, to define the problem correctly takes one more than halfway to its solution.

My method is as simple as my purpose. I have read the medical literature assiduously, looking for studies that test our current theory of AIDS. I have analyzed and synthesized this information and found that our theory of AIDS is full of glaring holes, confusing contradictions, and outright discrepancies. I am saying nothing more than what the medical literature itself says about AIDS. The only difference is that I am willing to say this in public, whereas most practitioners are not. But I am not willing to leave it at that. I am not interested in merely pointing out flaws in a laboriously constructed edifice erected by many dedicated and hard-working scientists, nor am I claiming that what they have discovered is wrong. On the contrary, my critique of AIDS theory assumes that most of the published experiments and clinical observations are accurate. The data are correct; our interpretation of them is not.

The problem is that a "fact" is not a "fact" until it is interpreted in the light of theory.

Consider, for example, standing in a lifeguard's tower at some beach and dropping a seagull feather and a lead ball. The feather will fall much more slowly than the ball. We all know this. So did Aristotle, hundreds of years before the birth of Christ. Aristotle therefore proposed that heavy things fall faster than light things. This makes sense and is confirmed by observation, but it is not scientifically correct. Galileo recognized during the Renaissance that the experiment is not acceptable for testing the law of falling bodies. There are at least two very different factors at work on the ball and the feather: the gravitational attraction of the earth and the retarding influence of the atmosphere. Galileo therefore argued that observations such as dropping a feather and a lead ball are totally irrelevant for testing physical laws, because the test is not performed under the controlled conditions of a vacuum in which the effects of the atmosphere can be eliminated. On the other hand, a modern aeronautical engineer trained in the Galilean tradition would nonetheless find the feather results fascinating—not as an experiment in the law of falling bodies but as an experiment in aerodynamic behavior. Here, in short, is one set of easily verified data, but three very different understandings of it. As the theoretical framework changes, the data stay the same, but their factual interpretations alter drastically—and so, too, our manipulative powers within nature. It is not enough, then, simply to measure. One must also interpret what is measured, spin out many possible explanations, and devise tests to validate or invalidate alternative theories and know their limitations.

Everything we know about AIDS may be like the feather-and-lead-ball experiment. The data are all easily validated by repeated observations and measurements, and yet they may still be misunderstood. A great deal of evidence suggests, for example, that we have attributed too much to HIV—the human immunodeficiency virus, or so-called AIDS virus—and too little to other causative agents in AIDS, just as Aristotle mistakenly attributed everything to the weight of falling bodies and nothing to the friction of the atmosphere. If this is the case, then we will understand AIDS only when we disentangle the effects of HIV from those that are due to other causative agents. AIDS may very well be as complicated as the physical description of a feather falling through a stiff and variable wind. Until we elaborate all of the possible interpretations of the AIDS data that we have and actively doubt both our eyes and our common sense, then we cannot know for certain that we have not followed Aristotle into a logical but mistaken conclusion. The cost in this case will not be the chagrin of having gotten it wrong, however; it will be measured in human lives.

It is imperative to rethink and research AIDS. This process will, I hope, change our conceptions of the syndrome and in altering our conceptions yield us the greater understanding of it that Galileo and other physicists provided in physics by rethinking the laws of falling bodies. Hundreds of thousands of people around the world have already died, millions more are threatened, and nothing, as yet, can be done to save any of them. The time required to reconsider what we know about AIDS and, more important, what we do not is a very small investment that may have an enormous yield.

ACKNOWLEDGMENTS

A book of this scope requires thanks to many people. Fred C. Westall helped me to formulate my early criticisms of the HIV-only hypothesis of AIDS and was both friend and collaborator during the development of our theory of autoimmunity. Sheila Hobbs made possible much of the implementation of that theory to AIDS. Peter Duesberg, Bryan Ellison, Maurizio Lucá Moretti, Eleni Papadopolous Eleopolus, and Geoffrey Hoffmann shared preprints and bibliographies. Scott Gilbert, Walter Gilbert, Martine Jaworski, Harry Rubin, and Mott Greene critiqued my AIDS writings and encouraged my heresy. Dr. Marlys and Dr. Charles Witte and Dr. Beverly Rubik provided scientific platforms for my ideas. Dr. Arnold Seid was a constant sounding board. Charlie Thomas, Jr. (Group for the Reevaluation of the HIV/AIDS Hypothesis) and Ed Vargas (HIV Connection?) have been valuable allies in the fight to rethink AIDS. William Bryan Coyle, Randy Peters, and Michael Callen have shared information and concerns that only those who are HIV positive can truly know. Heather Conlee and Chantel Lightner scoured the Michigan State University library for pre-1979 cases of AIDS-like infections, and my mother, Maurine Bernstein, dug up rare sources at UCLA. Linda Webster provided a fine index. I, however, am solely responsible for the contents of this book.

A special acknowledgment is due Dr. Joseph Sonnabend, who in 1982 began developing the approach to AIDS detailed in this book. He has been a guiding light. Thousands of lives might have been saved had the biomedical community and gay activists listened to him.

CHAPTER 1

ANOMALIES

OBSERVATIONS ABOUT AIDS THAT DO NOT FIT THE PICTURE

Most scientists believe that we understand AIDS and have trumpeted their belief to each other and to the public as well. The consensus is that HIV—the human immunodeficiency virus—causes AIDS and that when we learn how to vaccinate against HIV or develop an antibiotic that can treat HIV infection, then AIDS will be cured. This is the public face of AIDS research—the face that is meant to exude confidence, to reassure. Scientists are much more reticent about revealing their other face—the one that displays their ignorance, confusion, and puzzlement over the aspects of this disease that they do not understand. The best-kept secrets about AIDS are the questions unanswered, the puzzles unsolved, the contradictions unrecognized, and the paradoxes unformulated. Yet the degree of our ignorance must be the measure of our understanding. The existence of significant anomalies or departures from the regular expectations of the current theory must raise a red flag warning that our understanding of AIDS is not as profound as we might wish. Such anomalies do not mean that we have the facts of AIDS wrong but rather that we have not figured out how to explain all of the facts coherently and consistently within a single theoretical framework. The failure to explain AIDS accurately, in turn, raises the possibility that we are not addressing its causes and cures appropriately. The tragic cost of this failure is the constantly rising death toll of AIDS.

Significant anomalies to the current dominant theory of AIDS

1

exist, and they are important enough to warrant serious rethinking of the causes and nature of AIDS.

Anomaly 1: HIV and AIDS Are Older Than We Think: The Prehistory of AIDS, 1872–1979

One of the most important assumptions underlying our current understanding of AIDS is that it is a new disease caused by a new virus that emerged in Europe and the Americas only during the latter half of the 1970s. The recognition of AIDS as a distinct disease entity during the first years of the 1980s supports this assumption. Yet hundreds of AIDS-like cases were documented in medical journals for decades prior to the recognition of AIDS, some of whom have turned out to be infected with HIV.

Consider the following case. In December 1932, a forty-year-old black American male, identified only as W.J. in the medical literature, checked into a clinic at the University of Maryland Hospital. His physician, Dr. Francis Ellis, determined that he had never been out of the country and that his medical history was generally uninformative. Two possible exceptions were that W.J. reported having been treated for syphilis two years before (so we know that he was sexually active) and he had had a noncancerous "fatty tumor" removed from his thigh during the same year. Numerous, dark-red macules, which were now quite apparent and widespread, had appeared on his chest and abdomen at about the same time and he had similar spots on his ankles, which were unusually swollen, dry, and cracked. Upon admission to the hospital W.J. was found also to have impaired vision, low blood pressure, anemia, an infected anal fissure, and extensive internal hemorrhoids, causing bloody stools. Biopsy revealed that the man was suffering from an extremely widespread case of Kaposi's sarcoma—those dark-red patches on his skin. He was only the third black American ever diagnosed with this cancer. Although he survived for a year and a half following his visit to Dr. Ellis, no further information is available concerning his case. Whether he died, moved, or went to another physician is unknown.[1]

No one ever figured out why W.J. developed Kaposi's sarcoma, a cancer usually found only in elderly men of Jewish or Middle

European descent and so rare (four in a million people) that most physicians never see a case. Indeed, no one knows to this day what causes Kaposi's sarcoma. Moreover, no case had ever been reported in which the extremely unusual set of concurrent medical conditions such as anal fissures, extensive hemorrhoids, ocular impairment, and anemia were also present. Why this odd confluence of problems? No one, not even Dr. Ellis, dared to speculate. But he and the editors of the *Archives of Dermatology and Syphilology* found the case odd enough to want to publish its details. Notably, W.J. was not the first, nor would he be the last such patient, to puzzle his physicians either by developing Kaposi's sarcoma or the various other infections that plagued him. All of his symptoms would eventually be recognized as being typical of AIDS, but many decades and many more puzzling patients would have to intervene.

One of these patients was a twenty-five-year-old white male who checked into the Veterans Administration Hospital in Tucson, Arizona, in July 1948 with a cough producing blood-streaked sputum and severe chest pain. "Steve," as we shall call him, was married, worked as a trucker, and had served in the South Pacific during World War II, where he had contracted malaria. His past medical history revealed nothing unusual or contributory to his present symptoms, at least as far as his physicians were concerned. Laboratory tests failed to reveal the presence of any of the usual germs—tuberculosis, pneumococci, fungi—that are often associated with such symptoms, and chest X-rays revealed no apparent problems. After sixteen days Steve was sent home, mildly improved. He was back two weeks later. This time x-rays revealed abnormalities in both his lungs and heart. Once again, no specific disease organism could be isolated, but Steve was given a course of penicillin anyway. He developed an allergic reaction to it, and the treatment was discontinued. There was no improvement in his condition, but he was nonetheless sent home. Two months later Steve was readmitted to the hospital in what turned out to be the terminal stages of his illness. He had a low-grade fever, had lost over thirty pounds since July, and had developed enlarged lymph nodes on the left side of his neck. He still had the symptoms of an unspecified pneumonia, a tuberculosis test was now positive, a heart murmur had developed, and definite signs of ocular and brain damage had begun to appear. Cherry-red nodules had appeared on his scalp, which biopsy revealed to be Kaposi's sarcoma. Steve was transferred to the Veterans Administration Hospital at Sunmount, New York, for further treatment. Six months after his initial visit to the hospital in Tucson, Steve died of what the attending physicians, Martha Collins and Hyman Fisher, labeled an unusually fulminant course of Kaposi's sarcoma that had metastasized to the heart, lungs, lymph nodes, spleen, liver, pancreas, kidneys, bone marrow, and scalp. Once again, the cause of this unusual set of medical problems was not identified.[2] Once again, the symptoms, including prior malaria, are typical of AIDS, especially in sub-Saharan Africa and Asia.

A year later, a twenty-year-old woman—call her "Jane Doe"— died of another strange combination of symptoms. Jane had initially visited her physician, Dr. J. P. Wyatt at the Baptist Memorial Hospital in Memphis, Tennessee, and was found to have severe anemia. At that time, she had no other unusual medical conditions. The anemia failed to respond to iron supplements, so Dr. Wyatt decided to treat her with whole blood transfusions—a common practice during the middle decades of the century. Over a period of three years, Jane was given 135 whole blood tranfusions. Her white blood count dropped, she developed swollen lymph nodes in her neck and groin, she became anorexic, and she had a persistent low-grade fever. Eventually she developed pneumonia and died of combined respiratory and heart failure. Autopsy demonstrated a severe infection by the usually harmless infectious agent cytomegalovirus. The cytomegalovirus had infected virtually every organ in her body and caused sufficient cellular damage to kill her.[3] Jane's case, too, matches recent descriptions of AIDS.

Four years later, in 1953, Jane's physician, Dr. Wyatt, was the attending physician for a twenty-eight-year-old male who died in the aftermath of an atypical pneumonia. The man, known to us only as R.G., had a recent history of chronic, mild fevers and pneumonia, his lymph nodes were enlarged, and his white blood cell count was low, but there was no evidence of any other unusual medical problems. Extensive blood and sputum tests revealed that none of the typical germs could be responsible for his symptoms. After a few days in the hospital, R.G. was sent home. Two and half months later, he began to pass bloody stools, and his pneumonia returned. He was readmitted to the hospital. Again, no fungi or bacteria could be isolated despite a chest biopsy. Three months later R.G. died of a bacterial infection contracted during the biopsy. Autopsy revealed that, like Dr. Wyatt's earlier patient, R.G., too, had developed an uncontrolled cytomegalovirus infection that had spread throughout his body.[4] Again, a modern physician might suspect AIDS, since disseminated cytomegalovirus infection is one of the unusual symptoms of this modern scourge.

In 1959 in England, physicians encountered yet another puzzling case. "John," a former seaman in his mid-twenties, reported general good health up until November 1958. He subsequently developed odd skin lesions on his back and shoulders that were never definitively diagnosed. In December he began to experience problems breathing, night sweats, anorexia, weight loss, chronic fatigue, low-grade fever and a persistent cough. An ulcer developed around his anus and spread inexorably. Another appeared in one nostril and spread to the lip and tongue. John eventually died of pneumonia, and a postmortem demonstrated the presence of cytomegalovirus, *Pneumocystis carinii*—a protozoal infection that affects only immune-suppressed individuals—and staphylococci in his lungs. The skin lesions were found to contain cytomegalovirus and involved a combination of endothelial cells and capillaries such as those found in the early stages of Kaposi's sarcoma.[5] Notably, *Pneumocystis* pneumonia, like the fulminant Kaposi's sarcoma, is almost a dead giveaway for AIDS when found in a previously healthy individual.

Pneumocystis pneumonia also surfaced in 1959 in a forty-eight-year-old black male who had emigrated from Haiti to the United States. His physicians, Dr. H. A. Lyons and Dr. G. R. Hennigar, noted that the patient had clearly sustained a recent and significant weight loss, was anorexic, had night sweats and chronic fatigue of several months' duration, and displayed a chronic low-grade fever. AIDS patients, or those with AIDS-related complex, a pre-AIDS syndrome, will recognize these symptoms. Attempts to determine the cause of his pneumonia were unsuccessful, and the standard antibiotics were ineffective. The patient died within a few weeks of entering the hospital. Autopsy revealed the cause to be a severe *Pneumocystis* pneumonia infection. The two physicians published the case in two medical journals because of its apparent uniqueness. It was one of only a handful of cases of such pneumonia among apparently immunologically normal adults that they could find in the medical literature.[6] Once again, the similarity to present-day AIDS cases is striking, as is the fact that the patient was from Haiti—a risk factor that was identified during the first years of the AIDS epidemic.

Skip a decade, and we find yet another bizarre case. A sixteen-year-old black American known by the pseudonym Robert R. developed an extremely unusual panoply of symptoms that intrigued and dismayed his physicians, Memory Elvin-Lewis, A. Arthur Gottlieb, and Marlys and Charles Witte. Robert R. quickly developed, and within two years died from, a combination of disseminated *Chlamydial* infection, Kaposi's sarcoma, and a variety of other infections, including sexually transmitted diseases and chronic bowel infections. His case was so notable that it, like the other cases described, was written up and published in a leading medical journal in the hopes that one day someone would be able to identify the cause or causes of his unusual death.[7]

Three more cases of unusual and unexplained infections appeared in a Norwegian town at about the same time. Beginning in 1966, a Norwegian sailor developed lymphadenopathy, recurrent colds, and undiagnosed dark spots on his skin. He had been treated for venereal diseases twice. His wife also began to have recurrent medical problems beginning in 1967. She suffered repeated chest colds, disseminated candidiasis—a yeast infection—and recurrent, unexplained fevers. She eventually developed *Pneumocystis* pneumonia with neurological complications and dementia. Tests done on husband and wife during 1971 demonstrated T cell deficiencies of unexplained origins. Meanwhile, their nine-year-old daughter also developed chronic candidiasis of the bronchi, repeated severe infections, and finally a fatal case of disseminated chicken pox. She died in January 1976. Her father died four months later from Hodgkin's lymphoma complicated by *Pneumocystis* pneumonia. The wife died eight months after that. This cluster of cases was so odd that the attending physicians published an account of them, notable for the statement that the authors were unable to explain how or why all three of these apparently immunologically normal individuals should have developed immunodeficiencies of such magnitudes that they could acquire multiple opportunistic infections.[8]

The question of what caused all of these odd deaths must intrigue anyone who reads about these people. As different as the ten cases may appear upon first reading, all in fact share several characteristics. First, all of the primary diseases that afflicted these patients—Kaposi's sarcoma, *Pneumocystis* pneumonia, disseminated cytomegalovirus and *Candida* infections—are opportunistic diseases. They do not cause overt or serious disease in people whose immune systems are functioning normally. Instead, they are usually found only in individuals whose immune systems are deficient due to various genetically inherited disorders or who have had their immune system purposely suppressed by various drugs such as corticosteroids, cancer chemotherapeutic agents, or immunosuppressant agents used to prevent the rejection of organ transplants or to treat autoimmune or chronic inflammatory conditions. But only one of the ten cases (the Norwegian father) is described as having any of these immunosuppressive factors present, and it is unclear whether he developed his cancer before or after his other opportunistic infections. All of the other patients were reported to have been previously healthy, and therefore presumably immunologically normal, individuals. That is their second shared characteristic. Their third shared characteristic is that all of them satisfy the current definition of AIDS used by the medical community for diagnosis. Knowledge of AIDS is now sufficiently widespread that it does not take a physician to realize that all of these cases bear close resemblance to present-day cases of acquired immune deficiency syndrome.

AIDS first came to the attention of the medical community in 1981 when physicians in New York and Los Angeles reported small but significant groups of previously healthy young men suddenly afflicted with Kaposi's sarcoma and *Pneumocystis* pneumonia. Most of these patients experienced a very similar pattern of disease progression. A year or two before they were diagnosed with an opportunistic disease, they began to experience a low-grade fever that lasted many months, accompanied by night sweats that drenched them in perspiration. They began to lose weight for no apparent reason and developed anorexia (a distaste for eating), chronically swollen lymph nodes in their necks, groin, or armpits, and, often, chronic coughs as well. Then unusual infections began to afflict them. Various protozoa, sexually transmitted viruses, and other pathogens infected their intestines, resulting in chronic diarrhea that might last for months and finally necessitate surgical intervention. Funguses such as *Candida albi-*

cans, a yeast that commonly infects women's vaginas without significant long-term effects, spread uncontrollably into these men's mouths, nasal passages, bronchi, and esophagi. Cytomegalovirus, herpes simplex, and a dozen other viruses that are normally of little medical interest suddenly began to proliferate throughout their bodies. And, more frequently than almost anything else, Kaposi's sarcoma and *Pneumocystis* pneumonia appeared. Few of these men lived more than a year beyond their diagnosis of acquired immune deficiency, for the reason that their physicians had no idea how they had acquired their immune deficiency or what to do to try to correct it.

Perhaps the most unusual aspect of these cases was that initially every AIDS patient identified himself as being homosexual. Many reported being intravenous drug abusers as well. None of these men had any of the identified risks for opportunistic diseases, such as cancers, organ transplants, or inherited immune deficiencies. Thus, no reason appeared to exist for the failures of their immune systems. These patients represented anomalies, and, as such, they worried the medical community just as the pre-1981 cases described worried the individual physicians who had attempted to treat them. The difference was that the 1981 cases were not just a mere handful of randomly distributed, unexplained opportunistic infections without apparent connection. This time there was a clear-cut focus for the outbreak of opportunistic disease: gay, drug-abusing, promiscuous men, most of whom could be linked to one another through common sexual contacts. The original term for AIDS was GRID: gay-related immune deficiency. Physicians began to conjecture that a new sexually or needle-transmitted infectious agent might have entered the gay population and that it was only a matter of time before this deadly agent spread to heterosexuals as well.

Surprisingly little effort went into trying to trace the possible history of AIDS; of the ten cases just described, only one was identified as a possible AIDS-like case within the first two years of the epidemic, and even then it was generally discounted by the medical community at large. Indeed, all of the standard accounts of AIDS state quite explicitly that the earliest confirmed cases of AIDS have been traced back only to 1978 or 1979 and that HIV is not much older. Yet all ten of the pre-1981 cases fit the clinical picture of AIDS. All of the patients were young and previously healthy, and then they developed unusual and quickly fatal infections and cancers. Each had the persistent low-grade fever, fatigue, weight loss, and swollen lymph nodes typical of AIDS patients. All developed diseases that are known to affect only people with impaired immune function, yet none of them had a recognizable cause of immune impairment. Most of the cases were, like most AIDS patients today, male. One of the females, like many other females now contracting AIDS, apparently acquired her disease as a result of massive blood transfusions. The 1948 male case probably had repeated blood transfusions as a risk factor as well, since transfusions were a common treatment for the anemia associated with the malaria he had contracted as a soldier in the South Pacific. Unfortunately, no records exist describing his malaria treatment, so the transfusion risk must remain conjectural. The remaining two women apparently acquired their infections from a family member or from a life-style that exposed them to similar immunosuppressive risks, again typical of many heterosexual cases today.

Still, there are a few notable factors missing from these pre–AIDS era cases. Although the highest risk groups for AIDS today are gay and bisexual men and intravenous drug abusers, no mention of homosexuality or drug abuse is made in any of the original physicians' reports. These factors can by no means be ruled out, however. Bloody stools and anal ulcers, such as those described for the 1932, 1948, and 1958 cases, are common symptoms among promiscuous gay men who engage in receptive anal intercourse and are often associated with a syndrome now known as gay bowel syndrome, a panoply of chronic infections of the lower gut and anus. But gay men had not yet come out of the closet in the 1930s, 1940s and 1950s, and gay bowel syndrome was not recognized until the 1970s. Physicians would not have been able to recognize the symptoms as typifying a particular sexual preference.

The symptoms of intravenous drug abuse were also less obvious to physicians during the middle decades of this century than they are today. For one thing, addictions were then considered to be a mental illness, rather than a physical one, and were usually the private reserve of psychiatrists and psychologists rather than internists or general practitioners. Furthermore, the incidence of intravenous forms of drug addiction was much less common then: a grand total of twenty-four British opiate addicts were known to the Home Office in 1960 and only a few thousand in the United States. So the symptoms of drug addiction, too, would not have been readily recognizable to most physicians. Nonetheless, other AIDS-like cases were identified among drug abusers as long ago as the 1930s.

Other evidence serves to confirm the probability that these cases were truly AIDS cases despite significant diagnostic difficulties. Several patients had Kaposi's sarcoma or undiagnosed skin lesions that fit its description—a disease that is a marker for AIDS in patients lacking known causes of immune impairment. Definitive diagnosis of the questionable cases is not possible at this late date, but it is interesting to note that failure to recognize Kaposi's sarcoma several decades ago would not have been unusual. The incidence of the disease even among "high-risk" groups such as old men of Eastern European and Jewish descent is so low that most physicians had never seen a case until recently, and most textbooks stated that young adults did not get the disease. In consequence, Robert Gottlieb, one of the first physicians to diagnose AIDS, wrote in an early paper describing the disease that "clinicians may not entertain the possibility of Kaposi's sarcoma in persons who are still young and if lesions of that condition are in the patch or plaque stage only."[9] Indeed, the Centers for Disease Control, reporting on these early AIDS patients, stated that "although some of the patients have presented with the violaceous skin or mucous membrane lesions typical of Kaposi's sarcoma, many such lesions have been initially overlooked. . . . In many cases the histopathologic diagnosis from skin, lymph node, or visceral-lesion has been difficult even in specialized hands."[10] The problems for non-specialist were all the greater. Brownstein and his colleagues reported in 1973 that of one hundred pathology-verified cases of Kaposi's sarcoma, the correct diagnosis was made by the physician only 63 percent of the time. Thirty-five of the cases were misdiagnosed as pyogenic granulomas and seventeen cases were misdiagnosed as being melanomas.[11] Even physicians who may have considered Kaposi's sarcoma as a possible diagnosis for young, previously healthy men may have been led astray by their pathologist colleagues. Pathologists specialize in the origin, nature, and course of disease. Dr. James Ackerman reported in 1961 that approximately half of several hundred pathologists gathered at a conference in the United States were unable to identify a classic example of the disease.[12] We may therefore assume that many cases of AIDS-like Kaposi's sarcoma were overlooked prior to the recognition of AIDS.

The presence of *Pneumocystis carinii* pneumonia (PCP) is also a well-recognized and nearly universal symptom characteristic of AIDS. It may appear odd that only the most recent of the AIDS-like cases described were initially diagnosed with PCP, but historial and diagnostic factors can explain the apparent discrepancy. Diseases, like fashions, have their vogues. Just as pantyhose became widely available only after a durable nylon was invented, so PCP became a general subject of study outside the pathology laboratory only when techniques for identifying and isolating the organism in living people became available. The first definitive diagnosis of PCP in an adult prior to death was not made in the United States until 1965.[13] Such live diagnoses thereafter required open lung biopsy or needle aspiration, dangerous procedures that were carried out only on patients, such as premature infants and transplant patients, for whom a significant risk of PCP was already established. Pneumonia patients outside these risk groups— say young, previously healthy men or drug addicts—were never subjected to the necessary tests because they were not suspected to be at risk. Thus, diagnosis of PCP was a difficult task, dependent on technique and expectation, as one expert on the subject, Dr. Lowell S. Young of UCLA, has made clear. He wrote in 1984 that "clinical interest in *Pneumocystic carinii* pneumonia in the United States is less than two decades old. . . . In those days [c. 1970], the disease came to be known as 'transplant pneumonia' and it was often confused with cytomegalovirus."[14] One of Young's close friends and associates, Dr. Karl Western, "wagered me [at that time] that in any series of diffused pneumonias of undetermined etiology, appropriate sectioning and staining of pulmonary [lung] tissue would reveal *Pneumocystis* in a large proportion of cases. To diagnose *Pneumocystis* infection, one first had to consider that it might be present and use the right technique to demonstrate the presence of cyst or trophozoite forms in lung tissue. Unless one knew what to search for, one would not establish a diagnosis. . . . Many a patient with bizarre pulmonary processes that we did not understand turned out, sometimes too late for effective therapy, to have *Pneumocystis carinii* pneumonia."[15]

What Dr. Young did not say is that few physicians had his exper-
tise or had specialized in PCP. If he and his colleagues who knew
what to look for and how to look often had difficulty diagnosing
PCP, one can only imagine the difficulties that faced the average
practitioner.

The diagnostic difficulties were immense. Like Kaposi's sarcoma, PCP was not something most physicians saw or considered
in their diagnoses prior to the recognition of AIDS. Once again, I
quote Dr. Young: "*Pneumocystis* pneumonia is one of the most
difficult respiratory infections to diagnose and usually requires an
invasive procedure to obtain pulmonary secretions or direct samples of lung parenchyma."[16] *Pneumocystis* cysts are not secreted in
sputum and cannot be cultured by the normal techniques used to
isolate and identify other causes of pneumonia, such as bacteria.
Antibody tests for the presence of *Pneumocystis* are too inaccurate
to be useful in diagnosis. Compounding these difficulties is the fact
that most patients with PCP have concomitant infections present,
including cytomegalovirus (CMV), *Mycobacteria*, and *Mycoplasmas*, any or all of which may be mistakenly considered to be the
cause of the pneumonia. The clinical manifestations of PCP are
closely mimicked by reactions to tumor treatment, radiation therapy, and drug reactions. Finally, *Pneumocystis* pneumonia can
have very unusual manifestations on chest X rays that mimic other
diseases such as bacterial pneumonias, emphysema, and pneumo-
thorax.[17] More disconcerting are many reports of PCP is presenting without rales or other typical sounds and with a normal chest
X-rays.[18] The only sure way to diagnose PCP until 1982 was to
go into the lung and take a sample—a risky procedure that was
developed and disseminated among practitioners only during the
1970s. Even today, most cases of PCP in AIDS patients are pre-
sumptively diagnosed; the definitive tests are never performed to
demonstrate that the causative organism is present.[19] It is assumed
to be present due to the risk factors associated with the patient
and his or her symptoms.

Given the substantial difficulties associated with diagnosing
PCP, it is not surprising to find that no cases were diagnosed at all
in American adults until 1954—and that only at autopsy—and
subsequent cases were diagnosed only rarely in adults and usually
only postmortem until about 1970. Notably, during the 1950s Dr.
Henry Hamperl reviewed the case records and photomicrographs

of many cytomegalovirus cases and argued that Jane, Dr. Wyatt's 1949 case, probably had a PCP infection concomitant with her cytomegalovirus.[20] Both Dr. Hamperl and more recently Dr. Peter Nichols, a physician at UCLA, performed retrospective analyses of photomicrographs of lung tissue from Wyatt's 1953 case and concluded that R.G. not only suffered from disseminated cytomegalovirus infection but *Pneumocystis* pneumonia as well.[21] Thus, R.G. is the oldest definitive American case of PCP in an adult, and "Jane" may be his closest competition.

These findings are thought provoking. The presence of PCP, Kaposi's sarcoma in young patients, disseminated cytomegalovirus infections, the relatively rapid downward course of each patient toward death, their failure to respond to normal modes of treatment, their unusual combinations of symptoms, and in three cases, the development of vision impairment or other neurological symptoms are all typical of AIDS. Indeed, each of these cases meets all of the criteria developed by the Centers for Disease Control (CDC) in Atlanta, Georgia, and subsequently adapted by the World Health Organization for defining AIDS. AIDS, it may be postulated, was present in the United States and Europe fifty or more years before its epidemic form. One of the reasons that it was not recognized was that the diseases that make up the syndrome were so difficult to diagnose that it was almost impossible for anyone to find a pattern among the rare cases that did come to the light, and no sociological data, such as admitted homosexuality, existed to link the patients.

The case that AIDS was present in the Western world for at least a century prior to 1981 is actually much stronger than the meager instances just elaborated. Five of the people described above—the 1959 Englishman, Robert R. in 1968, and the Norwegian father, mother, and daughter during the early 1970s—were infected with HIV. Their cases so intrigued the attending physicians that frozen samples of blood and, in some cases, tissue as well were saved, in the hopes that subsequent medical advances would one day clarify the cause of their patients' deaths. These samples have recently been unfrozen to yield incontrovertible evidence that the putative cause of AIDS—the human immunodeficiency virus—was present in all five.[22] Indeed, James Moore and his colleagues, E. J. Cone and S. S. Alexander, Jr., reported in 1985 that they had tested 1,129 samples of blood collected from intravenous drug abusers in forty-three states during 1971 and 1972 and had found that fourteen of these samples (more than 1 percent) tested positive for the presence of HIV and antibody to HIV.[23] Forty-five samples (4 percent) had HIV antibodies present. None of a control group of eighty-nine blood samples from healthy individuals showed evidence of HIV or antibody to it. HIV, then, is much older and was much more widespread than most AIDS researchers generally admit. Just how old HIV is, we may never know, since relevant tissue and blood samples do not exist prior to 1959.

There is no doubt that AIDS itself, as distinct from HIV, is at least a century old. The ten cases described are only a small sample from a much larger pool of such examples. Several dozen cases of *Pneumocystis* pneumonia, cytomegalovirus infections, and other opportunistic diseases in patients matching the CDC surveillance definition of AIDS have been reported by Huminer and his colleagues in North America and Europe for the fifty-year period preceding the recognition of AIDS.[24] Katner and Pankey found twenty-eight cases of Kaposi's sarcoma in the United States, Canada, and Europe between 1902 and 1978 satisfying the definition of AIDS.[25] And Lars Breimer and I have recently reported the existence of several hundred cases of Kaposi's sarcoma dating back as far as Moritz Kaposi's first identification of the disease in 1872 that also satisfy the CDC surveillance definition of AIDS.[26] I have published extensive studies demonstrating that hundreds of AIDS-like cases of other opportunistic infections diagnostic for AIDS, including progressive multifocal leukoencephalopathy (a disease that progressively impairs brain function), disseminated cryptococcosis (a opportunistic fungal disease), and disseminated cytomegalovirus infections, also existed since the very first diagnoses of these diseases.[27] These are just the tip of an apparently unsuspected iceberg. Hundreds of cases of other opportunistic infections satisfying the CDC criteria also lie buried in the literature. Two specific diseases that are diagnostic for AIDS even in people without underlying causes of immune suppression and in the absence of a positive HIV test give some notion of the nature of the data: *Pneumocystis carinii* pneumonia and chronic, disseminated candidiasis.

Although *Pneumocystis carinii* pneumonia was first identified in 1911, the first unambiguous diagnosis of PCP in an adult was reported only in 1942 by van der Meer and Brug in Europe. The patient was a twenty-one-year-old male. No underlying disease or immunological abnormality is mentioned in their report on the patient, so he may have been the first of many such patients to have qualified as a presumptive AIDS case. Five years later, a sixty-year-old female developed PCP following prolonged vitamin A deficiency, a problem found all too often in drug addicts and chronic alcoholics as well as among starving people who develop AIDS in Third World nations.[28] The cases of Dr. Wyatt already summarized are the next two cases of PCP among previously healthy adults on record.

The number of PCP cases reminiscent of AIDS begins to grow during the 1960s. In 1960, Symmers reported the death of a forty-seven-year-old male from a combination of PCP, thrombocytopenic purpura, cytomegalovirus, and cryptococcus infections. Notably, thrombocytopenic purpura—an autoimmune disease in which antibodies destroy platelets and red blood cells, leading to intradermal bleeding and failure to form proper blood clots—has also been found as an unusually common concomitant of PCP in AIDS patients as well.[29] Also in 1960, a German doctor reported the death of a hemophiliac due to PCP.[30] About the same time, Sheldon detailed the first four of many other cases of PCP that were to be found over the next decade in older children who were not congenitally immunocompromised and had had no immunosuppressive therapy.[31] In 1960, Williams and his colleagues reported the young man in Birmingham, England, with PCP and disseminated cytomegalovirus who turned out to be HIV seropositive as well.[32] The same year, Anderson and Barrie reported a case of PCP in Canada in an adult whose only predisposing condition was cirrhosis of the liver, presumably associated with chronic alcoholism.[33] More than a dozen cases of PCP in previously healthy adults and children, and even an entire family, were reported over the next decade in North America and Europe, representing at least 5 percent of all PCP cases reported in the medical literature for that period.[34]

This astonishingly high incidence of AIDS-like PCP cases was confirmed in a review of 350 PCP cases made by Burke and Good in 1973. They found that 48 (14 percent) had no known congenital immunological deficiency or underlying immunosuppressive therapy, and in 11 cases (3 percent), the patients were explicitly stated to have been healthy prior to developing PCP.[35] In addition, Burke and Good themselves saw 46 PCP patients, of whom 7 (15 percent) were suffering from "lymphosarcomas," cancers now associated with AIDS. Notably, all but one of these "lymphosarcoma" patients was between the ages of fifteen and sixty-one and died within a year of diagnosis.[35] These may have been early, unrecognized cases of AIDS. Among non-immunosuppressive diseases correlated with PCP by the mid-1970s were a number now highly associated with AIDS: hemophilia, anemia, non-Hodgkin's lymphomas, hypoglycemia, cryptococcosis, disseminated tuberculosis, cytomegalovirus infection, protein-calorie malnutrition, venereal diseases, chronic fungal infections, and other parasitic infections.[35] These associated diseases accounted for approximately 10 percent of recorded PCP cases throughout the decades of the 1960s and 1970s.

Due to the difficulties inherent in diagnosing PCP, reported cases of the disease are almost certainly only a small fraction of the actual number that existed. Robinson found 2 cases of PCP in 203 random adult autopsies in 1961 (1 percent).[36] Only one of these patients had any known disease or treatment associated with immune suppression. Similarly, Esterly found 8 cases in 200 routine autopsies in 1968 (4 percent).[37] Reports in Europe during the early 1970s also demonstrated that 1 to 10 percent of all lungs autopsied had *Pneumocystic carinii* cysts present and that the presence of *Pneumocystis* antibody was much higher among people with cytomegalovirus and fungal infections and venereal diseases.[38] For example, one study of 200 consecutive routine autopsies at a Dutch hospital revealed that nearly 8 percent of the patients had *Pneumocystis* cysts present in their lungs. Two of these patients were particularly noteworthy; they were young men between the ages of twenty-one and forty who had died of complications of chronic bronchitis and bronchiectasis, persistent upper respiratory infections of undetermined cause that are also found in many AIDS patients today. No women in this age bracket had similar symptoms.[39] These figures suggest that, far from being a rare disease that almost never occurs in the general population, *Pneumocystis carinii* was a common, unrecognized infection. If it was overlooked, so was AIDS.

Other diseases diagnostic for AIDS were disregarded for similar reasons. Candidiasis, a fungal infection, is one of the diseases that can signify the presence of AIDS when it gets out of hand. The current surveillance definition of AIDS allows the syndrome to be diagnosed in the absence of an HIV test if the patient has chronic candidiasis of several months' duration, candidiasis disseminated to organs of the body other than the vaginal or mouth areas, or, in the case of a *Candida* infection of the esophagus, if the patient has no known cause of immune impairment.[40] According to this definition, cases fitting the description of AIDS have existed for at least half a century in the United States. For example, in 1946, a previously healthy twenty-eight-year-old man entered the hospital complaining of fever, severe headache, difficulty swallowing, and disorientation. Examination suggested meningitis, and a spinal tap revealed that his spinal fluid was infected with both *Mycobacterium tuberculosis*, the bacterium that causes tuberculosis, and with *Candida albicans*, the fungus that causes thrush in infants and occasional, transient vaginal infections in healthy women. The young man was also found to be suffering from *Candida* infections of the mouth and pharynx, the cavity between the mouth and the esophagus. Since this was some years prior to the discovery of the first antibiotic specific for fungi, nystatin, the man's physicians dosed him with iodides and swabbed his oral infections with gentian violet. When these treatments were ineffective, streptomycin therapy was begun; the young man responded favorably and was released from the hospital. Since streptomycin is an antibacterial that has no effect on *Candida*, one can only surmise that the young man was actually suffering from a combined infection. No follow-up was done in this case, so that it is unknown whether the patient recovered fully.[41]

At least ten more AIDS-like cases of esophageal candidiasis were reported in the next twenty-three years, including a thirty-year-old man with pneumonia,[42] three adults suffering from an illness of obscure origin that was characterized by a low-grade fever, enlarged spleen, deficiency of white blood cells, and esophageal candidiasis,[43] and a forty-three-year-old man with untreated lymphosarcoma (Kaposi's?), tuberculosis, pneumonia, and esophageal and bronchial candidiasis.[44] The patients did not survive their diseases. In 1962, a twenty-two-year-old black American female suffering from chronic anemia and iron deficiency due to sickle cell disease also developed esophageal candidiasis.[45] Both anemia and iron deficiency were identified as risk factors for chronic candidiasis and other fungal infections during the next decade.[46] She was lost to follow-up shortly after diagnosis, though whether through death or other reasons is not known. Three more well-documented cases of esophageal candidiasis in otherwise healthy individuals were reported between 1964 and 1969.[47]

The next unequivocal case of esophageal candidiasis in the absence of known causes of immune suppression occurred in September 1970, when "S.M.," a twenty-two-year-old male college student, appeared at the Wenatchee Valley Clinic in Wenatchee, Washington. He had an upper respiratory infection for which he was given oral tetracycline and a penicillin injection. Five days later, he found it difficult to swallow and had developed a pain under his sternum. *Candida* were cultured from the clumps of fungi found in his esophagus. His medical history suggested that this was the second time he had developed this combination of upper respiratory infection followed by *Candida* esophagitis. The first time had been two years earlier, while he was performing his military service. S.M. recovered from his infections following treatment with nystatin but was lost to follow up nine months later.[48] Whether he died, moved, or found another physician is unknown.

In analyzing the significance of these ten cases of esophageal candidiasis in previously healthy individuals, it is important to realize that by 1975, only about seventy-five cases of esophageal candidiasis had ever been reported in the entire medical literature. The vast majority of these cases had occurred in patients treated with steroids or with chemotherapy for various types of cancers.[49] Thus, while esophageal candidiasis was (and still is) quite rare, the ten AIDS-like cases listed represent a significant proportion (13 percent) of published cases. These published cases are, in turn, a very small proportion of actual cases.

Broad studies confirm a significant incidence of esophageal candidiasis in people without known causes of immune suppression during the decade prior to the recognition of AIDS, especially in cities such as New York where AIDS was to become so prevalent. For example, in 1976 Dr. Baboukh E. Kodsi and his colleagues at the Maimonides Medical Center of the State University of New York in Brooklyn reported that of 20,688 mainly white middle-class patients admitted to the hospital during 1973, 370 were ex-

amined for esophageal infections or other complications. Twenty-
seven of these had evidence of *Candida* esophagitis. Of these patients, only 8 had predisposing illnesses or treatments associated
with immune suppression. Nineteen of these cases of esophageal
candidiasis were, in other words, presumptive AIDS cases. Subsequent investigation of 1,065 additional patients with esophageal
complications between January 1974 and June 1975 by the same
physicians revealed an additional 61 patients with esophageal candidiasis, of whom about two-thirds were once again without identifiable predisposing factors for infection.[50] In other words, Kodsi
and his colleagues found about 60 people, most of them men, a
quarter of whom were under the age of forty-five, with symptoms
diagnostic for AIDS at a single hospital in New York City between
1973 and 1975. Similar figures were reported by other physicians
making broad surveys of patients in both the United States and
Britain.[51]

Dozens of reports of candidiasis disseminated to organs other
than the skin, mouth, and vagina also abound in the medical literature from the 1930s on, most of them satisfying the surveillance
definition of AIDS. The vast majority of these patients were under
the age of forty-five and male, and many were intravenous drug
addicts.[52] F.M., a forty-eight-year-old white man admitted on
April 24, 1939 to the surgical service of Cumberland Hospital in
Brooklyn, New York, is a typical case. He presented with a com-
plaint of severe abdominal pain accompanied by vomiting. He had
a moderate fever, his pulse rate was 114 beats per minute, and his
breathing was abnormally fast. He had an aortic murmur, suggesting some problem with his heart, and his liver was slightly
enlarged. His medical history was very revealing. He admitted to
a morphine and heroin addiction of twenty years' duration. He
had developed the heart murmur around 1932 and had survived
a case of unspecified pneumonia in 1936. He had begun to inject
drugs intravenously in July 1938 and had taken no precautions
whatsoever to sterilize the needles or syringes he used. In January
1939 he could no longer secure the drugs he craved due to lack of
money and entered a drug rehabilitation program. He had suffered withdrawal and now claimed to have been totally drug free
for six months. But the joint and muscle pains he had experienced
during withdrawal had not disappeared, and he had been progres-

sively losing weight and becoming weaker ever since. Recently, fever had developed, and abdominal pain and vomiting had appeared.

F.M.'s physicians, Henry Joachim and Silik Polayes, suspected septicemia, a systemic infection. They took blood, stool, and urine samples and a throat swab and sent them to Florence Bittman of the Bacteriology Laboratory of Bellevue Hospital for analysis. Bittman found nothing unusual from the throat swab, stool, or urine specimens, but she isolated huge numbers of the unusual, opportunistic yeast *Candida parakrusei* from F.M.'s blood. Ten additional blood tests revealed the presence of the same organism, and a culture taken from the container in which F.M. prepared his drugs was also contaminated with this yeast.

Dr. Joachim and Dr. Polayes were unable to help F.M.; antibiotics had not yet been invented. F.M. became weaker and weaker until he died on June 13, less than two months after he entered the hospital. The autopsy showed that F.M. had developed *Candida* infections of the respiratory system, urinary tract, central nervous system, spleen, and heart. His heart was found to be infected with *Streptococcus* and *Bacillus coli* bacteria, and he was diagnosed as having acute hepatitis as well, strongly suggesting an active hepatitis B infection.[53]

F.M. is the first case of systemic *Candida* infection in a drug addict that appears in the medical literature. He was far from the last, however. Within two years, five more drug addicts would be described as dying of the same infectious organism.[54] Hundreds more cases would be reported in the next two decades. *Candida* endocarditis—yeast infections of the heart and particularly heart valves—and *Candida* infections of the bone joints were also recognized as being frequent concomitants of intravenous drug abuse long before AIDS.[55]

These cases strongly suggest that the types of *Candida* infections that we associate with AIDS, and have become diagnostic of it, were not nearly as rare as has been thought. Indeed, the closer we get to the recognition of AIDS in 1981, the greater the number of relevant cases becomes. A few prophetic individuals such as Dr. Mark Orringer of the University of Michigan began to foresee the problem that became AIDS. In 1978 he presented a new technique for identifying esophageal candidiasis that was much simpler and more accurate than any previous technique, and he predicted, "We will be seeing more of these *Candida* esophageal infections because the infection is no longer restricted to patients with tumors."[56] How prescient Dr. Orringer was.

The crucial question is whether Dr. Orringer's prescience was due to better diagnostic methods, increasing numbers of patients, or both. The problem of improved diagnostic capabilities is not irrelevant to understanding AIDS. Notably, medical reports of almost all the opportunistic diseases associated with AIDS grow at an exponential rate from the time of their first observation until the recognition of AIDS. This observation suggests one of three things that are not necessarily mutually exclusive: our diagnostic ability has increased continuously during this century; as we conquer other infectious diseases, opportunistic ones become more common; or causes of acquired immune suppression have become ever more prevalent. Whatever the cause of this phenomenon, we must consider the possibility that at least part of the recognition of AIDS has resulted from an increased ability to diagnose the diseases that characterize the syndrome and to identify the exceptional cases.

One thing is certain: There were a large number of pre-1979 AIDS-like cases that have not been accounted for in our current theories of AIDS. No matter how we analyze these cases, they create significant problems for our understanding of AIDS. If HIV is a new and necessary cause of AIDS, as most AIDS researchers argue, what was the cause of these pre-1979 AIDS-like cases? Are there causes of acquired immune suppression other than HIV that may explain AIDS? What might these immunosuppressive agents be? Or is HIV much older than anyone has been willing to consider? If HIV is old, why has AIDS become epidemic only within the last decade? Have modes of transmission suddenly increased? No matter how one tries to answer these questions, the answers are disturbing. It is imperative that we explore their meaning.

Anomaly 2: HIV is Neither Necessary Nor Sufficient to Cause AIDS

The observation that AIDS is older than most people think is certainly odd, but it is not sufficient to make one question seriously the current theory that HIV is a relatively new virus or that it causes AIDS. At best, the existence of pre-1980 AIDS cases simply alters our notions of how AIDS may have evolved and raises questions about why it blossomed to epidemic proportions only during the 1980s. Expanded modes of transmission of HIV, rather than the sudden emergence of HIV itself, becomes the focus of interest.

A more serious threat to current notions of AIDS challenges the second tenet of AIDS dogma: that HIV is both necessary and sufficient to cause AIDS. The formal position taken by the U.S. government Health and Human Services Department, the National Institutes of Health, the CDC, and even the World Health Organization is that AIDS does not occur in the absence of an HIV infection and that HIV infection, in and of itself, is all that is necessary to result in AIDS. Infection with HIV is supposed to cause destruction of a specific type of immune system cell known as the T-helper or T4 cell. Because T-helper cells play an important role in promoting a wide range of immunological activities, destroying these cells cripples a broad spectrum of immune functions. This immune impairment leads to AIDS within an average of ten to twelve years in 50 to 100 percent of those infected. That is the official line espoused by all major health agencies and their spokesmen and by the vast majority of AIDS researchers. It is also the basis for almost all AIDS testing at present, and both physicians and the lay press have taken up calling HIV "the AIDS virus."

The announcement of the findings that would create these "facts" (for it is widely accepted by sociologists of science that scientific "facts" are created by the same sort of social consensus that creates "truths" in other belief systems) occurred on April 23, 1984. Secretary of Health and Human Services Margaret Heckler stood up before a huge audience of reporters to announce, "Today we add another miracle to the long honor roll of American medicine and science. Today's discovery represents the triumph of science over a dreaded disease. Those who have disparaged this scientific search—those who have said we weren't doing enough—have not understood how sound, solid, significant medical research proceeds." Heckler promised that a test for the AIDS virus would be available within six months and a vaccine within two years.[1] She then introduced the man who was going to make all of this possible: Robert Gallo.

Gallo explained how he and his co-workers, including Mikulas

Popovic, Zaki Salahuddin, and Elizabeth Read, had isolated a new retrovirus that could infect and kill the T-helper or T4 cells that are specifically destroyed during AIDS. Retroviruses encode their genes as ribonucleic acid (RNA) instead of the more usual deoxyribonucleic acid (DNA). Like other viruses, retroviruses infect cells, incorporate their genes into the chromosomes of the host cell, and then stage a coup in which they take over the cellular machinery to produce copies of themselves. Retroviruses must go through an extra step of back-transcribing their RNA into DNA sequences before they can incorporate their genes into the host's chromosomes, however. Then the copying process begins. The DNA is transcribed back into RNA; the RNA is translated into virus-specific proteins; and the RNA and proteins self-assemble into new copies of the retrovirus. This copying process can be so prolific that some viruses burst the host cell, releasing thousands of virus copies and simultaneously killing the cell.

The particular retrovirus that Gallo's group had discovered was actually the third one they had isolated in a decade. The first two, human T cell lymphotropic viruses I and II (HTLV I and HTLV II), were both associated with forms of leukemia (cancers of the white blood cells) in which leukocytes are over-produced rather than killed off, as in AIDS. Moreover, HTLV I had often been isolated from the white blood cells of AIDS patients. For a while, Gallo's group thought that HTLV I might actually be the cause of AIDS, but there was too much that HTLV I and II did not explain. One of the most important problems was that HTLVs immortalized the cells they infected. In other words, they caused the infected cells to divide endlessly (and often prematurely), so that each infected cell reproduced more infected cells ad infinitum. AIDS, however, was characterized by the selective destruction of T-helper cells. Thus, Gallo's group was quite excited to find that HTLV III killed the cells it infected, and it infected specifically the set of cells that are destroyed in AIDS. Subsequent tests were to show that the vast majority of AIDS patients had evidence of producing antibody to HTLV III, therefore corroborating its putative role as an AIDS-producing agent.

Gallo's announcement of the discovery of HTLV III was not without its controversial points. Luc Montagnier and his co-workers, Françoise Brun-Vezinet, Françoise Barre-Sinoussi, and Jean-Claude Chermann, had previously isolated a similar retrovirus from a patient with AIDS-related complex (ARC), or pre-AIDS, which they had called variously lymphadenopathy-associated virus (LAV) and T cell lymphotropic retrovirus.[2] Moreover, they had supplied Gallo's team with their virus on two separate occasions prior to Gallo's announcement of the discovery of HTLV III. Subsequent research has shown that LAV and HTLV III came from the same virus isolate. Gallo has even admitted that he and his team stole priority for the discovery from Montaginer's group using a virus isolate supplied to them by Montaginer himself. Moreover, the Gallo team has revealed other less than savory characteristics. Salahuddin has been found guilty of business fraud, and Gallo and several other team members are under investigation by the U.S. government for possible frauds of various types including improper business practices, creating false data, and claiming priority for a discovery that rightly belonged to the French. The full ramifications of these investigations will not be known until after this book is published. Two things only are certain. One is that the need to establish patent priority so as to allow profitable production of an HIV test for AIDS led to a quick resolution of the international controversy over priority for the discovery. Everything was smoothed over by renaming the new retrovirus human immunodeficiency virus, and patents for its discovery and tests for AIDS based on it were jointly issued to both the Americans and French, both of whom now collect a fee for the production and use of HIV tests.[3] The other thing that is certain is that one must look back with irony at Margaret Heckler's words when she said that "those who have said we weren't doing enough—have not understood how sound, solid, significant medical research proceeds."

In the meantime, the U.S. government endorsed Gallo's results, officially proclaiming HIV to be the cause of AIDS. Amazingly, these pronouncements not only came before other investigators had reproduced Gallo's and Montagnier's research; they came even before Gallo's paper had undergone peer review and publication.[4] Moreover, the Health and Human Services Department, which oversees the operations of the National Institutes of Health and the CDC and was spearheading the investigation of AIDS, also put its money where its pronouncements were. AIDS research became virtually synonymous with HIV research, and funding for investigating the roles of other infectious agents, drugs, and immunosuppressive risks associated with a gay life-style and intravenous drug abuse—all areas of vibrant research prior to 1984—quickly dried up as did most research in these areas.

Imagine the consternation created during 1990 by reports that HIV may be neither necessary nor sufficient to cause the syndrome. The key experiments have been performed in the laboratory of the discoverer of HIV himself, Luc Montagnier of the Institut Pasteur in Paris. If anyone has a stake in proclaiming that HIV is the sole cause of AIDS and that he is the primary discoverer of that cause, it is certainly Montagnier. Yet Montagnier has announced that HIV is not sufficient to cause AIDS.

Montagnier's case depends on standard virological methods. Most virologists studying HIV grow it in a special type of lymphocyte (white blood cell) that is itself grown in culture flasks. The recipe for growing HIV is readily available and reproducible. Montagnier's experiment consisted of simply adding one more ingredient to the mixture in which the HIV-infected lymphocytes are grown: tetracycline.[5]

Tetracycline is a commonly prescribed antibiotic whose mechanism of action is well understood. It is effective against a wide range of bacteria but not against any known virus, including HIV. Yet something very odd happened when the tetracycline was added to the HIV-infected lymphocytes: the HIV became inactive. It no longer replicated; it no longer caused cell death; it no longer infected other lymphocytes. Montaginer concluded that the tetracycline had killed off some bacterial contaminant present in his lymphocyte culture that had been activating the HIV. One of the implications of his finding was that everyone else's HIV-active lymphocyte cultures probably contained the same or a similar bacterial contaminant, a very likely scenario since cell cultures are easily contaminated and widely shared by researchers. Although no one has yet confirmed that their cell cultures are contaminated, no one has announced that Montagnier's experiments fail to work in their laboratories, either. The silence is deafening.

The next question for Montagnier was obvious: What was this bacterial contaminant? It has turned out to be a *Mycoplasma*, the smallest living thing that can replicate by itself. *Mycoplasmas* are bigger than viruses (which require the machinery of a living cell in order to reproduce themselves) but smaller than all other known bacteria. They also lack the usual cell wall that protects most bacteria. They are extremely difficult to isolate. They are also very common contaminants of laboratory cell cultures such as those used by AIDS researchers. And *Mycoplasma* infections are common in AIDS patients as well.

The response from most other HIV researchers has been to ignore Montagnier's data. A few, such as Robert Gallo, have responded by producing counter claims of their own. Gallo, for example, has published studies showing that both HTLV I (one of his earlier discoveries) and herpes simplex type 6, another virus, greatly increase HIV infectivity, replication, and cell killing.[6] He, in other words, thinks that Montagnier has picked the wrong cofactor.

Wrong or right, the necessity for cofactors to stimulate HIV activity raises the intriguing question of whether HIV is actually the cause of AIDS. In a 1989 article in *Discover* magazine, Gallo directly addressed this question. He wrote that the presence of other viruses such as herpes simplex or HTLV I "does not mean that HIV is not the true cause of AIDS, but that herpes virus (HHS-6) which is ubiquitous ['80 percent of adults have antibodies indicating chronic infection'] and which potentially kills T cells, may well accelerate its course. Ordinarily we think the immune system controls herpes—it's relatively innocuous, no big deal. But if the immune system is suppressed by HIV, the herpes virus will start to replicate, unleashing a vicious cycle of destruction. First it kills T cells itself; then it infects the same cells as the AIDS virus and awakens the virus from its latent state. Thus immune suppression caused by HIV activates the herpes virus, the herpes virus in turn activates the AIDS virus, and both kill T cells."[7] The only problem with this scenario is that it raises the question of which came first—the HIV or the cofactor? Which, therefore, is the "true cause" of AIDS? Or are both necessary and neither sufficient as Montagnier has been arguing?

Despite a general reticence among HIV researchers to address these thorny problems, Montagnier has at least one very active supporter, the American scientist Dr. Shyh-Ching Lo of the Armed Forces Institute of Pathology in Washington, D. C. Lo had isolated a new *Mycoplasma* from several AIDS patients as long ago as 1986. Despite being at first ignored, and then having his research subjected to the kind of scrutiny usually reserved for possible cases of fraud, Lo eventually demonstrated to the satisfaction of most experts in the field that his *Mycoplasma* actually exists and that it is both highly infectious and quite deadly in at least one strain of monkey.[8] It also causes immune suppression and may be an important initiator of a variety of autoimmune problems that afflict AIDS patients. Lo claims that this *Mycoplasma* has a causative role in AIDS: "I'm not saying that HIV plays no role in AIDS, but AIDS is much more complicated than HIV."[9]

There are, however, problems with any theory that asserts that HIV has a single cofactor. It seems highly unlikely that each AIDS patient would have exactly the same pair of infections as every other AIDS patient. The probability of actually contracting two necessary infections seems very unlikely to most investigators, and certainly existing data do not demonstrate that both HIV and *Mycoplasmas* (or HTLV or herpes simplex) are present in every AIDS patient. They are both present concurrently in many AIDS patients but not in all. And then there is the problem of proliferating cofactors. If Montagnier and Lo's *Mycoplasmas* and Gallo's HTLV I and herpes simplex virus type 6 can all stimulate HIV activity, why not other infectious agents as well? Is there, then, a single, universal cofactor for HIV, or does a varied set of cofactors exist? How large might that set of cofactors be? These questions are the subject of intensive research.

Experimental results and observations are accumulating suggesting that many cofactors may exist to promote the T cell death that occurs in AIDS. Some of these observations are bold enough that they may even lead us to question the dogma that HIV is the necessary and sufficient cause of AIDS. A number of investigators have reported that cytomegalovirus, adenoviruses, and hepatitis B virus, all of which are extremely common in AIDS patients, also potentiate HIV activity in laboratory cell cultures.[10] Noninfectious agents such as cocaine and heroin have also been reported to act as cofactors to speed the rate of HIV infection, increase replication, and stimulate viral cell killing. Perhaps HIV is transmitted and activated efficiently only in conjunction with ongoing stimulation associated with one or more of these additional agents.

Perhaps. But scientists are naturally skeptical of antidogmatic ideas. In fact, there is an old joke that circulates in various forms among scientists to the effect that whenever a new and profound idea comes along, it is met first with scoffing ("That's the most ridiculous idea I've ever heard! You can't be serious!"), followed by grudging admission that there might be something to the idea but not what the discoverer thinks ("What's new is not true, and what's true is not new"), culminating in total conversion to the new dogma accompanied by outright attempts to steal credit for it ("Of course, it's right; we've been saying that for years!"). The cofactor theory of AIDS is meeting just such a classic response. Mark Wainberg, one of the directors of the McGill AIDS Centre in Montreal, Canada, has told reporters, "It's nothing new that the progress of the disease can be accelerated by other agents. We've done work in our own labs to show that cytomegalovirus and herpes can have that effect." No doubt they have. But they have certainly not been among the investigators actively promoting and publishing cofactor research. Nor has another American AIDS researcher who, preferring to remain anonymous, has said, "I'd bet my professional reputation that something more than HIV is involved in this disease. But I wouldn't bet my grants, my ability to work." Nor will most of this man's colleagues. They remain silent, or they remain skeptics.[11]

A more recent set of studies may force these fence-sitters to choose sides. Not only does it appear that HIV may need one or more cofactors to initiate disease, there is also provocative evidence that some AIDS cases have never been infected with HIV at all. The CDC, which monitors AIDS cases in the United States, lists in its records more than three dozen HIV-negative homosexual men who have developed Kaposi's sarcoma sometimes complicated with other opportunistic infections.[12] Various studies have also found at least one HIV-negative hemophiliac who has developed Kaposi's sarcoma and some fifteen HIV-seronegative patients with *Mycobacterium avium intracellulare* (MAI) complex, another opportunistic disease diagnostic for AIDS.[13] Various neurological abnormalities typical of dementia, the first overt symptom of AIDS in about 10 percent of AIDS patients, have also been found in 7 percent of HIV-seronegative gay men—that is, men without evidence of HIV antibodies.[14] Several cases of chronic oral or esophageal candidiasis and immune thrombocytopenia accompanied by low T4 cell counts in HIV-negative patients without known causes of immunosuppression have been reported as well.[15] Even *Pneumocystis* pneumonia has been reported among HIV-seronegative adults without predisposing illnesses.[16] All of these patients fit the CDC definition of AIDS despite their failure to display HIV infection.

Notably, all of the cases just listed were known prior to 1991 but were ignored by all but a handful of AIDS researchers.[17] Most proclaimed with great assurance that, in the words of James Curran of the CDC, "There is not AIDS without HIV."[18] Then Luc Montagnier announced in May 1992 that he had three AIDS patients in whom he could find no evidence of HIV. Since his announcement came at an alternative AIDS conference attended mainly by critics of the HIV theory, its importance was at first lost. Two months later, however, not only Montagnier, but clinician after clinician rose to tell the audience at the internaitonal AIDS conference in Amsterdam that each had a handful of such HIV-free AIDS patients. Suddenly AIDS without HIV became big news because too many cases had surfaced to be ignored. There is no longer any doubt that HIV is not necessary to cause acquired immunodeficiency. The question is whether the causes of HIV-free AIDS are also at work in people with HIV, and therefore what role HIV plays in causing AIDS in anyone.

The observation of HIV-seronegative AIDS patients is not unique to the Western world. HIV-negative AIDS patients have also been reported in African countries. In one Ugandan study, for example, one of fifty patients (2 percent) fulfilling the World Health Organization (WHO) criteria for AIDS was HIV-antibody negative by an ELISA (Enzyme linked ImmunoSorbent Assay) test.[19] "ELISA" is an acronym for a particular type of test for the presence of HIV antibodies. These tests are notorious for giving false positives when validated with the more sophisticated tests now employed, so the presence of a negative test is all the more telling. Another study in Tanzania again employed WHO criteria for diagnosing AIDS and, using tests for both HIV antigen and antibody, found that one of eight AIDS patients (12 percent) was HIV negative on all tests.[20] A controlled study of 1,328 patients evaluated according to the WHO criteria in Uganda revealed that between 8 and 15 percent of all clinical diagnoses of AIDS were HIV negative.[21] In other words, these people had all of the symptoms of AIDS but not the virus that is supposed to be its cause.

How many such HIV-negative AIDS cases exist? Might these be flukes, without significance for understanding AIDS? Or might

they result from statistically insignificant failures of the test methods? In some cases, the lack of antibody and even of HIV virus
itself has been amply documented using the most sophisticated
techniques. In others, the data are inadequate. All that is known
for certain is that between 1 and 5 percent of AIDS patients who
are tested for HIV do not demonstrate the presence of antibody
to the virus nor can HIV infection be demonstrated directly.[22] The
CDC believes that most of these HIV-negative cases are not truly
AIDS cases but result from inaccurate tests or that these people
are too ill to make antibody to HIV that is actually present. The
CDC does, however, admit that some of these negative cases are
real.

The actual number of HIV-negative AIDS cases is irrelevant.
The existence of even a handful of HIV-negative AIDS cases is
sufficient logically to raise doubts concerning the necessity of HIV
as a cause of AIDS. From one perspective, it may mean that we
have attributed too much of the pathology associated with AIDS
to HIV. From another, it may mean that HIV is one cause of AIDS
but that other causes of AIDS exist as well. From yet a third perspective, HIV-negative cases may mean that there is a second epidemic masquerading under the guises of AIDS, which has yet to
have been detected and separated out from AIDS. If so, treating
every person with AIDS-like symptoms for the presence of HIV,
with or without evidence of infection, could be illogical and detrimental. Whatever the case, we need to rethink our theories of
AIDS. It is essential that the non-HIV immunosuppressive agents
be reevaluated as possible cofactors for HIV or, in some combination, as possible causes of AIDS in the absence of HIV.

Anomaly 3: Where Is the HIV Anyway?

It is axiomatic to the HIV theory of AIDS that the syndrome is
due to an infectious agent and that this infectious agent, HIV, is
transmitted from one individual to another either sexual contact
or by means of direct contact with infected blood or tissue. To be
transmitted sexually, HIV must be present in the reproductive
fluids. In most HIV-infected people, it is not.

Most people would imagine, I suspect, that an HIV-infected
AIDS patient who engaged in unprotected, high-risk forms of sex-

ual activity would almost inevitably infect his or her sexual partners. This is certainly the case for other sexually transmitted diseases such as syphilis and gonorrhea. A person whose sexual partner is infected with either of these diseases has a very high chance of contracting an infection. Syphilis has been found to be transmitted from one infected individual to an uninfected partner about 60 percent of the time if the infected person is a man, and about 40 percent of the time if she is a woman. Gonorrhea is transmitted from infected women to men at least 25 percent of the time, and from men to women 50 percent or more of the time. And about 25 percent of sexual partners of people infected with hepatitis B virus develop an infection.[1] In short, it takes, on the average, only two or three unprotected sexual contacts to infect the vast majority of people with most sexually transmitted diseases.

HIV, however, is anything but typical of sexually transmitted diseases. It can take hundreds of exposures to HIV for transmission to occur, or it may not occur at all.[2] Consider, for example, the demise of Rock Hudson and the unexpected fate of his lover, Marc Christian. Christian had met Hudson in November 1982 and become his live-in lover the next year. He stayed with Hudson until February, 1985. In the interim, Hudson began to look and feel less than healthy. He began to suffer from drenching night sweats, developed odd-looking purplish patches on his skin, often felt sick, had a persistent low-grade fever, and was losing weight. Marc Christian began to worry. Hudson and his personal secretary reassured Christian, however. They explained his drastic weight loss as being due to a new dieting program and then anorexia. When the skin lesions of Kaposi's sarcoma appeared, they were dismissed first as pimples and then as just a common form of skin cancer. Nothing to worry about. Still, during 1984, Hudson's personal secretary, Mark Miller, paid for a physical examination for Christian without telling him the reason and was vastly relieved to find that he was healthy. And it appears that sometime after June 1984, when Hudson was diagnosed as having AIDS and before Christian moved out in 1985, Hudson terminated sexual relations with Christian. Christian claims he did not know why.

He found out soon enough but in a most unexpected way. In July 1985 Hudson collapsed in his hotel room in the Ritz Hotel in Paris and was rushed to the hospital, where he was being treated for his HIV infection. Christian, watching the news when the story broke that Hudson had AIDS, felt betrayed. He had been engaging for three years, three to five times a week, in types of sex with Hudson that were known to be unsafe from the standpoint of contracting AIDS. Five weeks after Hudson died Christian sued Hudson's estate and Mark Miller for emotional distress. In February, 1989 he was awarded $21.75 million by a Los Angeles jury.

The most interesting aspect of this case from both the medical and legal perspectives was that Christian sued for emotional distress rather than a shortened life span or infection with a fatal disease. The reason for this line of legal attack was simple: Christian had been tested several times since 1985 for evidence of HIV infection, and none had been found. Christian was persistently HIV seronegative; he never produced antibodies against HIV at any time and was not, therefore, presumably infected with the virus. Thus, despite an estimated 600 high-risk sexual contacts with Hudson, he could not demonstrate that Hudson had in fact done him any bodily harm by exposing him repeatedly to his disease. He had not even developed antibody to HIV. The most he could do was to claim that Hudson had scared the wits out of him.[3]

Christian's experience in being repeatedly exposed to HIV without contracting it is surprisingly common. Earven "Magic" Johnson represents another highly publicized case. The great Laker basketball player retired late in 1991 after discovering that he had been HIV seropositive—infected with HIV sufficiently to produce antibodies to it—for an undetermined amount of time. The discovery had all the elements of a double tragedy since Johnson had just married and he and his wife, Earleatha, were expecting a child. Despite having unprotected sex (as the pregnancy amply testifies), Earleatha Johnson is not infected with HIV.[4] The wife of tennis star Arthur Ashe, who was recently diagnosed with transfusion-associated AIDS contracted nearly a decade ago, is also HIV free. Indeed, there are documented cases of some people having made love to their spouse thousands of times without the use of barrier methods of contraception and without the spouse's becoming infected with HIV. Transmission of HIV occurs only once in about 500 such unprotected heterosexual intercourses, on average.[5] These figures do not mean that it is safe to have sex with AIDS patients. Some partners do become infected after only a few contacts, and anal intercourse appears to increase the risk of HIV transmission greatly. But certainly these figures are vastly different from those documented for the transmission of syphilis, gonorrhea and hepatitis B virus.[6] Why is an HIV infection so terribly difficult to acquire as compared with other sexually transmitted diseases?

One of the reasons is that transmission may depend on the susceptibility of the sexual partner as well as the medical condition of the infected person. Another reason is that, unlike other sexually transmitted diseases agents, HIV is rarely present in semen in sufficient numbers to cause disease or is not present at all. That which is not present cannot be spread.

Despite early evidence that HIV was present in semen of HIV infected men, more recent studies have shown conclusively that the virus is only rarely there. Not surprisingly, the first laboratory to demonstrate the presence of HIV in semen was Robert Gallo's. He and investigators at several other hospitals demonstrated in September 1984 that HIV was present in the semen of two AIDS patients but not in the semen of three healthy heterosexuals.[7] A week later, David Ho and his colleagues at the Massachusetts General Hospital reported the isolation of HIV from both the semen and blood of a single healthy homosexual man.[8]

These reports created a plausible scenario for the transmission of HIV. The typical man ejaculates about a teaspoonful of semen during intercourse. This ejaculate contains, in addition to billions of sperm cells and the seminal fluid itself, millions of white blood cells. Many of these cells are either T-helper cells or macrophages that are susceptible to HIV infection.[9] As Jay Levy, a prominent AIDS researcher at the University of California, San Francisco, has written, if any of these cells becomes infected with HIV, they "can survive and serve as reservoirs (factories) for continual virus production and spread. When found in body fluids, they can transfer the infection to other individuals." Thus, Levy proposed in 1988 that cell-free HIV was unlikely to be a risk. The risk was conferred by HIV encapsulated within the protective armor of infected lymphocytes and macrophages. He therefore predicted that "the control and cure of AIDS will result only when the infected cell itself is eliminated," not simply by blocking free-virus entry into susceptible people.[10]

The early results demonstrating the presence of HIV in semen

are misleading, and several aspects of the studies deserve careful
scrutiny. First, the Gallo paper[11] clearly states that attempts to isolate HIV from semen resulted in extremely low levels of HIV that
appeared only transiently. As Jay Levy remarked several years
later, "T-helper lymphocytes generally die after HIV infection."[12]
They do not produce much or any HIV, nor do they infect other
lymphocytes either before or after their death. Thus, most HIV
researchers resort to a variety of tricks of the trade to verify the
presence of HIV in semen and blood. The most common is to co-
culture the semen cells with a lymphocyte cell line that is known
to be extraordinarily susceptible to HIV infection. In this way, the
HIV may be transmitted to cells that are much more likely to be
infected by it. These highly sensitive cells also have the unusual
property of rarely being killed by the HIV and so are able to produce very high levels of the virus. The need to co-culture the semen
in this way suggests, however, that little HIV is present and that
it cannot infect normal lymphocytes. Notably, Ho's group also
resorted to the same co-culture strategem to verify the HIV infection they reported. Moreover, careful reading of the Ho paper reveals that his group did attempt to isolate HIV from eleven other
HIV-seropositive men without success. Even with co-culture techniques, they could find no evidence of HIV in the semen of these
other men. Their single HIV-infected patient was, in the light of
their own data, an exception. Was the same true of Gallo's two
HIV-infected men? How many other semen samples did he test
without success?

More recent studies suggest that Gallo's choice of subjects must
have been very lucky. Once again Jay Levy's experience is instruc-
tive. He reported in 1988 that he could find HIV in only about
25 percent of the semen samples from AIDS patients whom his
group examined and the amount of virus was extremely low.[13]
Similar results have been reported by other laboratories as well.[14]
In all of the studies, less than a third of the infected men had any
HIV present in their semen and then generally less than one virus
genome per milliliter of semen, or perhaps one or two dozen virus-
infected cells per ejaculate, on average. Approximately the same
number of viruses are excreted in the saliva of HIV-infected individuals and in vaginal secretions. This amount of HIV is considered to be incapable of transmitting disease.[15]

Two provocative studies suggest that the actual incidence of HIV in semen may be significantly lower than many investigators report. Dr. Suraiya Rasheed of the University of Southern California reported in 1988 that she was able to detect HIV in only three of fifty (6 percent) semen samples from HIV-seropositive men.[16] These results were not associated with the presence or absence of retroviral therapy with AZT (zidovudine). Similarly, Dr. B. J. van Voorhis and his colleagues at the Brigham and Women's Hospital in Boston demonstrated in 1991 they could find evidence of HIV DNA in the semen lymphocytes of only one of twenty-five (4 percent) homosexual men with antibody to HIV, although they were able to demonstrate the presence of HIV in white blood cells taken from the blood of twenty-three of the twenty-five men. The importance of this study is underscored by the fact that the investigators used the most sophisticated techniques available for the identification of HIV and its DNA, including co-culture and a test for HIV genes known by the acronym PCR. Using controlled cultures, they demonstrated that their tests were sensitive enough to detect a single HIV-infected lymphocyte among 100,000 uninfected ones.[17]

A related phenomenon confirms the absence of free HIV in the genital and reproductive tracts of most AIDS patients. Many sexually transmitted organisms, including herpes viruses and various *Mycoplasmas*, are often found in the urine of infected people.[18] Although several independent studies have been performed to look for HIV in urine, no one that I am aware of has found any evidence of virus in any AIDS patient.[19] (Antibody to HIV is found in both semen and urine of most of these patients, however.)

The rarity of HIV in semen may explain not only why Marc Christian, Earleatha Johnson, and Jeanne Marie Ashe remained HIV free despite repeated unprotected sexual contacts with HIV-infected partners but also another, at first sight, unrelated set of anomalous results. One would expect that any woman inseminated with HIV-infected semen would have a reasonably high chance of developing an HIV infection and that children conceived during the process would also be at high risk for developing AIDS. This expectation is not confirmed. Less than 15 percent of women infected with HIV transmit their infection to their infants in Western nations.[20] Even more surprising, women who have been artificially inseminated with sperm from donors who were subsequently identified as being infected with HIV almost never become infected.

This result is puzzling. If our understanding of the way in which HIV-infected lymphocytes can transmit HIV is correct, then artificial insemination recipients should be at higher risk for HIV infection than women engaging in vaginal intercourse with an HIV-infected man. The reason for this increased risk is that it is common for the artificial insemination procedure to bypass the vagina so as to allow direct placement of the semen on the cervix, the narrow opening to the uterus, in order to increase the probability of fertilization. Since the uterus, but not the vagina, contains high concentrations of lymphocytes normally present to prevent infection, the probability of transmission of HIV from donor cells to the woman's immune system is generally considered to be greatly enhanced.[21] Nonetheless, no case of infant AIDS associated with artificial insemination exists, and the number of cases of HIV transmission to insemination recipients is so small that one must either conclude that HIV is virtually impossible to transmit to a healthy individual, or that it simply is not present to be transmitted in the first place.

Begin with the exceptions. In December 1990 "Jane Doe" filed a lawsuit against her doctor, "John Smith" (obviously not their real names) and the "ABC Company, Ltd." (a pseudonym) for "reduction of life expectancy." Her suit is based on the fact that she has been diagnosed as HIV antibody positive following artificial insemination with semen from a donor subsequently found to have been HIV positive. She was notified in January 1985 that she had been so exposed. Dr. Michael Rekart of the British Columbia division of the Canadian Centre for Disease Control reports that a total of thirty women were inseminated over a four-year period with semen from this donor and that only two of the twenty-four women who agreed to be tested are currently HIV seropositive. One of the two women conceived and carried her fetus to term. She, her husband, and her child are, as far as is known, healthy.[22] "Jane Doe," the other woman, denies having any risk factors that might have served as alternative sources of HIV infection, but her case remains to be demonstrated in court and no details are available.

"Jane Doe's" case is almost unique. As of fall 1990, Ian J. Wilson, who directs the semen donor program for the Mary Wilson Fertility Unit in Manhattan, could claim that there were "no cases of anyone getting AIDS from fresh sperm from a donor in the entire United States." This is a slight overstatement—two possible cases are on record—but it is still amazing that every year more than 80,000 women are inseminated artificially in this country and no more than a couple of HIV infections have resulted. Until very recently, none of the semen samples was tested for HIV. That means that more than a quarter of a million artificial inseminations were probably carried out between 1980 and 1985 without the benefit of screening for HIV and without resulting in any known cases of AIDS.[23]

No doubt some of these artificial inseminations utilized semen specimens from HIV-seropositive individuals. The AIDS Research Unit of the New York City Department of Health retrospectively identified six men who made regular semen donations in New York City between 1978 and 1985 and who subsequently developed AIDS. There are no known cases of AIDS associated with the semen donations made by these men despite the fact that the donated sperm was used to inseminate 176 women, most of whom agreed to be tested for HIV seropositivity after their exposure was discovered. Only one woman, who claims to have no other risk factors for HIV infection, has been found to be HIV positive, and she is otherwise healthy. She is listed by the New York Public Health Department as a "probable" case of insemination-acquired HIV infection since other risks cannot absolutely be discounted. Her husband is HIV negative. None of the children born to the 176 presumably HIV-exposed women is known to be HIV positive or to have developed AIDS.[24]

Only one other possible instance of HIV transmission by artificial insemination has been reported in the United States. An HIV-positive hemophiliac and his HIV-negative wife decided to try to conceive a child. Rather than risk transmitting the HIV infection during unprotected vaginal intercourse, they made use of an experimental procedure offered by their physician. This procedure involved processing the man's fresh ejaculate to attempt to separate out the sperm from the white blood cells that are a normal component of semen and that were the presumed carriers of HIV. Experiments carried out by the CDC suggest that these procedures may not have been very effective and that HIV-carrying lymphocytes had some probability of still being present in the semen. In the event, the woman underwent three uterine inseminations during 1989, and in January 1990 she became HIV seropositive. A review of this case by the CDC concludes, however, that "the mode of HIV-1 transmission to the woman described in this report cannot be determined definitively." Too many other risk factors, including possible condom failure, existed in this couple's life.[25] A second HIV-positive hemophiliac man and his HIV-negative wife underwent a similar procedure and the wife has remained HIV negative.[26]

HIV acquired by artificial insemination appears to be equally rare in other countries; a review of the medical literature has revealed only one other instance. Of eight Australian women inseminated prior to 1985 with semen from a man who subsequently developed AIDS, four eventually became HIV seropositive. Three of these women remain healthy to this day. One has developed AIDS-related complex, but it is reported that she has other identified risk factors for AIDS. None of the children or spouses of these women is HIV positive.[27]

In short, although HIV certainly can be transmitted through semen from one person to another, it is in fact transmitted so rarely to healthy sexual partners and is present at such low amounts in so few sperm samples from HIV-infected men that it is probable that those who become infected must be exposed repeatedly to many HIV carriers or have some unusual susceptibility for the virus. Clearly, this is not the picture of the cause of AIDS that the public has been given by the medical community or through highly publicized (and possibly misleading) cases such as that of Allison Gertz, a young woman who claims to have developed AIDS after a single sexual encounter with one HIV infected individual. Are cofactors such as concomitant infections or drug use responsible for enhancing the transmission of HIV? Is the mode of transmission—for example, anal intercourse as opposed to vaginal or oral—important? Is the crucial factor in transmission not HIV itself but the susceptibility of the recipient of HIV-infected semen? What regulates this susceptibility? These are the sorts of questions that are raised by the recognition that HIV is rarely present in semen and difficult to transmit sexually.

Anomaly 4: HIV Exposure Without Seroconversion or AIDS

Despite the difficulty of transmitting HIV to healthy people, some individuals are repeatedly exposed to HIV, in semen, blood, or tissue, and do become infected. A CDC study in 1988 found that "although most husbands and wives remained uninfected despite repeated sexual contact without protection, some acquired infection after only a few contacts. This is consistent with an as yet unexplained biologic variation in transmissibility or susceptibility."[1] Also consistent with such biologic variation is the fact that sexual partners of HIV-positive intravenous drug abusers have relatively high rates of infection as demonstrated by the fact that they are a recognized risk group.[2] Why? What controls transmissibility or susceptibility? Who are the people who resist both HIV infection and AIDS and how do they differ from those who become infected? Clearly, if we can answer these questions, we are well on the way to understanding the syndrome and its causes.

"Study says 61 will be infected by 1 AIDS carrier." This newspaper headline accompanied the 1988 publication by the *Journal of Sex Research* of a study by Steven Nahmias, a Santa Clara University statistician, on the role prostitutes were likely to have in spreading AIDS. Using the best available calculations of the number of sexual partners the average female prostitute has, the probability that HIV will be transmitted during a single sexual encounter, and the number of other female sexual partners acquired by the average male prostitution client, Nahmias concluded that a single prostitute with AIDS, whose clients failed to use condoms, would infect twenty men in a five-year period. Through these twenty men, forty other women would become infected and one infant.[3] Considering that the number of HIV-infected prostitutes was thought to be in the thousands in most of the major cities in the world by 1986[4], Nahmias's study sounded a sharp warning that prostitutes were likely to be the most devastating vectors for spreading AIDS. Moreover, Nahmias predicted what everyone has been fearing most in the AIDS epidemic: that AIDS would quickly become a predominantly heterosexual disease in Western nations.

The scenario he described has not taken place; it is not even close to being accurate. The number of American and European heterosexuals who have had sexual relations with a prostitute, who have no other admitted risk factors for AIDS (such as drug abuse), and who have subsequently developed antibody to HIV can be counted on the fingers of one hand. Sex with a prostitute is not even listed as a risk category by the American CDC.

There are two sides to the prostitute conundrum. The first is that female prostitutes themselves are perhaps the most surprising exception to the expectation that HIV exposure puts individuals at risk for AIDS. Prostitutes who do not abuse intravenous drugs almost never become infected with HIV. The other side to this coin is that even drug-abusing, HIV-positive prostitutes do not appear to transmit HIV to their drug-free patrons. Here is a surprise, indeed—a surprise strongly suggesting that susceptibility to infection may be more important in determining who becomes infected than is simple exposure to the retrovirus.

Begin by considering the ample evidence that prostitution is not, in and of itself, a risk factor for AIDS. M. Seidlin and his colleagues examined the prevalence of HIV infections in New York City call girls during 1987. They studied seventy-eight women who had been prostitutes for an average of five years each. Each woman had had an average of over 200 clients during the past year, or approximately 1,000 lifetime partners. Use of condoms was sporadic at best. Vaginal intercourse was common; anal, rare. Since it is estimated that nearly 5 percent of men in New York City are thought to be intravenous drug abusers and half of these are HIV seropositive, it is probable that each of these prostitutes had sexual relations with an average of twenty-five HIV-seropositive individuals.[5] Despite this unusual promiscuity and despite living in one of the AIDS capitals of the world, only one of the women was HIV seropositive. She admitted being an intravenous drug abuser. Her seventy-two non–drug abusing co-workers were all HIV negative.[6]

Another study carried out in New York City by Dr. Joyce Wallace and her co-workers between 1982 and 1988 found similar results. They surveyed several hundred streetwalkers (a lower class of prostitute than call girls) for a variety of measures of immunodeficiency. Excluding admitted intravenous drug users from their study, they found that only 4.5 percent of the prostitutes were HIV infected. The only statistically significant difference between those who were infected and those who were not was that the HIV-positive women had had a mean of 3,062 sexual partners during their lifetimes, whereas the HIV-seronegatives had had 1,047.[7] On the other hand, Wallace found an HIV seropositivity rate approaching 50 percent among drug-abusing prostitutes. These data clearly suggest that a history of drug abuse is a far greater risk for contracting HIV than is even the most outrageous heterosexual promiscuity. One can only wonder whether non-intravenous forms of drug use also promote HIV susceptibility among streetwalkers. No data are currently available.

Many other studies confirm that intravenous drug use, rather than prostitution, is the major risk factor for both HIV infection and AIDS among women in the United States and Europe. A 1988 American study concluded, for example, that "HIV infection in non–drug using prostitutes tends to be low or absent, implying that sexual activity alone does not place them at high risk, while prostitutes who use intravenous drugs are far more likely to be infected with HIV."[8] The same results were reported from Amsterdam, one of the world's centers for legalized prostitution. When several hundred prostitutes were studied, investigators found no evidence for HIV infections among non–drug abusers averaging more than 200 clients per year. High rates of sexually transmitted diseases (81 percent) and of HIV (30 percent) were found, however, among intravenous drug-abusing prostitutes.[9] In Zurich, Switzerland, a study of 123 prostitutes revealed that 18 were intravenous drug abusers. Fourteen of these drug abusers were HIV positive but only 1 of the 105 nonabusers.[10] Similarly, a year-long study of 839 registered Viennese prostitutes revealed that only 7 of them were HIV positive as of January, 1986, and that all seven of these cases were drug related.[11] Research conducted in London, Paris, Pardenone (Italy), and Athens has yielded virtually identical results.[12] Even more astounding, long-term studies in many of these cities have found no increase in the incidence of HIV seropositivity among non-drug abusing prostitutes since the early 1980s. In New York City, for example, the incidence was 4.5 percent in call girls who avoided intravenous drugs in 1982 and remained at that figure into 1989.[13] HIV is not ravaging drug-free female prostitutes.

The implications of these and related studies are astounding: HIV cannot be a sexually transmitted disease, in the usual sense of the term. The same studies cited demonstrate that most prostitutes, regardless of whether they are drug abusers and regardless of whether they regularly use condoms with their clients, eventually develop the standard sexually transmitted diseases of syphilis, gonorrhea, and herpes. But with only a handful of exceptions, only drug-abusing prostitutes contract HIV infections. The U.S. National Academy of Sciences even admits as much in their 1990 report, *AIDS: The Second Decade*, when they state that non–drug abusing prostitutes have no higher risk of AIDS than other women.

One strong hint about what may predispose some prostitutes to HIV infection comes from a a study made public by the Japanese physician Y. Shiokawa. He reported in 1988 that previously existing depression of normal T cell activity predisposes and is predictive of HIV infection among both prostitutes and male homosexuals. Shiokawa's work suggests, in short, that HIV does not *cause* immune suppression in prostitutes and gay men but rather that it *takes advantage* of prior immune suppression.[14] If Shiokawa is correct, there must be something related to intravenous drug abuse and a homosexual life-style that is conducive to immune dysfunction. Individuals with normal immune function should therefore be resistant to HIV.

Shiokawa's hypothesis brings us to the other side of the prostitution conundrum: Who contracts HIV from infected prostitutes? Is every "john" at risk? In fact, not only is heterosexual promiscuity unrelated to HIV risk, there is also very little evidence of transmission of HIV from female prostitutes, whether they are drug abusers or not, to non–drug abusing heterosexual men.

The most comprehensive study of HIV transmission from prostitutes to their clients was carried out in New York City. Of the 340 men subjects who regularly had sex with prostitutes and who denied ever having had sex with a man or ever having used intravenous drugs, only 3 had evidence of HIV infection.[15] Whether these three men had other immunosuppressive risks (nonintravenous drug use, chemotherapy, surgical procedures, and so forth) is not known, nor was their history of venereal diseases—a well-known risk for HIV transmission—studied. When such factors are taken into account, it has been found that prostitutes represent virtually no risk for spreading HIV to nonrisk heterosexuals. A German editorial in the medical journal *Sozial-und Präventativmedizin* (Social and Preventative Medicine) stated in 1988: "In Europe, there is as yet no evidence that female prostitutes are a source of HIV infection for the heterosexual population."[16] Several British studies have also concluded that prostitute-associated HIV transmission is almost always due to illicit drug use by the client and prior or concurrent venereal diseases of the client rather than of the prostitute.[17] The U.S. National Academy of Sciences, in *AIDS: The Second Decade*, argues that the concept of the female prostitute as a vector for spreading HIV to the heterosexual population has no basis in the United States either.

One clear implication of these studies is that the non–drug abusing heterosexual community should have little or no risk of HIV or AIDS. Indeed, testing for HIV-antibody positivity among new recruits by the U.S. Army, the U.S. Navy, and the Marine Corps[18] and among first-time blood donors of all ages by the Red Cross has been carried out since 1985.[19] All three sets of data demonstrate clearly that HIV infections among the general population are no more common in 1992 than they were in 1985. For some groups, such as white males, the incidence of infection actually seems to have decreased. The same general trends have been observed in Canada and Britain as well. HIV infections, and AIDS, are staying within the risk groups first identified for the disease: gay men, drug addicts, hemophiliacs, and blood transfusion patients.

HIV, in short, is spread by heterosexual contact only extremely rarely in the absence of predisposing factors. As revolutionary as this conclusion may sound, studies of other groups exposed to HIV support the idea that HIV infects only individuals with additional risk factors. For example, none of the husbands of four HIV-seropositive women exposed during artificial insemination developed antibody to HIV over a three-year period involving extensive and regular sexual relations without barrier methods of contraception.[20] In another study, no transmission of HIV was observed between twelve couples in which all of the women were HIV seropositive and in which at least one hundred sexual contacts had occurred.[21] Male-to-female transmission of AIDS in a similarly selected set of heterosexual couples yielded a seroconversion rate of 1 in 1,000 unprotected sexual intercourses.[22] As in homosexual men, anal intercourse was identified as the highest risk factor in this study.[23] Yet another study of fourteen wives of HIV-seropositive hemophiliacs similarly found that after a mean of three and a half years of unprotected intercourse, with an average of fifty sexual encounters per year, only one of the wives had become HIV seropositive. The wife who seroconverted had a recognized risk factor of immune suppression, however. She suffered from multiple sclerosis, which had repeatedly been treated with immunosuppressive drugs.[24] These and many similar studies all conclude, without exception, that in Europe and the Americas, the transmission of HIV through heterosexual intercourse is so rare that two heterosexuals without identified risks for AIDS have an equal probability of being struck by lightning, dying in a commercial airplane crash, or developing AIDS.

The same approximate risk seems to attend physicians and surgeons who are regularly exposed to HIV-infected individuals. So far, only four health care workers in the United States who have been exposed to HIV-contaminated blood or tissue since 1980 and who have no other acknowledged AIDS risk have developed symptoms of AIDS. There have, however, been more than 6,000 verified cases of health care workers' reporting subcutaneous exposure to HIV-infected blood or tissue as a result of needle-stick injuries, surgical cuts, broken glass, and so forth. This figure represents, without a doubt, a small proportion of actual exposure. In the first place, thousands of AIDS patients were treated prior to the identification of HIV and tests for it in 1984, and without extraordinary precautions on the part of health care workers. Moreover, since HIV testing has become available, none of my several health care friends who have had repeated subcutaneous contact with the blood or tissue of AIDS patients has so far been tested for HIV. Thus, the number of unrecorded exposures to HIV must be enormous. And yet only a few dozen health care workers are known to have become HIV seropositive during the entire decade of the 1980s in the United States.[25]

These figures are truly astounding when placed in the context of the frequency with which physicians become infected with other viruses and bacteria. Consider, for example, the transmission of hepatitis B virus in clinical settings. Several studies have found that surgeons puncture their gloves about once in every four operations and incur significant cuts or sticks about once in every forty operations. One percent of their patients are carrying hepatitis B virus. A surgeon who cuts or sticks himself or herself while working on such a patient has a 25 percent probability of becoming infected (if not previously vaccinated). Yet it has been documented that the rate of seroconversion among health care workers following skin puncture with HIV-infected tissue or blood is at most 1 in 1,000, and possibly as low as 3 in 10,000.[26] These studies of surgeons and other health care workers were not controlled for other immunosuppressive risks, such as concurrent disease or disease therapy, preexisting autoimmunity, prolonged exposure to anesthetics during surgery, or drug use, any of which may impair immune function. When one considers that HIV transmission in heterosexuals appears to be greatly heightened by concurrent drug abuse and that physicians are reputed to have the highest rate of drug abuse of any other group of professionals, one must also seriously raise the possibility that, among health care professionals exposed to HIV, the ones who seroconvert may well be those with other immunosuppressive risks. Unfortunately, there are no studies of risk factors for HIV seroconversion among health care professionals, and therefore no way to prove or disprove this conjecture at present.

What is clear from existing studies is that HIV is extremely difficult to transmit to a healthy individual. Indeed, Dr. Keith Henry and his colleagues in Minneapolis–St. Paul collaborated with John Sninsky and Shirley Kwok of the Cetus Corporation (now a subsidiary of Chiron, Inc.) in Emeryville, California, to demonstrate that even among health care workers with the greatest exposure to HIV, infection was virtually nonexistent. They took blood from twenty colleagues who had extensive laboratory exposure to HIV and tested for antibody to the virus. Only two people have weak signs of HIV exposure using the usually quite sensitive Western blot test, which tests for the presence of HIV proteins. In no blood sample was any sign of HIV viral genetic material found using the extremely sensitive polymerase chain reaction test. In short, Dr. Henry and his colleagues concluded that none of their highly exposed test subjects had become infected with HIV.[27]

Not only do health care workers not contract AIDS following workplace exposure to HIV; they do not transmit it during their professional activities either. One U.S. physician who became HIV seropositive in 1981 took part in over 400 surgical operations prior to his death in 1983. All of his patients were monitored; none became HIV positive. Similarly, a British surgeon who contracted AIDS prior to 1986 participated in over 300 surgical procedures before his death in 1988; none of his patients has developed AIDS or become HIV positive. Even more impressive is a study of 2,160 patients of another U.S. surgeon who died of AIDS in 1989. Only one case of AIDS was found among these patients, and the individual in question was an intravenous drug abuser whose surgery was for cervical lymphadenopathy complicated by tuberculosis—two indications that this patient had already developed AIDS prior to his operation. Moreover, although it is estimated that approximately 1,000 U.S. surgeons are currently HIV positive, no cases of HIV seropositivity or AIDS have been associated with any of these individuals or with any other surgeon or physician anywhere in the world.[28]

The only instance ever reported of HIV having apparently been transmitted from a health care professional suffering from AIDS to a patient is the widely publicized case of a dentist, David Acer. Approximately 1,700 people were identified as having been Dr. Acer's patients between 1984 and 1990, when Acer died of AIDS. Of these patients, 732 had been tested for HIV infection by January 1991 and found to be seronegative. Five of the dentist's patients, 2 men and 3 women, were found to be HIV seropositive. One of the men had identified risk factors for AIDS and probably did not acquire his infection through contact with the dentist. The second man may have had other risk factors and is known to have abused nonintravenous drugs to the point of overdose on at least one occasion. Information regarding possible risk factors other than visiting Dr. Acer has not been reported for 1 of the 3 women, subsequently identified as Lisa Shoemaker, thirty-five years of age. She is apparently still healthy.[29] The other two are explicitly stated to have had no identified risks for HIV other than their contact with Acer. One of these, Kimberly Bergalis, died of AIDS at the age of twenty-three. The other woman is identified as being elderly. She is still alive.

The argument that Acer was the source of their infections rests on a CDC study that found unusual similarities in the viral sequences isolated from the dentist, Kimberly Bergalis, the male drug abuser, and the elderly woman. It is possible that Acer infected each of these people while performing dental procedures. Lisa Shoemaker, for example, had surgery for the removal of two impacted wisdom teeth.[30] During her surgery, HIV could have been transmitted directly into her bloodstream. It is suspected that Dr. Acer may have failed to use proper sterilization procedures for his surgical tools between operations or may have cut himself and bled into the mouths of each of these patients during an operation. A close friend of Dr. Acer believes that he infected some of his patients on purpose to draw attention to heterosexual AIDS.[31]

Nothing about the Acer case is known for certain. Much of what the CDC has touted as fact about this case has recently been challenged by other scientists who claim that the viruses that infect the various patients and Dr. Acer are not nearly as similar as the CDC has maintained.[32] If they are not similar, then each patient may have acquired his or her HIV from independent sources. Also, many crucial questions, such as whether Kimberly Bergalis, Lisa Shoemaker, or the other patients had any sexually transmitted diseases, were on medications that might have been immune suppressive, were anemic, malnourished, or had any other immunosuppressive risks were never investigated by the CDC. We do not, therefore, know for certain that any of Dr. Acer's HIV-infected patients was genuinely free of factors that might have made them unusually susceptible to AIDS.

Still, even granted that HIV may have been transmitted from Dr. Acer to one or more of his patients, we are left with the observation that apparently 1,700 of his patients walked out of his office HIV free and that the tens of thousands of patients of other HIV-infected dentists have also remained HIV-negative. For example, none of the patients of three midwestern dentists who died of AIDS during 1991 has been found to be infected with the virus.[33] Thus, the possible exception represented by the Acer case underscores the fact that whatever may have occurred in his office, HIV is extremely difficult to transmit to a healthy individual and that the proper observation of sterilization and contamination containment procedures by physicians and dentists is extremely effective. The Acer case is not a reasonable model upon which to develop public policies for HIV containment.

Indeed, even if we take a worst-case scenario—one that was repeated thousands of times in the Western world prior to screening of blood transfusions for the presence of HIV antibodies and that continues to occur at significant rates in Third World nations—and consider what happens if a person receives several units of HIV-infected blood, the outcome is far less predictable than one might expect. Despite apparently widespread misconceptions among "experts" (who should know better) that "over 95 percent of the patients who were exposed to HIV through blood transfusions became infected," the actual figures are more encouraging.[34] Dr. John W. Ward and his colleagues at AIDS Program of the CDC report that of 765 people who received HIV-tainted blood, HIV tests for 257 were performed; 113 (55 percent) were found to be HIV seropositive within an average of five years following exposure to HIV; only 16 (6 percent) had developed AIDS.[35] Similarly, of 34 hemophiliacs inoculated an average of over a dozen times each with a single batch of HIV-contaminated blood clotting factor concentrate—the fraction of the blood they need to permit proper clotting—only 18 became infected.[36] So even among seriously ill people receiving very large amounts of blood or blood products containing HIV, the risk of developing AIDS is far lower than one might expect.

It is worth comparing these figures to the rate of infection following exposure to other viruses associated with AIDS, such as hepatitis B. Even with screening for hepatitis B viral antigens, it has been found that the vast majority of factor concentrates used to treat hemophilia are contaminated with sufficient hepatitis B virus to induce infection. Thus, almost every hemophiliac who has been treated with factor concentrate has developed hepatitis, abnormal liver function, and is hepatitis B virus antibody positive.[37] Moreover, up to 10 percent of such patients develop chronic hepatitis, which does not respond to treatment. Similarly, following transfusion, non-A, non-B hepatitis afflicts about 10 percent of all individuals who receive one to five units of blood, and 50 percent of those infected become chronic carriers of the infection.[38] In short, hepatitis viruses appear to be thousands of times as infectious in clinical settings as HIV and represent a much more prevalent medical problem.

To summarize, HIV is very difficult to transmit and equally difficult to acquire even after exposure. There are no other infectious agents that behave like HIV except opportunistic disease agents. These, like HIV, are widely disseminated but difficult to acquire and rarely result in disease unless a person has other causes of immune impairment. Does this mean that HIV requires prior immune suppression to become actively pathogenic? Or does it mean that HIV is so radically different from all other viruses that we cannot compare it to them? Regardless of how this question is finally answered, exceptions should always evoke skepticism in the mind of scientists and provoke them to question their assumptions. So far, HIV researchers have been far from skeptical.

Anomaly 5: Sometimes HIV Disappears

Several of the previously described anomalies suggest that HIV infection may not necessarily lead to AIDS if appropriate cofactors are lacking or if the infected individual remains healthy. Indeed, some people, such as health care workers and researchers with repeated subcutaneous exposure to HIV, never even develop antibody to the virus. These observations indicate that HIV may not be pathogenic (disease causing) under some conditions and that healthy individuals may be able to fight off infection successfully. The current dogma, however, states clearly that all, or nearly all, people infected with HIV will develop AIDS. Moreover, according to this dogma, the presence of antibody to HIV is tantamount to infection, so everyone who develops antibody to HIV can expect to die of the syndrome. In the light of this dogma, perhaps the most interesting, and certainly the most unexpected, anomaly that has appeared in recent AIDS research are cases of patients known to be infected with HIV who have successfully controlled or even eliminated the virus from their bodies without medical intervention.

One striking case has been reported by Gérard Dufoort and his co-workers in Paris. They identified nine recipients of blood donated by a Portuguese man infected with HIV before 1973. Of these, four died within a few months after transfusion from complications unrelated to AIDS. Two could not be located. Three were found. Two of these people, both women, submitted to tests for HIV. One, who was transfused in 1974, was determined to be HIV infected by two different types of tests. The other woman was transfused in 1982, but her tests were inconclusive (one was negative and the other borderline positive). Neither woman had developed any symptoms associated with AIDS as of 1988, nor have any reports of subsequent development of AIDS been published. These results strongly suggest that it is possible for a healthy individual to carry HIV antibodies for fourteen years or more without developing AIDS.[1]

Indeed, it has been demonstrated that repeated blood transfusions and transfusions of clotting factor concentrates used by hemophiliacs do not necessarily lead to HIV infection or the development of HIV antibodies. R. J. G. Cuthbert and his colleagues in Edinburgh have studied a group of forty-eight hemophiliacs who received blood clotting concentrates made from a single batch of HIV-infected blood. Of these individuals, only sixteen developed antibody. This is a particularly noteworthy observation, since some of the hemophiliacs who did not develop HIV infections used as many as thirty vials of contaminated factor concentrate, and some of those who did become infected used as few as nine vials. The only significant difference between the two groups was that those who developed HIV seropositivity had lower responses to tests of T cell activity in 1984 and 1985 than did those who remained seronegative. This finding suggests that the hemophiliacs who remained HIV free were healthier prior to HIV exposure than their less fortunate counterparts.[2]

Some people infected with HIV can apparently do even better than stave off AIDS; they can actually eliminate HIV from their bodies. In 1986, Dr. H. Burger reported that the wife of a hemophiliac with AIDS had developed and then lost HIV antibodies. Dr. Burger had utilized the then state-of-the-art test for the presence of HIV antibody, the so-called ELISA test. An ELISA test is performed by binding an antigen, or foreign compound, to the surface of a tiny well in a plastic plate. An antibody that is specific to the compound is introduced into the well and sticks to the antigen. The more antigen there is, the more antibody binds to it. The well is washed out to remove any unbound antibody. A second antibody is then added to the well, and the excess is again washed out. The second antibody binds to the first, but unlike the first antibody, this second one has an enzyme attached to it that can carry out a chemical reaction. This chemical reaction produces a measurable product, such as a colored compound, which is used to determine how much of the antigen was present in the well in the first place. To determine how much antibody an individual has made against HIV proteins, some blood is drawn and the antibodies purified. If these stick to a specially prepared well in an ELISA plate, a chemical reaction will occur, as outlined, that indicates the presence of the antibody. The initial screens for HIV seropositivity were all performed using various modifications of this ELISA technology, and most initial HIV screening still employs ELISA techniques.

Dr. Burger was naturally concerned that since the hemophiliac he was treating had developed antibody to HIV, his wife might also become infected. The woman in question agreed to be tested. An initial ELISA test had demonstrated the presence of antibody to HIV during 1984, and subsequent analysis of the woman's T cells revealed a number of abnormalities, including low numbers of T-helper cells in relation to T-suppressor (T8) cells, which were consistent with HIV infection. A later ELISA test found no HIV antibody, however, and repeated tests confirmed the loss of the antibody.[3] The woman's T cell counts also returned to normal, and she remained healthy[4]. The ELISA test for HIV antibody is known to give a reasonably high rate of false-positive results, however. As many as four out of five positive tests for HIV antibody reported using ELISA technology cannot be confirmed by more sophisticated techniques, such as Western blots, in which HIV-specific proteins are isolated.[5] Thus, many AIDS researchers found this case unconvincing, and we cannot know in retrospect whether this was the first clarion case on record of defeating HIV or whether it is nothing but an artifact.

The next set of cases was clearer. In 1987, Alfred Saah of the Johns Hopkins University School of Hygiene and Public Health reported that five men in a Baltimore-based AIDS study who had repeatedly tested positive for HIV infection had suddenly lost detectable levels of antibodies against HIV. Their condition persisted and in 1988 Saah and his colleagues published extensive evidence confirming that four of the men were antibody and HIV-antigen negative by a wide range of tests. Two of the four men retained evidence of inactive HIV infections in very low levels of T cells as measured by the extremely sensitive polymerase chain reaction test, while HIV genes could not be detected at all in the other two men. Active HIV could not be cultured from the cells of any of the four men, strongly suggesting that these men had effectively controlled or even eliminated their infections. The same phenomenon was reported in a preliminary form from several other hospitals participating in the Multicenter AIDS Cohort Study as well, although no publications on the outcome of these studies have ever appeared. C. R. Horsburgh and his colleagues at the CDC, however, reported a similar finding in four additional people in 1990.[6]

At least four other cases of loss of HIV positivity accompanied by normal T cell counts and continued good health are on record among adults and dozens of cases among infants. Two of the adult cases were reported by L. H. Perrin and his colleagues in Switzerland.[7] Their patients had had repeated contact with HIV-positive individuals and had themselves become HIV positive (they had developed antibodies to HIV). Their infections were confirmed by several tests, including the polymerase chain reaction test. Both patients then lost their antibody to HIV, retained normal T cell counts and activity, and remained healthy. They apparently retained a latent infection with HIV, however. Dr. A. Fribourg-Blanc also described two patients he had seen in France who presented almost identical case histories. In addition, Fribourg-Blanc reported a series of more doubtful cases of loss of HIV antibody and continued health that will not be described here. He suggested that the ability to eliminate HIV effectively was inversely correlated with the number of HIV exposures. The less frequently a patient was exposed to HIV, the greater was the probability that he or she might revert to seronegativity (a lack of HIV antibodies).[8]

This last conjecture by Fribourg-Blanc is particularly interesting since previous anomalies have already suggested that the differences between those who become infected with HIV or develop AIDS and those who do not may be the status of their immune system prior to infection. Certainly it is worth noting that most AIDS patients are not exposed to HIV or to any of the other agents that might act as cofactors in AIDS only once, twice, or even three times but encounter these agents repeatedly and sometimes nearly continuously for months or even years at a time through promiscuous sexual relationships, continuous exposure to a single sexual partner, or repeated sharing of contaminated needles and other forms of drug use. Hemophiliacs were repeatedly exposed to contaminated factor that contained not only HIV but hepatitis and many other viruses until very recently. Even blood transfusion patients who develop AIDS are unusual in often being exposed to other bloodborne viruses and having had many more transfusions than the average surgery patient. In other words, few people who develop AIDS are or were exposed to HIV only once or in isolation from other immunosuppressive agents.

But even individuals who are multiply exposed to HIV may not develop antibody to HIV or lasting HIV infection. For example, in 1985 Harold Jaffe and his colleagues identified twelve healthy homosexual men who had been HIV seropositive for between two and six years. When they attempted to isolate HIV from the lymphocytes of these men, they were able to do so for only eight; four of the men were apparently HIV free, although they still carried antibody to the virus.[9] Dr. David Imigawa and Dr. Roger Detels of UCLA reported in October 1991 that among thirty-one HIV-seronegative homosexual men from whom they had isolated HIV at least once during 1989, none developed HIV antibodies in 1991 and virus could be isolated from only a single individual that year. In other words, thirty of these men who had verified presence of HIV in their blood at one time or another had successfully eliminated HIV from their bodies. Notably, all of these men had altered their life-styles in the intervening two years from high-risk modes involving promiscuous anal intercourse to low risk modes involving safer sex practices and fewer partners. Their cases indicate that the elimination of immunosuppressive and infectious risks had immunologically beneficial results.[10]

The same phenomenon has been observed in hemophiliacs. One of the most surprising studies on record is a search for silent infections of HIV among hemophiliacs by Dr. J. Gibbons and his colleagues. They used the polymerase chain reaction technique to search for unexpressed copies of HIV genes in the lymphocytes of hemophiliacs who "had been exposed to contaminated blood products more than 3 years previously" but who were HIV seronegative. Perhaps such people were simply harboring latent or unexpressed HIV infections and were still at risk for AIDS. Gibbons studied fifty-seven hemophiliacs who had been so exposed but had remained completely healthy. They found evidence of HIV genes in the lymphocytes of only one individual, and that DNA was found only transiently. In other words, none of these hemophiliacs who had been exposed to HIV by direct, and in most cases repeated, transfusion of HIV-contaminated blood products developed any signs of infection; they did not have HIV antibody; and the sole individual in whom HIV genes were found subsequently eliminated the HIV genes from his system.[11] All of these individuals remained healthy.

Similar data have confused the picture of AIDS in infants. A number of investigators have reported instances of infants who were born with antibody to HIV and direct evidence of HIV infection and yet lost their seropositivity and have remained either completely healthy or have incurred only slight indications of transient immune dysfunction. In some cases, these children continue to harbor HIV in their cells in a latent and inactive form; in other cases it is not known whether the HIV infection persists.[12] What makes these children different from those who die of AIDS? No one knows. Yet rather than trying to find out so that clues to the survival of less fortunate children might be found, these infants are almost universally labeled as having "defective" or "paradoxical" antibody responses. Investigators are so sure that HIV must be present and must eventually cause AIDS that they cannot accept evidence that it, like every other infectious agent, can sometimes be vanquished.

Oddly, the ability of adults and infants to control or eliminate HIV infection in the absence of medical treatment is not seen by AIDS researchers as a source of hope for those at risk for AIDS but rather as a new public health threat: What if these apparently healthy individuals should pass their hidden HIV infections on to others through blood transfusions that would test negative for the presence of antibody to HIV and therefore be presumed safe?[13] Some investigators have gone as far as to invent a new type of infectious process they are calling "incomplete" or "abortive" infection to explain why homosexuals and hemophiliacs can apparently eliminate HIV from their systems.[14] Their claim is that such people were never *really* infected in the first place, although both virus and antibody to it were demonstrably present. While the fear of unrecognized HIV infections contaminating blood transfusions and organ transplants may be justifiable, it is not justifiable to ignore the positive implications for the HIV-infected individual who has become HIV antibody negative and even virus free.

I believe that data demonstrating loss of HIV and/or HIV antibody correlated with continued health suggest that even people in high-risk groups who may have initially had multiple contacts with HIV may successfully combat the viral infection. I believe that only people whose immune systems are already significantly impaired will develop and express HIV infections. Therefore, it is not just exposure to HIV that must be considered in determining risk for AIDS but the status of the immune system during and after HIV exposure. People who are healthy, or who adopt healthier life-styles after HIV infection, may successfully combat HIV.

I am not alone in this opinion. Alfred S. Evans, a professor of epidemiology and public health at Yale University and a member of the World Health Organization Serum Reference Bank and Section of International Epidemiology, has written, "In my view, many but not all persons infected with HIV will develop the full AIDS syndrome; others may develop AIDS-related complex, dementia, or some other clinical manifestation, but some will escape entirely (i.e., remain asymptomatic carriers). . . . My view is in contrast to that expressed by some others, who believe that probably all HIV-infected persons will develop AIDS or some end-stage manifestation. Only time will tell which view is correct, as determined through the several prospective studies now in progress. Some indicate that some homosexuals who have been HIV infected for as long as 10 years show no clinical manifestation of disease."[15]

Also, Gene Shearer and Mario Clerici of the National Cancer Institute recently studied gay men who have had repeated, unprotected anal intercourse with HIV-infected partners, laboratory workers repeatedly exposed subcutaneously to HIV in the workplace, uninfected infants of HIV-positive mothers, and intravenous drug users who have shared needles with HIV-positive addicts. They, too, have observed large numbers of these people who remain HIV-negative for several years according to all tests currently available. They have therefore "dared to think something that, to many scientists, has been unthinkable: that there might be a successful immune response to HIV."[16]

The longer the AIDS epidemic has lasted, the more people there are who are surviving HIV infections for ever-longer periods of time. One of the oddest observations that strikes a historian of the epidemic is that the latency period—the estimated time lag between HIV infection and the development of clinical AIDS—has expanded almost yearly. In 1986, the figure was less than two years; in 1987, it was raised to three; in 1988, it became five; in 1989, ten; and as of the beginning of 1992, the latency period was calculated to be between ten and fifteen years.[17] Is this relentless

increase because the most susceptible people died of their infections first, leaving the more immunologically hardy ones to follow? Is it because HIV is becoming less virulent? Is it because cofactors are necessary to trigger infection into full-blown AIDS and
those with the most cofactors died first, leaving people with lower-
risk life-styles to survive longer? Or is it that more and more
people are successfully fighting off the infection and remaining
healthy but inevitably carrying HIV antibody for years or perhaps
a lifetime as a memento mori of their near escape from death?
Only time and much more intensive studies will tell. But one curious and very striking thing is known: The median number of lifetime sexual partners for the first 100 gay men diagnosed with
AIDS was a wopping 1,120; all had had multiple and recurrent
venereal diseases and other chronic infections; and every single
one of these gay men abused a multitude of recreational and addictive drugs.[18] They were not typical of gay men then; they are not
typical of gay men contracting HIV today; and they are not typical
of most heterosexuals. Thus, it is quite plausible to hypothesize
that the people who died most quickly of AIDS and who were the
forerunners of the epidemic were those with the largest burden of
ongoing risk factors. If this is so, then HIV is only one actor in
the multitudinous company that performs the tragedy of AIDS.

But this is jumping the gun. Suffice it to say for the present that
attributing AIDS to nothing more than an infection by HIV is too
simplistic. It leaves too much unexplained and creates too many
anomalies to be a satisfying scientific explanation. HIV is not sufficient to explain the anomalies of AIDS. These anomalies represent the challenge of understanding AIDS. A more thorough and
skeptical analysis of the data is needed.

CHAPTER 2

THE ROLE OF HIV IN AIDS

Rereading the Script

Anomalies, problems, paradoxes, and contradictions are only the incentives for research. If no one pays attention to them, they are fruitless. Even when they are identified and scrutinized, they are only a beginning; they define the areas of our ignorance. Thus, the five sets of anomalies described in chapter 1 tell us nothing more than that something is wrong with our explanation of AIDS. Now the work begins.

This book will unfold according to the following logic. It began by defining some of the crucial, unsolved problems that remain to be explained. In order to understand why these problems are problems, we must look carefully at what the current theory of AIDS tells us to expect. I will not dwell at length on this subject, since many excellent books on the topic exist, including Robert Gallo's recent autobiography, *Virus Hunting*. Indeed, every school child and adult who can read or watch television is now bombarded with the basic "facts." I see no reason to rehash this material. Instead, I shall focus on how the basic "facts" of AIDS have been established and on the untested assumptions that underlie our theories, experiments, and clinical investigations of the syndrome. This chapter will explore the evolving definition of AIDS, the basic evidence used to argue that HIV is the cause of AIDS, and the standard criteria used by microbiologists and virologists for evaluating claims that a particular organism causes disease. It ends with a discussion of the difficulties inherent in proving that HIV is the cause of AIDS and a summary of alternatives to the notion that AIDS must be caused by a single disease organism such as HIV.

The upshot of the discussion will be that HIV has not satisfied any established criteria for demonstrating disease causation. Thus, although there is no doubt that HIV is an integral player in the drama of AIDS, we cannot say for certain that it is, beyond doubt, a solo actor doing a monologue. It may certainly be the lead player in its troupe, but existing evidence suggests, as Chapter 1 intimated, that HIV has a whole cast of supporting characters that foster its villainous work. Indeed, these henchmen may even act autonomously; there is some evidence that HIV is not even on stage when some of the "crimes" of AIDS are committed. The tragedy of AIDS has a more complex script than has yet been revealed.

So to begin, what, precisely, is AIDS, why has HIV been identified as the cause of the syndrome, and why have the arguments for its causative role been so compelling to so many scientists?

The Evolving Definition of AIDS

The first definition of AIDS appeared in the September 24, 1982, issue of *Morbidity and Mortality Weekly Report* published by the Centers for Disease Control:

> CDC defines a case of AIDS as a disease, at least moderately predictive of a defect in cell-mediated immunity, occurring in a person with no known cause for diminished resistance to that disease. Such diseases include KS [Kaposi's sarcoma], PCP [*Pneumocystis carinii* pneumonia], and serious OOI [other opportunistic infections]. These infections include pneumonia, meningitis, or encephalitis due to one or more of the following: aspergillosis, candidiasis, cryptococcosis, cytomegalovirus, norcardiosis, strongyloidosis, toxoplasmosis, zygomycosis, or atypical mycobacteriosis (species other than tuberculosis or lepra); esophagitis due to candidiasis, cytomegalovirus, or herpes simplex virus; progressive multifocal leukoencephalopathy, chronic enterocolitis (more than 4 weeks) due to cryptosporidiosis; or unusually extensive mucocutaneous herpes simplex of more than 5 weeks duration. Diagnoses are considered to fit the case definition only if based on sufficiently reliable methods (generally histology or culture). However, this case definition may not include the full spectrum of AIDS manifestations, which may range from absence of symptoms (despite laboratory evidence of immune deficiency) to non-specific symptoms (e.g., fe-

ver, weight loss, generalized, persistent lymphadenopathy) to specific diseases that are insufficiently predictive of cellular immunodeficiency to be included in incidence monitoring (e.g., tuberculosis,
oral candidiasis, herpes zoster) to malignant neoplasms that cause,
as well as result from, immunodeficiency.[1]

It is evident from this definition that the CDC was not sure what
AIDS was in 1982, other than that it appeared to be due to immune deficiencies of unknown cause that could be manifested by
any of fourteen different opportunistic diseases. Crucial to the definition, however, was the statement that diagnosis for AIDS could
be made only in people with these opportunistic diseases if they
had "no known cause for diminished resistance to that disease."
The reason for this caveat was that a number of groups of people
had previously been identified as having a significant risk for each
of these opportunistic diseases. Patients undergoing various cancer
chemotherapies, transplant patients, people treated with high or
chronic doses of corticosteroids to control inflammatory and autoimmune diseases, and people born with defective immune systems are prone to opportunistic infections of all kinds. They were
excluded as AIDS patients by definition, as were men over the age
of sixty who developed Kaposi's sarcoma, since such men were
already known to be at risk for this cancer. A diagnosis of AIDS
required no identified cause of immune suppression.

The definition of AIDS has evolved along with the disease itself.
Just how much it has evolved can be seen from the following example. In May 1991 a new and unexpected AIDS risk was broadcast to the world. "Organ recipients test positive for AIDS virus!"
screamed headlines. LifeNet Transplant Services of Virginia
Beach, Florida, announced that three people who had received organ transplants—one the heart and two others a kidney apiece—
from a man who had died of gunshot wounds in 1985 had developed AIDS and died. Three other recipients of the man's tissues
also tested positive for HIV. The frightening aspect of the cases
was that the gunshot victim had been tested twice for HIV prior
to the transplants and had been found to be HIV free. Subsequent
reanalysis suggested that the tests used during 1985 did not have
the sensitivity necessary to identify the man's very low level of
infection. On the other hand, it is equally possible that the patients
had latent HIV infections that were activated by the transplant

procedure. In either case, the cases raised the spectre, validated by similar instances,[2] of hidden HIV infections unwittingly being transmitted to or reactivated in a significant number of transplant and blood transfusion recipients. The story, coming as it did at the same time that a number of states were considering banning HIV-infected surgeons and dentists from performing surgery, added fuel to the hysteria that perhaps there is not, and never can be, any real protection against AIDS. Even the most scrupulous and clean-living individuals might, by chance and through no fault of their own, still contract this modern scourge through an improperly screened blood transfusion or an unwanted visit to the hospital.

No one seems to have realized that just seven years earlier, the same three organ transplant recipients could have died of exactly the same opportunistic infections without raising an eyebrow and without being diagnosed as having AIDS. They would have been in a group specifically excluded from being considered for a diagnosis of AIDS: transplant recipients. Their causes of immune suppression were known: the drugs they were treated with in order to prevent their immune systems from rejecting their new organs. These drugs, along with the rigors of surgery itself and the possibility of an immune system disorder called graft-versus-host disease in which the lymphocytes in the donated organ attempt to kill the recipient's body, result in very high rates of morbidity and mortality in organ recipients compared with the general populace or even with other surgery patients. *Morbidity* is the physician's term for sickness; *mortality* for death. Two of the transplant patients who died of "AIDS" received kidneys. Their probability of dying within three years of their operation was 20 percent if they developed no complications and 40 percent if they did. This figure rises to nearly 60 percent at five years for patients with complications.[3] Since the two patients who died clearly developed complications manifested as opportunistic infections, they were in the high-risk group. Thus, from a purely statistical point of view, each of these people was more likely to have died than to have been alive in 1991, no matter what their HIV status. The same approximate statistics apply to the unfortunate individual who received a heart transplant.

Chances are also good that the three would have died of the same symptoms and the same opportunistic infections whether they had contracted an HIV infection or not. HIV-negative transplant patients are prone to the same sets of opportunistic infections that characterize AIDS patients, including *Pneumocystis* pneumonia (originally known as "transplant lung"), cytomegalovirus, varicella-zoster virus, disseminated herpes simplex, and toxoplasmosis infections.[4] The only difference between the transplant patients who died of AIDS and those who die of the same symptoms but are not given a diagnosis of AIDS is the presence of antibody to HIV in the former group.

What, then, is AIDS? Why do we call a patient who dies of *Pneumocystis* pneumonia following a transplant operation unfortunate but one who dies of *Pneumocystis* pneumonia and HIV an AIDS tragedy? Ironically, this definitional problem has existed since the very beginning of the "epidemic." In the first report of GRID published by Michael S. Gottlieb and his colleagues at UCLA, one of the five patients was a twenty-nine-year-old male homosexual who had a known cause of immune suppression. He had been successfully treated with radiation therapy for Hodgkin's disease (a cancer of the white blood cells) three years earlier.[5] Radiation therapy is a well-recognized cause of immune impairment. Nonetheless, this case stands as one of the benchmark cases heralding the discovery of AIDS.

Beginning in 1984, the definition of AIDS was changed to make the Hodgkin's case less anomalous and eventually to include transplant patients and other immunosuppressed individuals under certain circumstances. The CDC revised its definition by adding to the list of diseases diagnostic for AIDS any lymphoma (cancer of the lymph system) limited to the brain.[6] The discovery of HIV and its identification as "the cause of AIDS" during 1984 caused a second revision in June 1985.[7] To the previous set of fourteen diseases predictive of cellular immune suppression, the CDC added seven more diseases. If a person was found to be HIV seropositive by any test and had histoplasmosis (a fungus) disseminated beyond the lungs or lymph nodes; isosporiasis (a protozoal infection) causing chronic diarrhea for more than a month; bronchial or pulmonary candidiasis; many types of non-Hodgkin's lymphomas; Kaposi's sarcoma over the age of sixty; chronic lymphoid interstitial pneumonitis if a child; or any cancer of the lymph system diagnosed three or more months after a diagnosis of any opportunistic infection, then he or she was an AIDS patient. Thus, a number of groups that had previously been excluded from diagnoses of AIDS, such as certain cancer patients and elderly men with Kaposi's sarcoma, were suddenly potential AIDS patients despite previously demonstrated risks for opportunistic diseases. The crucial question was whether they had become infected with HIV as well.

Even more important in the light of recent questions concerning the necessity of HIV for causing AIDS, the 1985 revision of the AIDS definition also stated that some opportunistic diseases previously diagnostic for AIDS would be diagnostic in the future only if HIV was present: "To increase the specificity of the case definition, patients will be excluded as AIDS cases if they have a negative result on testing for serum antibody to [HIV], have no other type of [HIV] test with a positive result, and do not have a low number of T-helper lymohocytes or a low ratio of T-helper to T-suppressor lymphocytes."[8] In other words, people with the same clinical symptoms as an HIV-infected person (for example, disseminated tuberculosis) but without evidence of HIV or obvious immune impairment were not AIDS patients. This alteration causes problems. Twelve of fourteen cases of Kaposi's sarcoma diagnosed in individuals without identified risk factors for AIDS during 1981 and 1982 had normal immunologic results and were not tested for HIV (since HIV had not yet been discovered).[9] According to the 1985 definition, they might not have been diagnosed as AIDS patients. Even more interesting are the more than twenty HIV-negative cases of Kaposi's sarcoma among homosexual men with normal immunologic results that have been reported in the medical literature during the last two years. Do these people have AIDS? If not, is there a second epidemic of Karposi's sarcoma (and perhaps other opportunistic diseases) superimposed upon the so-called AIDS epidemic and appearing in the same risk group? How are these two diseases, if they are two, to be distinguished? What do they tell us about the necessity of HIV in AIDS?

These issues become more confused in the light of the next set of alterations announced by the CDC in August 1987. According to this set of revisions, the list of opportunistic infections indicative of AIDS grew to twenty-four, again enlarging the pool of potential AIDS patients. One set of twelve opportunistic diseases, including *Pneumocystis* pneumonia, Kaposi's sarcoma, disseminated cytomegalovirus infection, and esophageal candidiasis, were diagnostic for AIDS regardless of whether there was any evidence of HIV infection. Twelve other diseases were diagnostic for AIDS only in conjunction with a positive HIV antibody test. This meant that a large number of AIDS patients (45 percent of all cases diagnosed in the United States during the past decade and 1 percent of patients specifically tested for HIV seropositivity continued to be diagnosed as having AIDS in the absence of evidence of HIV infection.[10] By far the most important of the changes made in 1987 was the statement that "regardless of the presence of other causes of immunodeficiency, in the presence of laboratory evidence for HIV, any disease listed . . . indicates a diagnosis of AIDS."[11] In other words, *acquired* immune deficiency syndrome attributed to HIV infection is now diagnosed even among people who were born with congenital immune deficiencies; who have demonstrable, preexisting, or coexisting causes of immune suppression due to chemotherapy, radiation treatment, or corticosteroid use; among transplant patients who are on regimens of immunosuppressive drugs for life; and so forth.

AIDS, in short, has become a schizophrenic disease. Some people with diseases identical to those classically used to define the syndrome, such as disseminated tuberculosis, are not AIDS patients in the absence of HIV. Some people are AIDS patients if they develop opportunistic infections even in the absence of evidence of HIV. And in the presence of HIV, almost any rare disease is diagnostic for AIDS regardless of whether the person has other, more fundamental causes of immune suppression.

The definition changes are apparently not over. In 1992, the CDC proposed altering the definition of AIDS to include any person who had developed a significant loss of a particular type of white blood cell called T-helper lymphocytes.[12] Normally, a healthy person has a T-helper lymphocyte count of around 1,000 cells per cubic millimeter of blood. AIDS may now be diagnosed when the number of these T-helper cells falls below 200 per cubic millimeter of blood if the individual is HIV seropositive and even if he or she has no opportunistic infections. In other words, the primary criterion that allowed the identification of AIDS in the first place—that a person have an opportunistic disease in the absence of an identified cause of immune suppression—may be abandoned completely. People may be diagnosed as having AIDS even if they have no infections typical of AIDS, as long as they have a significantly low number of T-helper cells and antibody to HIV.

This latest proposed definition change has little, if any, scientific merit. Indeed, the CDC itself has been fighting against the definition change, and Dr. James O. Mason, assistant secretary for health in the Department of Health and Human Services, says forthrightly that changing the definition "messes up the baseline for comparison from past to future" and that it "will make interpretation of trends in incidence and characteristics of cases more difficult."[13] Then why alter the definition?

The reason for this latest definitional alteration is social and economic, not scientific. AIDS activists are now dictating how AIDS is to be diagnosed and who is to be included in the count.[14] For them, the issue is not one of correct diagnosis or elucidating the cause of AIDS; it is the understandable desire to increase access to health care. As Erik Eckholm has written in the *New York Times*, "The definition [of AIDS] has become a political as well as a medical question as people infected with the human immune deficiency virus, HIV, compete for treatment. For years, people weren't considered to have AIDS until they showed symptoms of certain infections and cancers that invade the body once the immune system breaks down. But after complaints that many ailing people were being excluded from the count, the Federal Centers for Disease Control has begun revising its definition. . . . It has been estimated that the broader definition . . . will add 160,000 people to the current caseload of 200,000 classified as having AIDS."[15] In other words, the number of AIDS cases may double with one fell swoop, not because AIDS has suddenly spread to new risk groups or even because it has spread within acknowledged risk groups but by definitional fiat.

It is worth putting these developments in historical perspective. Mirko Grmek, a French physician and historian of medicine, notes in his *History of AIDS* that AIDS "is not a disease in the old sense of the word, inasmuch as the virus is immunopathogenic, that it affects the immune system and produces symptoms only through the expedient of opportunistic infection or malignancy. . . . Its pathological manifestations could not even have been understood as a disease before the advent of new concepts resulting from recent developments in the life sciences. In the past, a disease was defined either by clinical symptoms or by pathological lesions, which are morphological changes in organs, tissues, or cells. Nothing of the sort, neither clinical symptoms nor lesions, observable by the old means, characterizes AIDS. It is not a disease in the sense given to the term before the mid-twentieth century. Persons affected by HIV virus suffer and die with the signs and lesions that are typical of other diseases. As recently as twenty years ago, these opportunistic disorders were the only reality that physicians could observe and conceptualize."[16] In other words, AIDS is new not only in the sense that it was only recently recognized; AIDS is also new in the way that biomedical researchers have defined it. These are important points to remember when we try to determine what AIDS is, what causes it, and whether its causes are in fact new. After all, if the biomedical tools and concepts did not, as Grmek asserts, exist twenty years ago for recognizing AIDS, how could it have been observed even if it had existed?

The schizophrenic and metamorphic nature of the definition of AIDS are of considerable importance in evaluating the possible cause or causes of the syndrome. Consider an analogy. A man drowns. The pathologist finds that he has much too much carbon dioxide in his blood. From a purely factual standpoint, we know that too great a percentage of carbon dioxide in the air one breathes can be fatal. This is the point of the rebreathers that divers sometimes use; they absorb the carbon dioxide from the air supply, allowing prolonged reuse of the air. We also know that when people drown, the level of carbon dioxide in their blood increases dramatically since their cells continue to respire even when their lungs cease to exhale. Yet it does gross injustice to logic to maintain that the level of carbon dioxide in a drowned man's blood is his cause of death. One must take a step back and ask why the man's carbon dioxide level became so high; that reason, quite clearly, is that he could not breath; he could neither exhale nor inhale. Thus, the high level of carbon dioxide in his blood is what is known to pathologists and philosophers of science alike as an *epiphenomenon*—a secondary or additional symptom or complication arising during the course of a malady, treatment, or experiment. Clearly the drowned man had many problems besides this buildup of carbon dioxide. For instance, he also ran out of available oxygen, a problem at least as severe as the increase in carbon dioxide levels that he experienced. Yet neither the buildup of carbon dioxide nor the lack of oxygen is, in a purely logical sense, the primary "cause" of death. Indeed, there is no single cause of drowning, no matter how similar the outcome. At the most fundamental level, the man drowned because he could not swim, because he got a cramp that incapacitated him, because he had a heart attack, because he struck his head on something and passed out, because someone held his head under the water until he was unconscious, or any number of other reasons. In short, the existence of high levels of carbon dioxide in the man's blood is factually correct, it is a finding invariably present in drowning victims and extremely rare in other people, but it is most definitely not the primary cause of death.

The drowned-man analogy is highly relevant to understanding AIDS. We must be absolutely certain that HIV is not an epiphenomenon of AIDS before we assert that it is the primary cause. The fact that it is an extremely frequent finding in AIDS patients is not logically compelling. It is only suggestive. Other active infections, such as cytomegalovirus, are also nearly universal among AIDS patients. If both are correlated with AIDS, which is the cause? Or are both viruses reactivated by previous and perhaps more diverse causes of immune suppression? How do we know what is cause and what is effect?

The existence of the full range of AIDS symptoms and opportunistic infections in both HIV-free and HIV-infected transplant and cancer patients warns us that this logical caveat is one that must be acknowledged in AIDS. HIV infection may be an epiphenomenon of immune suppression rather than a necessary cause. Immune suppression may predispose people to HIV infection (just as it predisposes them to other opportunistic infections) rather than resulting from such an infection. I argue in the following chapters, in fact, that HIV may be just such an epiphenomenon. Every AIDS patient has multiple causes of immune suppression other than HIV, many of which precede HIV infection and some of which occur in the total absence of HIV. The existence of these largely unrecognized immunosuppressive agents in AIDS not only requires a rethinking of the definition of the syndrome as occurring mainly in people without previously identified causes of immune suppression but also necessitates a critical look at the role of HIV as a causative agent in AIDS.

Before turning to the adequacy of the arguments supporting HIV as the sole, necessary cause of AIDS, two final comments are necessary concerning the definition of AIDS. The effects of the definition changes go far beyond mere questions of who has AIDS or how it is to be diagnosed. Much of our public health policy rests upon calculations of how fast AIDS is growing and into what groups it seems to be spreading. Each time the definition of AIDS changes, all of these calculations change as well. Previously excluded people suddenly qualify as AIDS patients. Diagnoses skyrocket. The 1985 definition change resulted in about a 4 percent increase in the number of diagnoses, a small enough fraction that translates into 2,000 additional cases a year in the U.S. The 1987 revision resulted in about a 30 percent increase in diagnoses, or some 10,000 cases in 1988 and some 15,000 additional cases during 1991. The proposed 1992 definition may double the the number of diagnoses overnight. In consequence, a significant proportion of the continued explosive growth of AIDS throughout the past decade has been fueled not by the transmission of AIDS to new groups of people but rather by the inclusion of previously excluded groups of people into the category of AIDS. People fitting these revised definitions of AIDS had always existed, but they were not counted as AIDS cases. Indeed, prior to 1981, they were not even recognized. Thus, despite claims that AIDS is the worst plague since the Black Death of the Middle Ages, despite the fact that AIDS is now the tenth most common cause of death in the United States, and despite the fact that there are no new miracle cures for the most common causes of death—heart disease, cancers, diabetes, stroke, and accidents—life expectancy for people in the U.S. has increased every year since 1980 at an almost constant rate.[17] One could justifiably argue that the AIDS epidemic is due at least partially to the grouping of two dozen causes of death under one rubric rather than to a new disease.

Finally, it is imperative that one gaping lacuna in the AIDS definition be pointed out: There are no criteria listed in any definition of AIDS that allow for a person to fight off AIDS or to be cured of it. Once a person is diagnosed, he or she will have AIDS forever after, regardless of any improvement in state of health and regardless of whether death results from a non–AIDS associated disease (for example, heart disease or diabetes). This is another way in which the definition of AIDs is a medical novelty. A person has pneumonia as long as he or she is symptomatic and the germ causing the disease is present. Destroy the germ and eradicate the clinical symptoms, and the person is cured, regardless of the fact that both antibody to the germ and scarring of the lungs may persist for their lifetime. Even in slowly progressing diseases such as cancer or heart disease, five-year survival is often taken as tantamount to a cure if disease symptoms are essentially absent. No such criteria exist for AIDS, despite the fact that some AIDS patients are still alive a dozen years after diagnosis with Kaposi's sarcoma, *Pneumocystis* pneumonia, and other opportunistic diseases. As AIDS survivor Michael Callen writes in his inspirational book, *Surviving AIDS*,[18] long-term AIDS survival does occur, but no one, once diagnosed definitively with AIDS, has ever been taken off the lists kept by the CDC except at death. This makes AIDS the first disease that no one can survive, by definition. Not only is this description of AIDS logically bankrupt, it sends the demoralizing and inaccurate message to people with HIV or AIDS that they have a disease that is not worth fighting. A more legitimate, and more hopeful, definition must be devised.

The Case for HIV as the Cause of AIDS

Despite continued modifications of the definition of AIDS, most AIDS researchers "know" its cause. "HIV causes AIDS," wrote William Blattner, Robert Gallo, and Howard Temin in 1988.[1] In their opinion there is no room for doubt. This statement, and others so similar that the differences do not bear examination, appears almost daily in the press, in textbooks, in lectures, and in medical journals around the world. HIV is now often called the "AIDS virus." Clearly, the case for HIV's causing AIDS must be extremely powerful. What constitutes the case, and why is it so convincing?

The evidence that HIV causes AIDS has been summarized in key papers by Blattner, Gallo, and Temin in the journal *Science*, by Robin Weiss and Harold Jaffe in *Nature*, by Manfred Eigen in *Naturwissenschaften*, by Jonathan Weber in *New Scientist*, and by Alfred Evans and Harold Ginsburg in the *Journal of Acquired Immunodeficiency Syndromes*. Most of these men are leaders in the field of AIDS research, and all have eminent qualifications. Interestingly, all of their papers were written in response to critiques leveled at the HIV theory by the maverick virologist from Berkeley, Peter Duesberg, who does not believe that HIV can cause AIDS.[2] These refutations of Duesberg's criticisms represent the clearest and, in some cases, the only complete analysis of the arguments that have convinced most scientists that HIV is the sole cause of AIDS.

The primary evidence that HIV causes AIDS is that the virus and the disease were first found in temporal proximity. Dr. Jonathan Weber, a senior lecturer (the American equivalent of a full professor) at the Royal Postgraduate Medical School in London, has put the argument this way: "If an infectious agent caused AIDS, then scientists had to show that the existence of the agent in a given population predates the emergence of clinical AIDS in those people. In other words, the effect should not come before the cause. Furthermore, they had to show that the agent was newly introduced into these populations, and not something that has been around for some time."[3] Weber himself is convinced that these criteria have been satisfied in the case of AIDS by the demonstration that "studies of homosexual men and haemophiliacs revealed that antibodies to HIV began to appear in 1978, just before the first cases of AIDS appeared in the US. The 'cause' indeed predates the 'effect.'"[4]

The data seem quite convincing on this point. Retrospective studies of HIV antibody in Europe and North America have found it only in very recent samples. For example, Dr. David Madden and his associates tested 310 blood samples from pregnant women and other hospital patients collected between 1959 and 1964 in the United States. None was found to be HIV seropositive.[5] Similarly, the CDC performed tests on blood samples stored between 1972 and 1976 from 585 former intravenous drug abusers tested for hepatitis at Veterans Adminstration clinics on the East Coast. None was HIV positive. Significant numbers of HIV-positive blood samples do not begin to appear until 1978 in gay men and intravenous drug abusers and thereafter in hemophiliacs.[6]

The figures are similar for Europe. In a series of 716 blood samples from Italian drug users collected between 1978 and 1985 in Milan, Italy, none of the 15 samples from 1978 was HIV seropositive. Two of 28 samples from 1979 were positive; 5 of 79 from 1980, 10 of 121 in 1981, and 28 of 123 in 1982. By 1985, 2 of every three Milanese drug addicts were HIV positive.[7] Serial blood samples from 12 Scottish and Danish hemophiliacs provide corroborating evidence. Blood samples available for 4 of the individuals prior to 1980 contained no HIV antibody. All of the individuals were HIV seropositive by 1984.[8]

The recognition and spread of AIDS followed hard on the heels of the spread of HIV. D. M. Auerbach and his associates at the CDC Task Force on Kaposi's Sarcoma and Opportunistic Infections reviewed tumor registries across the United States and found that the incidence of Kaposi's sarcoma had shown a marked increase only in 1979 and that only one or two cases of *Pneumocystis* pneumonia fitting the description of AIDS could be located in the CDC's registry prior to 1979. The CDC officially concluded, therefore, that AIDS was a new disease dating from about 1979.[9] This conclusion was strongly supported by retrospective studies of the incidence of *Pneumocystis* pneumonia and Kaposi's sarcoma among various risk groups, such as hemophiliacs and drug addicts, during the 1970s.[10] Perhaps most convincing was evidence that AIDS was not present in hemophiliacs until several years after HIV and AIDS were first found in homosexuals, suggesting that HIV was spread to hemophiliacs through blood donated by infected individuals. For example, Dr. J. M. Mason and his associates performed an extensive search for *Pneumocystis carinii* pneumonia cases among hemophiliacs in CDC records and the published medical literature for the period prior to 1983 and found only one relevant case.[11] The CDC itself listed no cases of AIDS among hemophiliacs until the last quarter of 1981 (a single case) and a mere seven cases in 1982.[12]

The correlation between the first observations of HIV and the first observations of AIDS is bolstered by analysis of the evolution of HIV itself. Intensive studies by virologists of the gene sequences of hundreds of isolates of HIV have shown that it mutates extremely rapidly—perhaps ten to a hundred times faster than influenza viruses. By comparing the mutations, these virologists can put together family trees of how they are related. These family trees suggest that if the current rates of mutation have been taking place for several decades, then HIV 1, the basic type of retrovirus found in Europe and the Americas, probably diverged from some prototype virus no more than twenty to thirty years ago. HIV 2, the retrovirus type found most frequently in Africa, diverged from its prototype somewhat earlier, possibly explaining why AIDS is older in Africa than elsewhere in the world. Similarities between both HIV 1 and HIV 2 to simian forms of HIV found in Africa argue strongly for an African origin for AIDS. These simian forms of HIV also cause an AIDS-like disease in some species of monkeys. HIV may therefore be nothing more than a new form of simian HIV that has a particular affinity for human beings.[13]

A third very strong argument touted by all HIV-proponents is that AIDS occurs only in populations in which HIV is present. To quote Blattner, Gallo, and Temin: "The strongest evidence that HIV causes AIDS comes from prospective epidemiological studies that document the absolute requirement for HIV infection for the development of AIDS. . . . AIDS and HIV infection are clustered in the same population groups and in specific geographic locations and in time. Numerous studies have shown that in countries with no persons with HIV antibodies there is no AIDS and in countries with many persons with HIV antibodies there is much AIDS. Additionally, the time of occurrence of AIDS in each country is correlated with the time of introduction of HIV into that country; first HIV is introduced, then AIDS appears."[14]

According to those who follow the HIV dogma, there is no doubt that AIDS is transmissible and that it must be caused by a microorganism such as HIV. "The infectious, transmissible nature of the underlying cause of AIDS has been apparent from early contact-tracing studies."[15] Early studies by the CDC demonstrated, for example, that AIDS developed in clusters of men associated by sexual ties. Thus, of the first nineteen patients, nine were patients in southern California who had had contact with at least one other person who developed AIDS, and one resident of another state was sexually linked to eight additional cases.[16] The subsequent discovery of hemophilia and transfusion cases bolstered the argument for a transmissible agent. It is often asserted that every case of transfusion- or hemophilia-associated AIDS has been traced to HIV-contaminated blood donated by someone who either had AIDS or subsequently developed it. (In fact, as we have seen in some of the organ transplant cases, there are exceptions in which the source of the HIV can never be documented.)

Some investigators, including Peter Fischinger, Dani Bolognesi, and Jonas Salk, cite inadvertent transmission of HIV to blood transfusion patients and hemophiliacs, followed by the development of AIDS, as proof of the causative nature of HIV. They claim that people exposed to HIV develop AIDS; people not exposed do not.[17] The same argument is often made *mutatis mutandis* for infants who develop AIDS following infection with HIV acquired from their mothers. Such infants, it is argued, have no risk for AIDS other than HIV, and therefore their development of the syndrome following HIV infection is proof that the retrovirus is sufficient to cause AIDS. Moreover, "HIV is the singular common factor that is shared between AIDS cases in gay men in San Francisco, well nourished young women in Uganda, haemophiliacs in Japan and children in Romanian orphanages."[18]

"The tight association of HIV and AIDS in risk groups is not the only causal evidence. Case-control studies show that within risk groups it is only the HIV infected who develop AIDS (for example among haemophiliacs and their wives, or in the infected infants compared to the uninfected siblings of HIV-positive mothers)," according to Weiss and Jaffe.[19] Similarly, Roger Detels reported that a study of 5,000 homosexual and bisexual men at four centers in the United States demonstrated that HIV-seronegative men never developed AIDS, whereas 5 percent of the HIV-seropositive men developed AIDS each year.[20] Thus, AIDS appears where HIV infects and is never found where HIV is absent, according to these experts.

HIV can also explain how the immune system is destroyed as AIDS progresses. The immune system consists of several layers. First, there is the skin, mucous membranes, gastrointestinal tract, and respiratory tract, which function to prevent foreign material from entering the body. Any foreign material that does get in, and can elicit an immune response, is called an antigen. Antigens can be processed in several ways. First, cell scavengers called macrophages (literally, "big eaters") may engulf the antigen through a process called phagocytosis ("cellular eating"). Second, cell-mediated immunity may be induced, in which T lymphocytes (the "T" is for "thymus activated") may directly attack the foreign material. Finally, B cells, so-called because they were isolated first from the bursa of Fabricius in chickens and are activated in bone marrow in human beings, may be stimulated to produce antibodies. Antibodies are specialized proteins that are secreted into the blood and lymph to bind specifically to the foreign material. Once the antibody binds to the antigens, a complex chemical reaction involving the complement system is initiated to destroy the antigen. AIDS involves the relatively specific destruction of macrophages and a special type of T cell called a T-helper cell. HIV, not coincidentally, infects macrophages and T-helper cells.

The destruction of macrophages and T-helper cells by HIV is considered to be the primary problem underlying AIDS. The reason why can quickly be grasped when the normal function of the immune system is examined. To ensure health, T cells, B cells, macrophages, and other types of immune system cells participate in a process as complex as a security check at a super secret military installation. The immune system must differentiate cells, proteins, and other chemicals that are produced by the body itself ("self") and those that are introduced from outside the body ("nonself"). The distinction between "self" and "nonself" is much the same as having a security pass or not having a security pass (Figure 1) The process of evaluation usually begins when macrophages or other cells ingest the material. Viruses and bacteria will be engulfed and enzymatically broken down. Proteins will be chopped into small fragments called peptides. These fragments—the antigens—will then induce various chemical changes in the

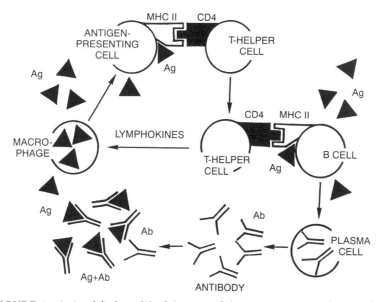

FIGURE 1. A simplified model of the part of the immune system that regulates T and B lymphocyte activity. Ag = antigen; MHC II = MHC class II proteins; CD4 = a protein characteristic of T-helper lymphocytes and macrophages that is used by HIV to recognize and invade these cells; Ab = antibody.

macrophage that transform it into an antigen-presenting cell (APC). The antigen is literally presented on the surface of the cell complex to a protein aggregate called a major histocompatibility complex (MHC). The MHC system is responsible for determining whether the antigen is "self" or "nonself." It is, in a sense, like a magnetic card reader or fingerprint identifier in a security system. If the antigen has a security clearance, no further action is taken; the antigen is allowed to be on the premises, as it were. If, however, the antigen is found to be without an appropriate security pass (it is foreign), then the MHC system will initiate an immune response to eliminate it.

The MHC system initiates an immune response by calling in T or B cells—the immune system's equivalent of security guards. Often the first type of cell activated is the T-helper cell, which mediates all the rest of the immune system interactions. Initiation of T-helper cell activity occurs when a chemical interaction occurs between the antigen, a specific type of MHC protein called MHC class II (or HLA-D), and a protein called CD4 on the T-helper cell. In other words, the alarm system has to be activated, the foreigner identified, and the security guards properly briefed before any action can be taken. When MHC class II, antigen, and CD4 protein form a complex, the CD4-bearing T-helper cell begins to clone itself (that is, it makes many identical copies of itself). In essence, an army of identical cells is created, all specifically trained to kill the newly identified invader. As these clones are produced, they secrete special proteins called lymphokines that stimulate the production and activity of more macrophages (security-identifying devices). Security is beefed up. The more antigen there is, the more APCs are formed, and the more T-helper cells are activated.

A similar interaction between T-helper cells, the foreign antigen, and B cells (which also have MHC proteins on their cell surfaces) can initiate a second phase of the immune response in which the B cells are transformed into plasma cells. These plasma cells clone themselves, and each clone produces antibody specifically targeted at the foreign antigen. Unlike the T cells that must directly attack the foreign invader in hand-to-hand combat, as it were, antibodies are secreted into the blood and lymph, where they can act independently. They are like long-range, target-identifying bullets. One other type of cell, the "natural killer"

(NK) cell, may also participate in eliminating the antigen. NK cells are apparently like special forces commandos that operate independent of the MHC security system. This part of the system stimulates a quick and effective immune response.

To prevent the immune system from overreacting to an antigen, the network of interactions just described can also control itself. As antibody is secreted, the amount of antigen present in the body drops; the number of macrophages that encounter antigen decreases; the number of APCs decreases correspondingly; T-helper cell cloning is slowed; B cell proliferation ceases; and antibody levels decline. Thus, the degree of immune response is regulated by the amount of antigen present. A second set of T cells, called T-suppressor lymphocytes, are also involved in regulating the immune response. The more T-suppressors that are activated, the less actively the immune system can attack a potential antigen. T-suppressor cells are therefore thought to protect us from accidentally attacking our own bodies. If they are overly stimulated, they can interfere with a proper immune response. Such overstimulation may occur in AIDS.

Once the security network has eliminated the antigen and disengaged from battle, a few cloned cells are retained forever after as a sort of physical memory of the successful elimination of the invader. They are the seasoned security guards who remain at the installation after the rest of the force is decommissioned. These memory cells "remember" what the invading antigen "looked like" and can mount a quick, massive immune response if it returns. Thus, these memory cells are responsible for the protection against infection afforded by vaccination or previous infection.

One of the reasons that HIV is such a good candidate for explaining AIDS is that it infects just those cells that are depleted most significantly during AIDS. HIV actually uses the same CD4 protein that is involved in the MHC-antigen recognition system to invade T-helper cells. In other words, HIV masquerades as the part of the immunological security system that identifies "self" from "nonself." To return to our analogy, if MHC is a key that is necessary to turn the CD4 lock that initiates cloning of T-helper cells, then HIV uses the same key. HIV, however, does not turn on T-helper cells; it invades them and eventually kills them.

The results of interfering with T-helper cell activity are devastating to the immune system. In a typical healthy human being, the T lymphocytes in blood and lymph will consist of approximately 30 percent T-suppressor cells (which are present to regulate immune responses), 60 percent T-helper cells, 2 percent natural killer cells, and a variety of other cell types. Each of these cell types can be identified by the presence of some particular type of protein on its cell surface, just as different parts of the armed forces have different uniforms or badges. T-helper cells have a protein called CD4, T-suppressor cells have a protein called CD8, and natural killer cells have a protein called CD11 (Figure 2). Scientists have made antibodies to each of these proteins. Each antibody binds only to its particular protein antigen but not to the others. Thus, it is possible to extract the white blood cells from a sample of a person's blood and to count the number of lymphocytes attacked by each type of antibody. The typical healthy human being has been found to have about twice as many T-helper

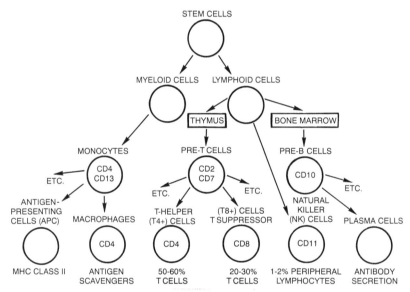

FIGURE 2. A simplified chart of the development of selected white blood cells (leukocytes) from stem cells in bone marrow. Some of these cells are activated in the bone marrow to become antibody-producing B lymphocytes. Others are activated in the thymus to become T lymphocytes. Others, the myeloid cells, are inmvolved in antigen scavenging and presentation. Each type of leukocyte is characterized by specific proteins found on their cell surfaces: CD4, CD8, CD13, MHC class II, and so forth. Some of these proteins are used by viruses to recognize and invade these cells. Some are also targets for autoimmune processes in which B cells secrete antibodies that attack T cells.

(or T4) cells as T-suppressor (or T8) cells, so that a typical T-helper/T-suppressor ratio is around two. In contrast, AIDS patients almost invariably have ratios of less than one. In other words, instead of having twice as many T-helper cells as T-suppressor cells, they have half as many T-helper cells as T-suppressor cells. The T-helper cells have been destroyed (and in some cases, T-suppressor cells augmented)—a clear diagnostic sign of AIDS or AIDS-related complex. HIV also invades other cells besides T-helper cells that share the CD4 protein on their cell surfaces. These include macrophages, dendritic cells, and other monocytes involved in antigen scavanging and presentation.[21]

The devastation caused by the loss of two-thirds or more of the body's T-helper cells and macrophages can be imagined by tracing the effects on the immunologic security system. First, there are far fewer macrophages and therefore fewer APCs to stimulate T-helper cells. Simultaneously, T-helper cells are being destroyed, and there are fewer of them to clone. The fewer T-helper cells there are, the less lymphokine is produced to stimulate the remaining macrophages and the fewer interactions are possible with B cells. Thus, less antigen scavenging and less antigen-specific antibody production will result. The number of HIV-infected cells and the amount of HIV circulating in the body will continue to increase. As it does, the immune system becomes less and less able to react to other infections. Finally, even diseases that a healthy immune system can easily handle—opportunistic diseases—will be able to establish themselves in the body with impunity. This is AIDS.

Problems with the Evidence That HIV Causes AIDS

As seductive and compelling as are the arguments supporting HIV, they reveal a very different face when the flattering makeup of overgeneralization and dogma are removed and the bare facts revealed. The problems begin with the set of five anomalies outlined in Chapter 1. If HIV is new, then why is there evidence of HIV in people in Europe and the Americas with all of the symptoms of AIDS back to at least 1959? What are the causes of AIDS-like syndromes in people recorded in the medical literature for nearly a century prior to that? How can an epidemic be reconciled with a disease agent that is so difficult to transmit? Why do healthy people seem to be less susceptible to HIV infection, and resistant to the subsequent development of AIDS, while drug abusers, some homosexual men, and other risk groups seem to be very susceptible? In short, why has AIDS remained among identified risk groups? When these anomalies are contrasted with the conclusions reached by those investigators proposing that both HIV and AIDS are new and that anyone can get AIDS, then outright contradictions result. Either the anomalous data are false, or the studies claiming that HIV and AIDS are new are flawed. I argue the latter.

The evidence that HIV is new looks impressive until it is analyzed closely. The retrospective study that has looked back the furthest is that of 310 blood samples collected over a period of six years from American women pregnant between 1959 and 1964. None had evidence of HIV. These studies, however, averaged only about 50 blood samples per year. The incidence of HIV among heterosexual women today is estimated to be about 1 per 1,000 per year in the U.S. Thus, the probability that one would find an HIV-positive blood sample from a pregnant woman in a batch of 50 is extremely small (about 1 in 20). The sample size is too small to be meaningful. If HIV were less prevalent two decades ago (as is highly probable), then many more blood specimens would have to be tested to be statistically significant.

The data on Milanese drug abusers are also flawed. Recall that the numbers of HIV positives grew from zero in 1978 to two of every three in 1985. Another factor changed as well, however: none of the fifteen drug abusers tested from 1978 was an intravenous abuser, whereas all of the 1985 group were. Thus, rather than demonstrating that HIV is a new virus, these data may document instead profound changes in how drugs were abused that affected the transmission of HIV.[1] The Milanese data also suffer from problems of sample size. The investigators found 2 HIV seropositives out of 28 addicts in 1979, for a seroprevalence of 7 percent. Five HIV positives in 1980 appears to be a significant increase until you realize that they tested 79 blood samples from that year. That equals 6 percent. Ten of 121 in 1981 equals 8 percent. Not until 1982, when the percentage jumps to 23, is there a significant change in the incidence of HIV seropositivity. In other words, despite the fact that the number of positive samples grew steadily from year to year, the rate of seropositivity among these drug addicts was constant between 1979 and 1981. This figure may therefore represent a baseline rate of HIV infection prior to the AIDS epidemic. If so, then only one of the fifteen drug addicts from 1978 would be expected to have been seropositive even if all had been intravenous abusers. This individual could easily have been missed. As Mirko Grmek notes in his *History of AIDS*, negative results cannot prove that HIV was not present— only that it was not found.[2] Lack of evidence is, logically, never a basis for arguing anything except the fact of our own ignorance. The only data that count are the positive reports that indicate that HIV is not new.

Consideration of sample size suggests that a valid study of random blood samples must have many hundreds or, preferably, thousands of samples. The only study that meets these requirements was conducted by James Moore and his associates, who found that between 1 and 4 percent (depending on the test employed) of 1,129 blood samples drawn from drug addicts in the United States during 1971 and 1972 were HIV positive.[3] This study, plus the individual HIV-positive cases from the 1950s and 1960s that were discussed in Chapter 1 convince me that HIV is not a new virus but rather one that is being spread by means of new modes of transmission that dramatically increased its prevalence beginning a decade or so ago.

But what about the many studies that purport to demonstrate that AIDS itself (as distinct from HIV) is new? Not only has Huminer's, Katner and Pankey's, and my historical search through the medical literature cast doubt upon this conclusion, flaws also invalidate claims that AIDS is new based on retrospective studies of the incidence of *Pneumocystis carinii* pneumonia (PCP) and Kaposi's sarcoma.

All retrospective analyses of AIDS-like PCP incidence, save for the one summarized in Chapter 1 of this book, have relied upon Centers for Disease Control records of pentamidine requests. Pentamidine is a drug used to treat PCP. The CDC has had complete control over its prescription and therefore has all available records. These records show that there was only one AIDS-like patient between 1967 and 1970[4] and one patient between 1976 and 1980.[5] The period between 1970 and 1976 has apparently not been studied, but at least one AIDS-like case treated with pentamidine was reported in the medical literature in 1973.[6] Thus, all that is known for certain is that at least three cases of AIDS-like PCP were treated with pentamidine between 1967 and 1979. Since this number is so obviously small, the sudden jump in the number of requests for pentamidine in previously healthy young adults to forty-one in 1981 is often cited as strong evidence that AIDS did not exist prior to that time.[7]

Unfortunately, this conclusion is highly suspect. Pentamidine-treated cases represent a small minority of PCP cases. Physicians preferred (and many still prefer) to prescribe trimethoprim (TMP) combined with the sulfa drug, sulphamethoxazole (SMX).[8] No records are kept of TMP-SMX prescriptions, but some sense of the relative numbers compared with pentamidine can be gleaned from the fact that only one of the first fourteen AIDS patients who developed PCP in 1980 and 1981 was treated with pentamidine. The other thirteen were treated with TMP-SMX.[9] Thus, utilizing CDC records of pentamidine requests is not only inaccurate for the period preceding 1980, it is also grossly inaccurate for the early AIDS period. For every pentamidine-treated AIDS-like case, there may have been between ten and twenty TMP-SMX cases. Certainly, a significant proportion of AIDS-like PCP existed in the pre-AIDS era, as I have demonstrated in Chapter 1.

The Kaposi's sarcoma (KS) story is almost identical to the PCP story. Retrospective studies of UCLA records by Dr. Fredrick W. Gilkey,[10] Danish cases,[11] Haitian cases,[12] and CDC records[13] all concluded that KS among young, previously healthy men was strikingly new. These studies, along with textbook pronouncements and review articles stating that KS was a normally non-fatal disease of elderly Jewish men or those of Eastern-European descent and almost never seen among young adults or women, helped cement the belief that AIDS itself was novel.[14]

Apparently no one bothered to look at the medical literature itself, or they would have found the hundreds of cases of KS satisfying the criteria for AIDS in the century prior to 1980 that I, Katner and Pankey, and other investigators have resurrected (Chapter 1). I estimate that no less than 10 percent of all cases of KS reported in the medical literature prior to AIDS satisfy the criteria for AIDS. The Surveillance, Epidemiology, and End Results (SEER) Program of the National Cancer Institute demonstrated that the incidence of KS for the years 1973 to 1980 was 36 per 10,000,000 people in the U. S.[15] or about 750 cases per year. Seventy-five or more of these may have qualified as AIDS cases each year—sufficient numbers to explain all the AIDS cases diagnosed during the first three years of the "epidemic." Johannes Clemmesen, a former head of the Danish national cancer registry, has warned that retrospective studies of such data banks are likely to lead to a significant underestimation of cases, due to changes in how KS is classified and diagnosed, so that the true number of cases may be even higher.[16]

Ample evidence exists that an unrecognized pool of KS cases did, indeed, exist. The SEER program data mentioned above was reanalyzed by Dr. Robert J. Biggar and his colleagues in 1984 to determine the incidence of KS prior to 1980 in various participating cities such as San Francisco, Denver, and Atlanta. They found that by 1973—the earliest year studied—the incidence of KS in San Francisco, where AIDS was soon to be identified, was already five to ten times that of cities such as Atlanta and Denver and mid-western cities where AIDS was and is relatively rare. The vast majority of these San Francisco cases were among middle-aged and older, single men, many of whom identified themselves as gay. In other words, the incidence of Kaposi's in homosexual males was already significantly elevated for at least a decade prior to the recognition of AIDS, and a clear correlation between sexual preference and age (and therefore probable number of homosexual partners) was evident. Unfortunately, neither New York nor Los Angeles were included in the SEER program, and no data are available for these cities.

AIDS-like KS, in other words, was clearly emerging in San Francisco, if not elsewhere, at least a decade prior to its recognition. Now, this is not to say that the incidence of KS did not increase dramatically after 1981 and with the recognition of AIDS as a reportable disease by the CDC. It certainly did. The SEER data show a sudden and dramatic increase in diagnoses of KS beginning in 1981 that are mirrored in CDC data. But data can be misleading if not placed in the proper context. In this case, the same study that analyzed the SEER program for the incidence of KS also analyzed the data concerning another cancer of the skin called mycosis fungoides. Mycosis fungoides (MF) is unrelated to AIDS and therefore provides a useful control for interpreting the increase in diagnoses of KS after 1981. Amazingly, diagnoses of MF in San Francisco increased at the same rate, at the same time, and in the same cities as diagnoses of KS. There were parallel "epidemics" of KS and MF in exactly the same years. Or were there?

There is an alternative explanation of the KS-MF data. When a new disease is first identified, the disease-causing agent is not the only thing that can spread quickly: so does information about how to recognize and diagnose its effects. There is absolutely no doubt that news of the correlation between homosexuality, KS, and PCP spread extremely rapidly within the medical and gay communities in 1981. It is therefore a reasonable conjecture that both physicians and homosexual men suddenly found intense interest in previously unexceptional and uninteresting skin lesions. Gay men probably consulted their physicians in unusual numbers. Taking no chances, physicians probably ordered a biopsy of every such lesion from every young or single man, so that the rate of diagnosis of every type of skin cancer rose tremendously. As a result, many cases of KS, and incidentally MF too, that had previously been ignored or overlooked, were diagnosed in epidemic proportions. All this because new diagnostic information about who was at risk for KS had been disseminated.[17] The same story appears to help explain the sudden rise in diagnoses of other opportunistic diseases associated with AIDS. Physicians told to look for PCP in homosexuals and drug abusers began looking for it for the first time, and not surprisingly, found it.

Even the studies showing that AIDS in hemophiliacs only appeared after AIDS in other risk groups, as clean and convincing as it seems, may be grossly inaccurate. During January of 1983, more than a year before HIV was announced as the cause of AIDS, Robin Henig of the *New York Times* interviewed Dr. David L. Aronson about AIDS in hemophiliacs. Dr. Aronson was the director of the coagulation branch Division of Blood and Blood Products of the National Center for Drugs and Biologics, a part of the Food and Drug Administration in Bethesda, Maryland. When asked if some new virus or other infectious agent had invaded the blood supply of the nation, Dr. Aronson replied, "I can't see anything different in the death patterns of hemophiliacs now from four years ago [1979]."[18] He explained that during the 1970s five to ten hemophiliacs were dying each year of opportunistic infections and rare cancers that would now qualify the patients as having had AIDS.

This was a shocking statement, since at that time the CDC listed the first case of hemophilia-associated AIDS as occuring at the end of 1981, and did not record more than ten cases per year until 1983. Dr. Aronson, however, published a retrospective study of the causes of death among hemophiliacs between 1968 and 1979 that argued otherwise.[19] He found that of 949 deaths recorded during that period, most were attributed to "hemophilia," but two were due to fatal *Candida* infections, eight to hepatitis, and sixty-six to forms of pneumonia due to unidentified infectious agents. Some or all of these may have been due to undiagnosed *Pneumocystis* pneumonia. The cause of death was not known for an additional ninety individuals, who may have had extremely unusual diseases. The sixty-six cases of pneumonia are very striking, since they are far in excess of the incidence of pneumonia among healthy Americans, and most of these hemophilia deaths occurred in patients under the age of forty-five, whereas the majority of pneumonia deaths in the general population occur in individuals over the age of forty-five. If all of these cases were actually AIDS, then as many as seventeen American hemophiliacs were dying each year of AIDS by the beginning of the 1970s. Even if the number is only a quarter of this, it is still more AIDS cases than were actually diagnosed among hemophiliacs until 1983. Similar retrospective results were reported by Dr. C. R. Rizza and Dr. Rosemary Spooner in the United Kingdom for the years 1976 to 1980.[20]

How were these cases missed? Easily. Even in 1985, the incidence of AIDS-like diseases was only one in a thousand hemophiliacs in both the United States and Europe.[21] AIDS is—and was— a very minor problem compared with the much more common life-threatening complications that many hemophiliacs face. Even an expert on hemophilia would not have been likely to see more than a couple of AIDS-like cases in his or her lifetime and, in the absence of clear diagnostic criteria, could not have recognized their significance. Thus, we cannot be sure that either AIDS or HIV is nearly as new as most investigators insist. Notably, then, a retrospective study of possible pre-1979 AIDS cases carried out by the Study Group on the Epidemiology of AIDS in France, under the director general of health, reached a quite different conclusion from its American counterparts. Nine of the first twenty-nine cases of AIDS identified by this group were diagnosed between 1974 and 1980. "This suggests," they concluded, "that the illness and the supposed infectious agent(s) are not new in France. Only its recent epidemiological features, in the USA and by the same token in France, are new."[22]

Other fundamental issues are also less certain than they at first seem. For example, Blattner, Gallo, Temin, Weiss, and Jaffe all agree on "the absolute requirement for HIV infection for the development of AIDS" (to quote Blattner, Gallo, and Temin) or, in Weiss and Jaffe's words, "only the HIV infected . . . develop AIDS." And yet both groups have to admit that these absolute statements may not be true. A few paragraphs after stating that HIV is absolutely necessary, Blattner and his colleagues note that even in Gallo's own laboratory, only slightly over 90 percent of AIDS patients can be shown to be actively infected with HIV.[23] Weiss and Jaffe similarly hedge: "Because other causes of immune suppression also give rise to similar symptoms, there will be HIV-negative individuals who fall within the CDC definition of AIDS. The relative risk of AIDS among HIV-infected people, however, compared with HIV-negative people in each risk group is much greater than the relative risk of lung cancer in smokers [versus nonsmokers]."[24] Their analogy begs the question. Physicians know that lung cancer has many causes other then smoking (such as asbestos exposure). Why is it so difficult for them to admit, similarly, that AIDS may have more than one cause?

Blattner, Gallo, and Temin would undoubtedly respond that other causes have been investigated and demonstrated to be irrelevant to AIDS. They say, for example, that one of the major arguments in favor of HIV as the sole cause of AIDS is "that HIV infection, and not infection with any other infectious agent, is linked to blood transfusion-associated AIDS." They document as their source for this information a paper by their colleagues at the CDC, Thomas A. Peterman, Rand Stoneburner, James R. Allen, Harold Jaffe, and James W. Curran.[25] Not only is this paper not a study of transfusion-associated cases (it reports on transmission of HIV from people who acquired AIDS through blood transfusions to their spouses), but there is nothing in the article indicating that Peterman and his colleagues performed any tests or analyzed any data for any infectious agent other than HIV. Moreover, no paper on blood transfusion patients produced from the CDC by these authors during the previous three years had studied the possible role of non-HIV immunosuppressive factors in AIDS. In fact, as we will see in subsequent chapters, there is very good evidence from studies by other laboratories for other factors being involved in susceptibility to AIDS in these and all other AIDS risk groups.

Another argument proposed by Blattner, Gallo, and Temin to bolster the plausibility that HIV causes AIDS also fails due to their inaccurate reporting of data. These three scientists contend, correctly if somewhat simplistically, that "scientists conclude that a virus causes a disease if the virus is consistently associated with the disease and if disruption of transmission of the virus prevents occurence of the disease."[26] This line of reasoning discards many other criteria that most scientists believe are necessary to demonstrate disease causation, as we shall see shortly, but the basic logic is correct. Gallo and his colleagues claimed in 1988 that just such interruption had been demonstrated for HIV: "As a result of the decrease in blood transfusion–associated transmission of HIV [due to HIV screening beginning in 1984], the incidence of blood transfusion-associated AIDS among U.S. newborns showed a decline. This decline in pediatric AIDS became evident before that in adult AIDS because of the shorter latent period for AIDS in infants."[27] If this were true, it might represent evidence supporting HIV as a cause of AIDS, if it could also be demonstrated that no other virus (such as cytomegalovirus) was eliminated from the blood supply by the same screening procedures. (In fact, screening for cytomegalovirus and Epstein-Barr virus—the primary cause of infectious mononucleosis—has been implemented.)

Again, the CDC data available to Blattner, Gallo, and Temin at that time belie their claim. Adult cases of transfusion-associated AIDS doubled to 752 cases in the year ending May 1988, and pediatric cases tripled to 63 over the same period from a year before.[28] Between January 1 and December 31, 1988, a total of 66 transfusion cases in infants were recorded, and another 40 the following year. Adult cases increased to 777 during 1989.[29] These figures increased for adults to 866 in 1990 and stayed the same (39 cases) among children. Thus, Blattner, Gallo, and Temin had no basis for claiming in 1988 that screening blood for HIV had dramatically lowered the rate of AIDS in transfusion-associated cases. This is not to say that screening blood for HIV (and hepatitis B virus, cytomegalovirus, and Epstein-Barr virus) is either useless or ineffective, for the numbers of such cases might have been much higher today had such testing not been instituted, but the data certainly do not support the contention that HIV screening has ended the risk of AIDS for transfusion patients.

Even the epidemiologic studies claiming that AIDS is sexually transmissible are not nearly as strong as the HIV proponents claim. At the same time that the CDC and other groups studying homosexual men were claiming that contact tracing studies and the incidence of AIDS proved that a sexually transmissible agent must be the cause, Dr. Joan Kreiss and her collaborators at the Wadsworth Veterans Administration Hospital in Los Angeles reported that they could find no evidence of transmission of T-lymphocyte abnormalities from hemophiliacs with AIDS to their spouses.[30] Even today, despite an estimated 15,000 HIV-infected hemophiliacs in the United States, such cases are extremely rare. By July, 1991 only ninety-three cases of AIDS among spouses of hemophiliacs had been identified during the entire AIDS epidemic in the U.S.[31] As we shall see later in this book, most of these cases are associated with known immunosuppressive risks for the sexual partner, including old age, anal intercourse, and corticosteroid treatments. These spouses also have unusual access to needles, syringes, and various drugs.

Even the data demonstrating homosexual cases linked by sexual contact do not stand up to scrutiny. The majority of men with AIDS in southern California and New York City diagnosed by the end of 1982 could not be linked by sexual contact (40 linked versus 208 nonlinked). Sixty-five percent of the linked men had had more than 1,000 lifetime sexual contacts and lived near one another, making the probability of random linkage (as opposed to actual infection with an AIDS-causing agent) reasonably high. In fact, the average number of lifetime sexual partners for all AIDS patients at that time was over 600, the majority had engaged in rectal intercourse, most had used inhalant nitrites more than 1,000 times during their lives as well as other drugs, and all had had multiple venereal diseases. The CDC therefore concluded that although there was presumptive evidence of an infectious agent, "the cluster [of linked AIDS cases] may represent a group of homosexual men who were brought together by a common interest in sexual relations with many different partners or in specific sexual practices, such as manual-rectal intercourse ["fisting"]. Frequent social contacts among some patients enabled them to identify other patients by name. Although these men were sexual partners of each other, nonsexual activities, such as drug use, may have contributed to the development of AIDS."[32]

Other epidemiological problems also exist. No one has explained, for example, why AIDS has remained almost totally within defined high-risk groups, and particularly why female prostitutes are not vectors for spreading HIV. If AIDS is a simple, sexually transmitted virus, then it should be running rampant in the heterosexual community by now. Even among HIV-infected people, vast differences in AIDS risk have been identified. Berkeley virologist and member of the U.S. National Academy of Sciences Harry Rubin summarized some of the relevant data in 1988: "Although the case for HIV causation of AIDS is largely based on guilt by epidemiological association, that same epidemiology raises serious questions about the rigor of the association. The rate of onset of clinical AIDS in the high-risk groups is 5 percent per year. The Public Health Service estimates that there are 1 to 1.5 million HIV-positive individuals in the country. If they were at equal risk of developing the disease as the HIV-positive individuals of the high-risk groups, there should be 50,000 to 75,000 new cases per year, but the estimated number for 1986 was 16,000. The three- to fivefold shortfall of cases implies that the probability of developing the disease in HIV-positive individuals varies greatly with the behavior of those individuals. It is not inconceivable that the probability approaches zero in heterosexuals with pure HIV infection who do not abuse drugs."[33] Rubin's analysis has been seconded and extended by other HIV skeptics, including British epidemiologist Gordon Stewart and American professor of preventative medicine, Dr. Steven Jonas.[34]

Even investigators well entrenched in the HIV camp have pointed out the same phenomenon. For example, in 1988, Dr. T. A Peterman, Dr. R. L. Stoneburner, and their colleagues at the CDC investigated the risk of HIV transmission from heterosexual adults infected through blood transfusions to their sexual partners. They found that some partners developed antibody against HIV after only a few unprotected sexual encounters, whereas the vast majority were HIV-seronegative after hundreds of instances of unprotected sex. They concluded that there must be some "as yet unexplained biologic variation in transmissibility or susceptibility" to HIV and AIDS.[35] Does this wide variation mean that HIV requires cofactors, as the French discoverer of HIV, Luc Montagnier, believes? Does HIV take advantage of a previously impaired immune system? Or does this data indicate that some forms of HIV are pathogenic and others are not? No matter how one answers these questions, it turns out that the HIV = AIDS equation is too simple.

Problems such as these have led other eminent scientists to caution their colleagues not to jump to too many conclusions about the role of HIV in AIDS. Three of the most notable of these cautionaries are Walter Gilbert and Manfred Eigen, both Nobel laureates, and Beverly E. Griffen, Director of the Department of Virology at the Royal Postgraduate Medical School, Hammersmith Hospital, London. Neither Gilbert nor Eigen is directly involved in AIDS research, but both have extensively reviewed the available literature and warn that premature consensus can have deleterious effects on scientific research. Eigen, in particular, concluded a 1989 article with these words: "The correlation between HIV and AIDS is undeniable. Epidemiology and evolutionary arguments substantiate the claim that HIV is necessary for AIDS. . . . What remains unsettled is the etiological [causation] argument. . . . It is only fair, as far as causation is concerned, to conclude that we need to know more about the pathogenetic mechanism [how HIV causes AIDS] before we can decide whether HIV is not only necessary but also sufficient to cause AIDS."[36] Dr. Griffen argues similarly that the case for HIV has not yet been clearly demonstrated and that the burden of proof still lies with those claiming that HIV is sufficient.[37]

Rather than receding as further research is performed, this skepticism has been growing. During 1991, Charles A. Thomas, Jr., a former professor of biochemistry at Harvard University and now president of the Helicon Foundation in San Diego, organized the Group for the Scientific Reappraisal of the HIV/AIDS Hypothesis. Gilbert, Griffen, and other AIDS mavericks such as Peter Duesberg and the inventor of the polymerase chain reaction technique used in AIDS testing, Kary Mullis, are members. A similar group, comprising health professionals, researchers, and government officials and called HIV Connection? was formed by Edward Vargas in San Francisco in 1992. And the first international conference on the subject, An Alternative View of AIDS, was held in Amsterdam under the leadership of the Dutch-based Foundation for Alternative AIDS Research during May 1992. Luc Montagnier was one of the keynote speakers. Questioning the role of HIV in

AIDS may not yet be a mainstream activity, but it is no longer done only behind closed doors.

One of the reasons that skepticism is mounting is the failure of HIV researchers to explain the most elementary facts of AIDS. Some of these mysteries are based on unexpected clinical findings. Immunologists and virologists expected to find that HIV causes its massive destruction by avoiding immunologic recognition, sneaking into T cells, and then slowly eating the immune sytem away from the inside out. On the contrary, recent studies have discovered that the antibody response to HIV is very effective at controlling infection and driving the virus into a latent form.[38] No one can explain why this antibody response is not sufficient subsequently to keep HIV in check.

A related problem is that the time between infection and the onset of AIDS seems to be increasing exponentially. As noted previously, the average length of time between HIV infection and the first opportunistic infection diagnostic for AIDS was about a year, and is now more than a dozen.[39] These are averages; some people still contract HIV and die within a year. Others apparently will be healthy two or more decades from now. The question is, Why is there such a huge and medically unprecedented variation? No theory based solely on HIV can explain this phenomenon.

Finally, and most important, the account that I gave about how HIV destroys T cells and macrophages—standard textbook fare—turns out to be based largely on speculation. As Dr. John Coffin of Tufts University said at the Sixth International Conference on AIDS in June 1990, "We do not yet know *how* HIV causes AIDS."[40] Similarly, Martin A. Nowak and his colleagues in Europe wrote in the November 15, 1991, issue of *Science*: "Much uncertainty still surrounds the processes governing the development of acquired immunodeficiency syndrome (AIDS), after an individual is infected with the human immunodeficiency viruses (HIV-1 and HIV-2)."[41] No one has corrected these investigators, or begged to be allowed to purge them of their ignorance. Indeed, one can find dozens of similar statements appear throughout the medical literature on AIDS by some of the most respected experts on HIV, including Robert Gallo, Anthony Fauci, and Luc Montagnier.[42] If anything, the problems seem to be getting worse with the passing years. An article in the November 1991 *Journal of NIH Research*, for example, states outright that studies of AIDS dementia (a loss of brain function that affects most people with AIDS) are foundering: "From the outset, the study of the neurological manifestations of infection by the human immunodeficiency virus-type 1 (HIV-1) has been beset with practical problems, experimental complexities, and theoretical improbabilities. As the years wear on, the problems seem to be growing more complicated."[43] Thus, the discovery of HIV, far from leading to the quick and deep insights into the nature of AIDS promised by Margaret Heckler and Robert Gallo in 1984, has led instead to increasing confusion.

Mysteries are, of course, the main course of science. Scientists dig into them with gusto, and none more so than HIV researchers who are well aware of the problems I have just summarized. Indeed, without these problems, there would be nothing for HIV researchers to do. The fact that we do not know how HIV creates such havoc in the immune system does not necessarily mean that it does not. It may simply mean that HIV is more clever than we have yet imagined. Thus, scientists keep formulating ingenious new possibilities to explain how HIV may cause AIDS. Most of these new theories have in common the notion that HIV does not actually cause T cell death directly, as was originally presumed, but that it somehow triggers indirect mechanisms of T cell death. For example, HIV may preferentially infect the stem cells that mature into T cells, thereby preventing T cells from developing in the first place. This is a plausible scenario—but with no evidence to back it. Another possibility is that many HIV-infected cells can form what are known as syncytia with other uninfected cells. These syncytia are connections to other T cells, which can result in clumps of dozens or even hundreds of cells. These clumps cannot perform their proper immune functions, and thus one or two HIV-infected cells may be able to inactivate large numbers of healthy cells. Why and how syncytia form is still a mystery, and why they form in the presence of some forms of HIV but not others remains a major puzzle.

Another indirect method of HIV activity has been proposed by a group at the University of Brescia in Italy. They claim that HIV may have "superantigens"—proteins that result in the selective elimination of particular types of immune system cells when inoculated into suckling mice or rats.[44] If HIV were a superantigen it might explain how it eliminates T-helper cells specifically. But superantigen activity has never been demonstrated in an adult animal, let alone in a human being; it will take extraordinary efforts to prove such a function for HIV.

Yet another recent attempt to explain how HIV actually works is diametrically opposed to the superantigen theory. Martin Nowak, Robert May, and several of their colleagues in England and the Netherlands have suggested that the incredibly high rate of mutation of HIV is what makes it so deadly. HIV mutates faster than influenza and cold viruses, so that every HIV-infected person is actually infected with a family of what may be dozens of different forms of HIV, some of them very active and some apparently harmless. Nowak and his colleagues developed a mathematical model that suggests that when a sufficient number of dissimilar families of HIV have been produced in an individual and each destroys a specific T-helper cell clone, the immune system collapses.[45] In other words, the longer a person has been infected with HIV, the more different versions of it mutate and the greater the number of T-helper cell clones that are destroyed. This scenario can explain why AIDS generally takes a long time to develop in most people and why the time from infection to full-blown AIDS varies from person to person. Depending on how quickly the HIV mutates and on the specific forms of the mutations, AIDS may develop more or less quickly. Nowak's hypothesis, however, is just that—a hypothesis. Moreover, Nowak has addressed the possibility that other infectious cofactors may promote the effects of HIV in his model.[46]

The hypothetical nature of how HIV is supposed to cause AIDS would not normally be of any consequence, since we know how only a small handful of microbes cause disease, but for two further problems. One, which I deal with at length in the next section, is that HIV has not satisfied the etiological standards laid down in the past for identifying the cause of a disease. There is no direct evidence that HIV causes AIDS—only correlations. The other is that the HIV theory is based on a false assumption, made explicitly by Blattner, Gallo, Temin, Weiss, Jaffe, and their colleagues, that HIV is the only agent present in AIDS patients that can explain T cell depletion. Obviously, any virus or other immunosuppressive agent that can infect or kill T4 (T-helper) cells and macrophages, or which can interfere with the network of interactions necessary for proper immune function, can produce the same ef-

fects as HIV. If such agents exist—and Chapters 3 through 5 will
show that they do—then it is not enough to demonstrate that HIV
is present and highly correlated with AIDS. It is also necessary to
demonstrate that these other agents are not present in AIDS patients. That is impossible. Some subset of these other immunosuppressive agents are very much present in every AIDS patient and
do have demonstrable effects on their progression to AIDS. Particularly surprising, at least in retrospect, is that evidence for the
presence, and indeed necessity, for some of these other agents was
available from the beginning of research on HIV. Researchers in
all of the major laboratories, including Gallo's and Montagnier's,
were responsible for finding this evidence, and it should have
warned them long ago that HIV was only part of the AIDS story.

Evidence of the necessity of cofactors for HIV was found at the
outset. Virologists attempting to isolate the virus in 1983 initially
used procedures developed previously for the isolation of other
human retroviruses. They were only marginally successful. Blood
from AIDS and AIDS-related complex patients was used as a
source of the presumed virus, and efforts to infect lymphocytes
and macrophages in cell culture were attempted. Transmission of
the virus to the whole blood, bone marrow, or umbilical cord
blood of fetuses was also tried. At best, virus production was observed "from time to time, but only transiently," noted one leading
group of researchers.[47] Even attempts to mimic co-infection with
other viruses or bacteria by adding immunologically stimulatory
compounds rarely caused HIV to be expressed in the blood of
healthy individuals.[48] If this was so in the laboratory, what reason
do we have to believe that it is not also so in living human beings?
Recall that only a fraction of those people such as blood transfusion patients and hemophiliacs who are repeatedly exposed to
HIV-contaminated blood or blood products actually become infected, and these people are not even healthy.

The breakthrough that has made all HIV research possible was
the discovery that the virus could be transferred to lymphocyte cell
lines derived from a variety of leukemia (cancerous white blood)
cells that could grow in laboratory culture flasks.[49] Even then,
according to Mangalasseril Sarngadharan and Phillip Markham,
two of the investigators who pioneered this research, "careful
examination of cultured cells subsequently suggested that activation of [HIV] viral synthesis required immune stimulation of in-

fected cells. Stimulation *in vitro* could be provided by mitogen or added cells (allogeneic antigens). . . . Certain manipulations of culture conditions were found to improve the outcome, e.g., co-cultivation of patient cells with mitogen stimulated peripheral blood leukocytes from uninfected donors. Isolation of virus from cultured cells also was substantially facilitated by inclusion of hydrocortisone in the culture media."[50] In other words, HIV replicates much better in cancerous lymphocytes than in healthy ones, and even in the cancerous ones, antigenic stimulation is generally necessary to induce virus expression. The list of antigenic stimulators should have been a tip-off as to the types of cofactors necessary in AIDS: foreign antigenic substances stimulating T-cell cloning (mitogens) such as often accompany infection; foreign cells, such as red blood cells or activated lymphocytes from another individual (allogeneic stimulation); and hydrocortisone, an immune-suppressing-drug.[51]

Here, in the laboratory, as early as 1984, was evidence that concurrent bacterial or viral infections might be necessary to stimulate HIV, that allogeneic stimulation with material such as blood transfusions and semen (which also contains a high proportion of leukocytes) might trigger HIV replication, and that immunosuppressive drugs could also activate the virus. All of these agents were, after all, highly associated with modes of transmission of HIV or were found with unusually high prevalence in AIDS patients. However, none of these specific agents was tested in any laboratory at the time, nor did a single scientist take the logical step of proposing that if infectious, allogeneic, and immunosuppressive cofactors were necessary to promote HIV infection in the laboratory, then they might also be necessary to cause AIDS in living human beings. As we shall see, there is good reason to think now that just such cofactors are necessary to cause AIDS.

The reason no one took this logical step, it seems, is that everyone was locked into a simplistic, though venerable, medical dogma. It has been taken for granted since the formulation of the germ theory of disease by Louis Pasteur and Robert Koch a century ago that every infectious disease has one, and only one, causative agent. The one germ–one disease dogma is so pervasive and so powerful that having isolated HIV and found that it infects T cells, investigators believed that they need not look any further. Had any of them carefully studied the history of medicine or followed the criteria laid down by other investigators of human disease over the prior century, AIDS research might have taken a quite different turn. It has been known since at least 1930 that there are classes of diseases, including various diseases associated with viral and bacterial infections, for which the one germ–one disease dogma does not apply. AIDS may be one of these.

Koch's Postulates, Etiological Criteria, and AIDS

The most interesting flaw in the arguments supporting the HIV = AIDS equation is what HIV proponents universally leave out: Koch's postulates, a set of criteria developed a century ago by the German physician Robert Koch in order to evaluate claims that a particular microorganism was the cause of a specific disease. These postulates have been the focus of intense controversy among AIDS researchers because HIV has not satisfied their standard criteria for causation. Scientists, such as virologist Peter Duesberg, argue that HIV does not satisfy Koch's postulates and therefore cannot be the cause of AIDS. Some of those who accept HIV as the cause of AIDS, such as Jonas Salk and Dani Bolognesi, reply that Koch's postulates have been satisfied by indirect means, such as transmission of HIV via transfusion, followed by the development of AIDS. Other scientists who believe that HIV is the cause of AIDS, such as Rockefeller University president David Baltimore and British virologist Robin Weiss, agree that Koch's postulates have not been satisfied but argue that they are irrelevant for testing the HIV theory. Much hinges on which, if any, of these positions are correct. The one thing that is certain is that where there is so much heat, there is fire. Some burning issue has yet to be resolved.

Robert Koch developed his postulates to provide criteria for unambiguously determining etiologies. Etiology is the study of the causes or origins of disease. Its own origins are usually traced to 1876 when Koch, a thirty-three-year-old country doctor, set out to identify the cause of anthrax, a deadly infection characterized by ulcerating nodules on the skin and lesions in the lungs, that was decimating herds of sheep all over Europe. Months of work finally yielded an agent common to every case of anthrax Koch examined: the tiny, spore-forming bacterium now known as *Bacil-*

lus anthracis. Koch's discovery, however, was just the beginning of his problems. How was he to prove that this bacterium was the cause of anthrax, and not just some common correlate of the disease or, worse, an epiphenomenon like high carbon dioxide levels in drowning victims? Moreover, how was he, an unknown young man with no academic affiliation or important connections, to prove to the medical and veterinary communities that he had discovered the cause of a disease that had stumped the greatest investigators of Europe?

Koch planned his experiments carefully, drawing on the training he had received from his teacher, Jacob Henle. He articulated in the process a strategy that became the medical standard in bacteriology. His rules were simple: (1) find a microbe that is present in all (or nearly all) diseased animals but that is not found as a concomitant to any other disease; (2) isolate the microbe in pure culture so that it, uncontaminated by any other microbe, can be inoculated into test animals; (3) produce the original disease in healthy test animals by inoculating them with the pure microbe; and (4) prove that the inoculated microbe actually caused the disease by demonstrating that the disease process was accompanied by growth of the offending organism and by reisolating it in pure culture from the newly diseased animal. These four rules are known as Koch's postulates. Koch's use of his postulates to explore anthrax and tuberculosis was sufficient to earn him a Nobel Prize.

The logic of Koch's postulates is straightforward: Demonstrate that one, and only one, organism is associated both with the occurrence of a specific disease and with its onset by isolating and controlling its transmission independent of all other factors. The etiology of virtually every controllable infectious disease known to medical science—anthrax, rabies, poliomyelitis, and tuberculosis, to name but a few—has been solved by following Koch's postulates. Diseases whose causes are thought to be understood but have not yet satisfied Koch's postulates are often described as having probable causes that have not satisfied these postulates. A few diseases, such as hepatitis due to hepatitis B infection and mononucleosis due to Epstein-Barr virus infection, have not satisfied Koch's postulates since no animal model has been developed. As with yellow fever a century ago, these diseases have been studied by direct transmission from a sick human being to a healthy one.

Consider HIV in the light of Koch's postulates. His first postulate is that the disease agent must be correlated highly with incidence of the disease. There is little doubt that this postulate is satisfied for HIV and AIDS. Certainly the vast majority of AIDS patients demonstrate antibody to HIV, and most of these also have signs of active infection. Two issues remain unresolved. First, are patients who appear to be HIV free throughout the course of their disease the exceptions demonstrating that HIV is not necessary to AIDS? Second, is HIV the only factor correlated with immune suppression in AIDS patients? The definitive demonstration that AIDS can develop in HIV-negative individuals or that other immunosuppressive agents are as highly correlated with AIDS as is HIV would raise serious problems for the theory that HIV is the sole, necessary cause of AIDS. I will demonstrate in Chapters 3 to 6 that other immunosuppressive agents are as highly correlated with AIDS as is HIV and that they do cause demonstrable immune suppression in its absence. The fact that HIV is virtually synonymous with AIDS does not prove its causative role.

Koch's second criterion—that the disease agent be isolated in pure culture—is difficult to interpret for HIV. At one end of the spectrum, Peter Duesberg has argued that there is no such thing as a pure culture of HIV. In a literal sense, this is true. There is no such thing as a pure culture of any virus, for the simple reason that viruses can be cultured only in other living cells. This problem is not inconsequential, as people working on vaccines against HIV have recently learned. E. James Stott and his colleagues at the National Institute for Biological Standards and Control in Hertfordshire, England, reported in the September 26, 1991, issue of *Nature* that chimpanzees inoculated with uninfected human T lymphocytes developed immune responses that protected them against HIV infection nearly as well as chimpanzees inoculated with human T lymphocytes infected with HIV. In other words, the T lymphocytes were creating an immune response nearly identical to that created by the HIV itself. Since nearly all vaccine work has involved cellular components, it is no longer clear whether it is the virus or the lymphocytes that are responsible for the observed protection. The same may be true in the natural process of HIV infection, since HIV is rarely, if ever, transmitted from one individual to another except through direct contact between lymphocytes.[1] If the lymphocytes are necessary for disease transmission, and they produce an immune response, how can there be any certainty what disease symptoms are due to exposure to foreign lymphocytes and what are due to HIV?

This problem has been exacerbated by a recent report by microbiologists Tracy Kion and Geoffrey Hoffmann of the University of British Columbia in Canada.[2] They report that mice with autoimmune diseases sometimes develop antibodies against HIV even though they have never been exposed to HIV. Since the usual mode of demonstrating an HIV infection is to demonstrate the presence of HIV antibody in the blood of the suspected AIDS patient, these findings throw into serious question whether such antibodies necessarily demonstrate the presence of HIV. Antigens other than HIV—cellular components such as lymphocytes or semen have been suggested—are sufficient to produce HIV seropositivity in the absence of HIV.[3]

On the other hand, failure to satisfy the letter of Koch's second postulate has not created an impediment to the demonstration that viruses such as influenza and polio are the causes of specific diseases. Virus grown in laboratory cells can be transferred back and forth to animals such as chimpanzees that also develop the disease, and vaccines and drugs to prevent or cure the disease can be tested. The inability to work with pure virus cultures therefore does not necessarily represent an insurmountable roadblock. The obstacle may be Montagnier's demonstration that he cannot get HIV to infect or replicate in cells in the absence of an infectious cofactor. His discovery suggests that Duesberg's charge is true in more than the merely literal sense. It may be the case that all cell cultures (and human beings) that are actively reproducing and spreading HIV virus are contaminated by other microorganisms, and thus that a pure culture with disease-causing activity is impossible.

There is no doubt that Koch's third postulate is not satisfied for the theory that HIV causes AIDS. Although it has been demonstrated that chimpanzees and macaques alone of all nonhuman mammals can be infected by the human form of HIV, no monkey has yet developed AIDS. The basic facts are these: As soon as HIV was identified as the putative cause of AIDS, chimpanzees at a number of research institutions were inoculated with HIV-infected blood or concentrated HIV cultures. Researchers were able to demonstrate that the HIV inserted its genes into the genome of

T-helper cells and macrophages in these monkeys, that the virus replicated, and that the chimpanzees produced an active antibody response to the HIV. There are now several dozen chimpanzees who were infected with HIV in 1983 and 1984. None of these chimpanzees has yet developed AIDS-related complex or AIDS. Even given a ten-year average latency period from infection to AIDS, this result is unexpected. At the very least, it demonstrates that the latency period of HIV in chimpanzees is even longer than that predicted for the average HIV-infected human being. Alternatively, since HIV can infect chimpanzees, some other mechanism that sets our nearest evolutionary cousins apart from us (perhaps a genetic difference) may prevent the development of AIDS (a clue that, if true, is worth investigation). These data also raise the possibility that HIV is incapable of causing AIDS by itself or perhaps at all.

Since Koch's third postulate is not satisfied for HIV, his fourth postulate—reisolation and transmission of the infectious agent resulting in disease—is not satisfied. Arguments about transmission of HIV in blood transfusions, in blood clotting factor concentrates, and from HIV-infected mothers to infants do not satisfy this criterion since none of the recipients of the HIV is healthy at the time of infection. Moreover, the HIV is definitely not pure in these cases. Each of these recipients is exposed to blood components, lymphocytes, foreign antibodies, and often several other infectious agents simultaneously, as I will document in the next chapter. In short, the standard criteria of proof that have been utilized in the past to test etiologies have not been satisfied for the theory that HIV causes AIDS.

Interestingly, the failure of HIV to satisfy Koch's postulates was admitted as long ago as 1988 at a meeting of the American Federation of AIDS Research. Present at that meeting were a Who's Who of AIDS researchers, including Dr. William Haseltine of Harvard, Anthony Fauci, director of the National Institute of Allergy and Infectious Diseases of NIH, and several other luminaries. At the meeting, Dr. Marcel Baluda of UCLA noted that the fact that chimpanzees can be infected with HIV but do not develop AIDS "may argue against HIV being the etiological agent."[4] He noted further that the only direct evidence of HIV's role would have to come from accidental transmission of HIV in laboratory or medical settings to truly healthy people. Thus, the report that a labora-

tory worker was infected with a cultured strain of HIV and shown
to be infected by it was viewed with anticipation: Would this individual go on to develop AIDS?[5] Four years later, the answer appears to be a provisional "no": No laboratory researcher exposed
to HIV has yet been reported to have developed AIDS. In consequence, most HIV researchers admit that Koch's postulates are
not satisfied for HIV.

The fact that HIV does not satisfy Koch's postulates does not
convince HIV proponents that it is not the cause of AIDS. On the
contrary, "knowing" that HIV causes AIDS, most researchers reject Koch's postulates.[6] The argument that Koch's postulates are
irrelevant to understanding AIDS may look, at first glance, like a
mere rhetorical device for squirming out from underneath some
uncomfortable facts. It is not. The issue of whether Koch's postulates are relevant to virology has been debated for more than fifty
years, and it is a historical fact that viruses have been at the center
of that debate.[7] Tom Rivers, one of the grand old men of virology,
argued in his presidential address to the American Society of Immunology in 1937, for example, that Koch's postulates could not
be applied to any virus because no virus could be procured in pure
culture. He proposed that a modified form of the postulates be
used by virologists: "1) the virus must be isolated in association
with the disease with some regularity; 2) the virus must occur in
the sick individual not as an incidental or accidental finding, but
as a cause of the disease; 3) antibody to the virus must be absent
prior to the onset of illness and make their appearance during the
period of recovery."[8] Unfortunately, Rivers's second criterion
rather begs the question of how one is to demonstrate that something is a cause rather than an accidental correlate of disease.

Rivers's criteria were not universally accepted. In 1957, Robert
J. Heubner, chief of the Laboratory of RNA Tumor Viruses at
the National Cancer Institute in Bethesda, Maryland, suggested a
much more specific set of criteria to augment Rivers's overly simplistic ones.[9] Heubner's prescription contained eight steps. (1)
Verify that the virus is real by establishing it in cell culture or in
live animals. (2) Demonstrate that it is not a contaminant or passenger isolated along with some other microorganism. (3) Demonstrate that antibody results from active infection and was not present previously. (4) Demonstrate that the virus is unlike any other
virus on record. (5) Demonstrate that the virus is constantly asso-

ciated with a unique disease state and that the virus can be isolated from diseased tissue. (6) Either establish an animal model for the viral disease or organize human volunteers for a double-blind study to reproduce the disease. (7) Undertake epidemiological studies to identify patterns of infection as compared with patterns of disease. Disease should occur in some subset of the infected population but not outside it. (8) Prevent the disease by vaccinating against the viral agent, or otherwise demonstrate that immunity against the virus results in protection from the disease state. Heubner, tongue in cheek, also added an additional requirement for demonstrating disease causation: (9) "financial support." This requirement may not be quite so amusing if HIV is not the whole explanation of AIDS, since there is essentially no funding anywhere for exploration of alternatives.

Neither Heubner's nor Rivers's criteria have been addressed by HIV proponents. Either they do not know them, or they know that HIV does not satisfy them any more than it satisfies Koch's postulates. Clearly Rivers' third criterion—that antibody be absent during the disease and appear during convalescence—is not typical of AIDS (or of several other viral diseases, such as paralytic polio). On the contrary, presence of antibody to HIV is interpreted as a failure of the immune system in AIDS. Similarly, several of Heubner's criteria are unsatisfied. There is no animal model in which human HIV causes disease. No human volunteers have stepped forward to act as guinea pigs, as occurred among investigators attempting to determine the cause of yellow fever a century ago. Exceptions to the epidemiology of HIV seem to exist, in which people not infected with HIV develop AIDS. And no vaccine yet exists that can interfere with HIV infection or the development of AIDS. Even if a vaccine is available in the next five years, it may take another five to ten years to determine whether it is effective because of the ever-increasing period between HIV infection and the development of AIDS. Thus, none of the last four of Heubner's criteria is satisfied or is likely to be satisfied in the near future.

AIDS researchers have ignored previous criteria for establishing disease causation in favor of ad hoc inventions of their own. The earliest set, which became the model for the arguments summarized in the first section of this chapter, were outlined by Robert Gallo, George Shaw and Phillip Markham—three of the American isolators of HIV—in a 1984 article, "The Etiology of AIDS." This article makes no reference to Koch's or anyone else's criteria for determining etiologies. Instead, Gallo and his colleagues listed seven lines of evidence as support for their contention that HIV causes AIDS; (1) AIDS is new. (2) AIDS spread from a limited region. (3) AIDS appeared in distinct groups that shared only a proneness to communicable diseases. (4) Clusters of disease were found to be linked by common contacts. (5) AIDS appeared to be spread by contact with blood and must therefore be infectious. (6) The infectious agent isolated (HIV) was associated with selective depletion of a particular type of T4 lymphocytes (T-helper cells) in all of the patients they had studied. "We knew of no agents, aside from a family of human T-lymphotropic retroviruses [viruses that are attracted to a particular type of white blood cell and insert themselves into their DNA] that we had discovered three years earlier . . . that demonstrated such tropism to a subset of lymphocytes."[10] Finally, (7) HIV is rare among non-AIDS patients. What these arguments amount to is the demonstration that there is a good correlation among HIV, the incidence of AIDS, and one of its specific symptoms (T-helper cell loss). Such a correlation satisfies only two of Koch's postulates and four of Heubner's.

What is somewhat astonishing is that in 1984, when Gallo first championed HIV as the cause of AIDS, the correlation between HIV and AIDS was not even particularly convincing. At that time, Gallo's group had isolated HIV from only 50 percent of AIDS patients, 80 percent of patients with AIDS-related complex, and 20 percent of clinically healthy high-risk patients.[11] Although isolation rates are now above 90 percent in most controlled studies, we must also realize that since 1984 the definition of AIDS has changed to incorporate HIV seropositivity as a diagnostic feature. Thus, not only have techniques for isolation improved, a regimen of preselecting patients for the presence of HIV has also been instituted. AIDS is defined by the presence of HIV, and HIV is defined as the cause of AIDS because it is so prevalent among AIDS patients. Under these conditions, it would indeed be surprising if HIV could not be isolated from the majority of people with antibody against the retrovirus. But correlation, no matter how good, is never grounds for asserting causation. One must also also have experimental control over the disease.

Some HIV proponents have recognized these logical flaws. Two

notable attempts to modify Koch's postulates to fit HIV were subsequently made by Alfred Evans of Yale University and Warren
Winkelstein of the University of California, Berkeley. Winkelstein
proposed in 1988 that five criteria suffice to demonstrate that HIV
is the cause of AIDS: "Criterion 1: Prevalence of disease should be
significantly higher in those exposed to the factor than in unex-
posed controls. . . . Criterion 2: Incidence of the disease, i.e., new
cases, should be significantly higher in those exposed to the causal
factor than in those not exposed. . . . Criterion 3: A spectrum of
host responses should follow exposure to the hypothesized agent
along a logical biological gradient. . . . Criterion 4: Temporally
the disease should follow exposure to the causal agent. . . . Criterion 5: Elimination or modification of the hypothesized agent
should decrease the incidence of the disease."[12] HIV, Winkelstein
claims, satisfies these criteria and is therefore the cause of AIDS.

A year after Winkelstein made his proposal to modify Koch's
postulates, Alfred Evans proposed a very different list of five criteria, which he claims are also satisfied by HIV and suffice to demonstrate its causal role in AIDS. The fact that these criteria are
so different from both Winkelstein's and Gallo's, as well as from
Heubner's and Rivers's, shows how deep the problem of demonstrating causality truly is. Evans suggested that an infectious agent
be said to cause a disease when: "(i) Antibody to the agent is regularly absent prior to the disease . . . (ii) Antibody to the agent
regularly appears during illness . . . (iii) The presence of antibody
prior to exposure to the agent indicates immunity to the clinical
disease . . . (iv) The absence of antibody prior to exposure to the
agent indicates susceptibility to infection and disease . . . (v) Antibody to no other agent should be similarly associated with the
disease."[13]

Careful reading of both Winkelstein's and Evans's criteria
would, I suspect, shock Koch if he were still alive. Neither set
retains the basic logic inherent in Koch's postulates, which insist
on isolation of the disease agent, control of its transmission, and
demonstration of disease causation in its presence. Winkelstein's
criteria amount to nothing more than establishing correlations and
explicitly allow AIDS to be diagnosed even when HIV is not present. Evans does not even require isolation of a disease agent; the
presence of antibody is sufficient. Such evidence can be very mis-

leading. As we have already seen, Stott and his colleagues and Kion and Hoffman have demonstrated the presence of HIV-like antibodies in animals that have never been exposed to HIV. What, then, do antibody studies in the absence of virus isolation mean? Moreover, antibody against HIV regularly appears a decade or more prior to overt illness and, far from being protective, is predictive of AIDS—the opposite of what one would expect from Evans's criteria. This phenomenon of antibody predicting disease instead of protecting against disease is not generally true of any other type of disease save for autoimmune diseases, which are not infectious. Moreover, people who are HIV seronegative, far from being the most susceptible to AIDS as one would infer from Evans' criteria, are known to be the least susceptible.[14] And antibodies to HIV are far from being the only ones highly correlated with AIDS. As I shall demonstrate in Chapter 4, antibody to two other immunosuppressive viruses, cytomegalovirus and Epstein-Barr virus, are also synonymous with AIDS. Thus, there is no way to distinguish clearly the effects of HIV from the effects of these other active infections in AIDS patients. Antibody studies, if anything, only confuse the issue of what the causative agent in AIDS may be.

In short, HIV does not satisfy any of the etiologic criteria that existed prior to its discovery, and the etiologic criteria that have been developed since are all logically flawed. This means either that HIV is not the true cause of AIDS[15] or that we still do not understand how such viruses produce disease and are therefore utilizing the wrong criteria. In either case, we clearly need clarification of the issues and a new, higher standard of etiological criteria applicable to viruses and retroviruses.

In the meantime, HIV remains a microbiological anomaly. Such uniqueness is worrisome. Although Evans has argued that "the fact that an etiologic agent presents new findings that do not fit old dogma should not be interpreted as implying a lack of causality," he and his colleagues ignore the fact that extraordinary proofs are required to overthrow well-established dogma.[16] Far from having extraordinary proofs of HIV's causal role in AIDS, we have less than the usual set of proofs required in establishing previous disease etiologies. Given this state of affairs, attempts to modify Koch's postulates after the assertion that the causative agent has been identified smack of a posteriori reasoning. Such reasoning is always suspect to logicians and should be equally suspect to physicians and scientists as well.

Multifactorial, Synergistic Disease Models of AIDS

How is the dilemma of HIV's role in AIDS to be resolved? One way is to be even more radical than the HIV proponents in modifications of Koch's postulates. Perhaps the reason that AIDS has not yet yielded to the standard etiological approaches developed over the past century is that the model itself is wrong. Perhaps the assumption that every disease is caused by a single disease agent is invalid. Perhaps HIV is necessary but it is not sufficient to cause AIDS, for example, and therefore all attempts to demonstrate HIV causality, or the causality of any other single disease agent, are doomed from the start. If they are, then what alternatives exist?

Many diseases, in fact, break the mold of one infectious microbe producing one disease manifestation. Disease processes are much more multifarious than that. In general, diseases can be categorized into at least five groups: (1) those caused by failure of the organism to develop properly; (2) those caused by lack of essential nutrients; (3) those caused by a buildup of toxins or poisons; (4) those due to a single infectious agent; and (5) those that result from some confluence of several factors. The first group consists of genetically inherited diseases and drug-induced fetal malformations such as those associated with thalidomide and alcoholism. The second group of diseases results from chronic insufficiency of nutrients (scurvy, ricketts, osteoporosis) or acute insufficiency of necessary elements such as oxygen (resulting in brain damage, for example). The third group of diseases may be due to causes such as toxic levels of mercury or lead in water or food and paints or chronic alcholism. The fourth group consists of those caused by single infectious agents: measles, tuberculosis, smallpox, polio, and so forth. Finally, there are the diseases that are caused by some complex of interacting factors: cancer, heart disease, late-onset diabetes and so on.[1]

The multifactorial diseases are the ones most likely to be a useful model in considering AIDS since all AIDS patients, as I will demonstrate at length in the next three chapters, have multiple infectious and noninfectious causes of immune suppression. There are at least two subcategories of diseases in this fifth, multifactorial group. First are the synergistic diseases. Synergism is defined as the joint action of agents that increase each other's effects. For example, neither alcohol nor barbiturates are likely to be fatal in low to moderate doses, but their combination can prove fatal. Synergistic actions require that all of the factors be active simultaneously.

Another subset of multifactorial diseases also requires several, interactive factors, but these act sequentially, or in a stepwise fashion, rather than synergistically. Most models of cancer, for example, now assume at least two factors, one called the "initiator" and the other the "promoter." The initiator induces a genetic mutation in a gene that can lead to cancer but does not itself, cause cancer. The mutation remains "silent"—or, like several of the viruses relevant to AIDS, "latent"—until a promoter activates the mutated gene. When this mutated gene is turned on, it results in uncontrolled cell growth and differentiation—cancer. The promoter, however, cannot induce cancer by itself any more than can the inducer, since, in the absence of an appropriate mutated gene, there is nothing for the promoter to turn on. A similar model seems to apply to late-onset (or noninsulin-dependent) diabetes, in which a genetic predisposition for diabetes can be triggered by pregancy or a large weight gain in older age. Unlike cancer, which seems to involve an irreversible step, late-onset diabetes seems to be reversible, or at least controllable, by eliminating the promoter. Weight loss and control of food intake can be sufficient to treat such diabetics.

Notably, Evans himself has recently focused attention on infectious diseases that behave in synergistic or multistep fashions, although he makes little of these ideas in his writings on AIDS.[2] Evans believes that the reason "most persons infected with most viruses such as poliomyelitis, hepatitis, and Epstein-Barr virus develop mild or inapparent infections, while only a few develop the characteristic clinical manifestations of the disease [is] the presence of certain 'clinical promotion factors.'"[3] In other words, Evans believes that susceptibility to infection, nutritional status, the presence of other infections, and one's history of previous infections may all play some role in who gets ill and who does not. There is, in fact, nothing new about Evans's ideas except the label he gives them. Koch noticed at the end of the nineteenth century that alcoholics had a much higher incidence of cholera than anyone else,[4] and Louis Pasteur, who first proposed the germ theory of disease, supposedly admitted on his deathbed that "the terrain is more important than the germ."[5] He certainly wrote, as microbiologist René Dubos constantly reminds his readers, passages such as: "I'm convinced that when a wound becomes infected and festers, the course that wound takes depends upon the patient's general condition and even his mental condition."[6] If Pasteur, Koch, Dubos, and Evans are correct, then etiologies cannot be established merely by isolating a disease organism; one must also specify the conditions under which that disease organism is pathogenic to its hosts.

In some cases, the conditions require the presence of two or more synergistic microbes. One of the earliest examples of a synergistic infectious disease was elucidated during 1931. R. E. Shope of the Rockefeller Institute (then located in Princeton, New Jersey) demonstrated that swine influenza was not caused by either a virus or a bacterium but rather by a specific combination of a virus and bacterium. The proof was quite rigorous. Shope isolated a bacterium called *Hemophilus influenzae* from every pig that developed the symptoms of swine flu. Infecting unimmunized pigs with this bacterium did not, however, result in swine flu or in any disease manifestations at all. Thinking that they had isolated the wrong organism, Shope and his colleague, Paul Lewis, started over again and isolated an unidentified virus. (This virus was later identified as a parainfluenza or Sendai virus.)[7] This virus also failed to produce the symptoms of swine flu. Finally, since both microbes were highly associated with the disease but neither seemed sufficient to cause it alone, Shope tried mixing the virus with the bacterium. Lo and behold, the pigs came down with all the symptoms of swine influenza.[8] The synergism was not a mere question of facilitating infection. Shope could demonstrate that the individual agents were just as infective as the combination, because each resulted in a good antibody response. In fact, the antibody response to the individual microbes was capable of protecting previously infected pigs against the fatal results of subsequent infection with the *Hemophilus influenzae*–virus combination.[9] The combination of virus and bacteria had effects that neither microbe had by itself.

Many additional examples of such synergistic infections are now on record. Sherwood I. Gorbach and John G. Bartlett of the Brentwood Veterans Adminstration Hospital and the UCLA School of Medicine in Los Angeles reviewed information concerning a similar phenomenon involving infections with anaerobic bacteria during the early 1970s. Anaerobic bacteria grow only in the absence of oxygen and do not normally cause disease. In the presence of other bacteria that can use oxygen, however, anaerobic bacteria can become pathogenic. Thus, various combinations of anaerobic streptococci with staphylococci, bacterioides, and diphtheria can cause gangrene, septic shock, and thrombophlebitis (an infection of the veins). As with Shope's swine flu, only the combinations of agents, not the pure cultures of the individual germs, satisfy Koch's postulates.[10]

Other human diseases with multiagent causes have also been documented. For example, adenoviruses, which are common among people at risk for AIDS, rarely cause anything more serious than influenza-like symptoms in healthy adults. In July 1970, however, three U.S. Army recruits died of adenovirus pneumonia. At least two of the three also had concomitant bacterial infections, one of which was identified as *Haemophilus influenzae*.[11] These recruits may represent the human equivalents of Shope's swine influenza–stricken pigs.

These and many other experiments and clinical observations that will be detailed later in this book must lead us to question whether the very basis of the HIV theory of AIDS—that a single virus is responsible for AIDS—is necessary or even tenable. Clearly the work of Montagnier and Lo on *Mycoplasmas* strongly suggests that HIV alone is not sufficient to cause AIDS-like effects in cell culture, let alone in human beings. Even Gallo and many of his colleagues have published evidence that HIV can certainly be triggered by the presence of other co-infections and drugs associated with AIDS.

We have reached a deep and important set of medical questions about the very nature of disease causation. Dr. Gorbach and Dr. Bartlett have summarized the nature of these questions clearly: "Our philosophic models of infection are based on concepts of microbial monoetiology [single causes]. Pasteur demonstrated that certain microorganisms were responsible for specific syndromes. This concept was formalized into its present liturgical style by

Robert Koch in his famous 'Postulates.' To complete the trilogy, [Paul] Erhlich created a 'magic bullet,' the drug designed for [killing] a specific infection. In its final form, this principle reads: *one microbe—one disease—one drug.* The concept of monoetiology applies to infections such as lobar pneumonia, typhoid fever, diphtheria, and cholera. But this classic design does not fit most infections associated with anaerobic bacteria. . . . Models of bacterial synergy are perhaps more applicable to such mixed infections."[12] Perhaps models of microbial synergy are more applicable to AIDS as well.

If AIDS turns out to be a synergistic or stepwise multifactorial disease, then the reason that Koch's postulates, or any modification of them, have not been satisfied is clear: we made an incorrect assumption of monoetiology at the outset. We need a new set of criteria for evaluating claims of causality for the more complex cases of synergy or multifactorial disease. I therefore emphasize something very important at this point. I have no quibble with those who believe in the need to revise Koch's postulates. I am, in fact, among these revisionists. Several other investigators, (most notably Ernest Witebsky) and I have proposed that autoimmune diseases (in which the immune system turns upon its host) require rather different criteria for their elucidation than are embodied in Koch's postulates or in the virologists' revisions of them.[13] In particular, I have proposed that autoimmune diseases result from immunological responses to sets of concurrent infections but never to single disease agents. This observation, which is quite controversial, may be important in understanding at least some aspects of AIDS, since autoimmunity is universal to it. If I am right, then no single disease agent, HIV included, is sufficient to induce AIDS or the autoimmune complications that accompany it. That is why Koch's postulates and its revisions have not been satisfied.

AIDS may belong to a new category of diseases whose causative rules have yet to be fully elaborated. These new diseases will need new criteria that maintain the basic logic of Koch's postulates but also include the identification of specific sets of disease agents rather than individual ones. Such diseases will fit an enumeration of criteria like the following: (1) two or more infections or immunosuppressive agents will be active simultaneously in individuals affected with a particular disease; (2) people with only one of these agents will either have no symptoms of the disease or symptoms of a different disease that is associated with the single agent; (3) these multiple agents will be isolatable or identifiable individually; (4) no single one of these agents will be capable of inducing the disease by itself when introduced into healthy human beings or animals (although they may cause other disease symptoms); (5) an immune response to each agent will be demonstrable during disease induction and will be significantly altered by the combination of agents as compared with any agent individually; (6) the disease will be transmissible from one animal or human being to another by means of an appropriate combination of the putative causative agents. In the case of autoimmune diseases, the transmissible factors may include tissue or purified proteins, immune serum, lymphocytes, or antibodies.[14] According to these criteria, the causative agents need not be transmissible or infectious as a general rule but may be of a noninfectious nature, like the synergistic combination of barbiturates and alcohol. In the case of AIDS, however, since significant evidence exists supporting transmissibility, the primary agents responsible are presumably transmissible, although cofactors for these need not be.

The criteria for evaluating synergistic and autoimmune diseases that I have presented may, or may not, be accurate for AIDS or may, or may not, be acceptable to the scientific community. This is a matter for ongoing research. My point in presenting them here is not to argue for their validity but more simply to argue that there is a well-established set of diseases that have many of the characteristics of AIDS—multiple, concurrent infections, autoimmune processes, and multiple disease-causing agents—that may provide an as yet untested model for AIDS. Since AIDS has thus far been refractory to the one agent–one disease–one drug approach, it is certainly a valid scientific position to consider the possibility that AIDS may be due to synergistic or multifactorial causes until proved otherwise.[15] One thing is certain: The case that HIV causes AIDS is still open, and surprises are still possible.

CHAPTER 3

IMMUNOSUPPRESSION AND AIDS

Noninfectious Immunosuppressive Agents Associated with AIDS

The theory that HIV is the sole agent responsible for the immune suppression in AIDS has been brought into question by a series of anomalies concerning who develops AIDS, whether HIV can cause immune suppression on its own, by the failure of HIV to satisfy any of the widely accepted criteria of causation, and by apparent exceptions in which AIDS develops in the absence of HIV infection. These outstanding problems may be resolved in favor of HIV at some time in the near future, but in the meantime, their existence forces us to consider the possibility that some assumptions about HIV or AIDS are inaccurate.

The role of unexamined assumptions in science, as in logic, is crucial; they can make or break the most rigorously derived conclusions. Even schoolchildren are aware that if the assumptions underlying an argument are incorrect, no matter how rigorous the reasoning or how detailed the subsequent research, the conclusions will be false. Who, after all, has not been taught such simple syllogisms as, "Jim is a man; all men are mortal; therefore Jim is mortal." But who cannot see the fallacy in the similarly constructed syllogism: "Logic is infallible; men use logic; therefore men are infallible." The reasoning is precisely the same. The assumptions in the second argument—that logic is infallible and that (even if it were) logic is always used properly—are, however, incorrect. The argument fails despite the rigor of the logic itself because the information that went into the logical construct was inaccurate.

Our arguments concerning HIV and AIDS are similarly flawed by inaccurate assumptions. The most important of these assumptions is that HIV is the only immunosuppressive agent present in those at risk for AIDS and the only agent necessary to explain the immune suppression that characterizes the syndrome. As Dr. James R. Allen, chief of the Surveillance Section on AIDS Activity for the CDC, wrote in 1984: "The surveillance definition of AIDS . . . demands two criteria: (1) The presence of a reliably diagnosed disease at least moderately predictive of cellular immune deficiency; and, (2) the absence of an underlying cause for the immune deficiency or of any defined cause for reduced resistance to the disease."[1] The specific causes of immunodeficiency that disqualify diseases as indicators of AIDS—that is, the known causes of reduced immunological resistance, are clearly defined: "1. high-dose or long-term systemic corticosteroid therapy or other immunosuppressive/cytotoxic therapy less than 3 months before the onset of the indicator disease. 2. any of the following diseases diagnosed less than 3 months after the diagnosis of the indicator disease: Hodgkin's disease, non-Hodgkin's lymphoma (other than primary brain lymphoma), lymphocytic leukemia, multiple myeloma, any other cancer of the lymphoreticular or histiocytic tissue, or angioimmunoblastic lymphadenopathy. 3. a genetic (congenital) immunodeficiency syndrome or an acquired immunodeficiency syndrome atypical of HIV infection, such as one involving hypogammaglobulinemia [that is, deficiency of B cells or antibody, rather than a deficiency of T cells]."[2] According to the logic of diagnosis laid out by the CDC, anyone with any of these three categories of immune suppression was not an AIDS patient because he or she had a recognized cause of immune suppression. Anyone with similar symptoms but who did not fit into one of these categories was an AIDS patient because they had no known cause of immune suppression. In other words, such people were assumed to be immunologically normal prior to contracting HIV, and HIV alone was responsible for causing their immune deficiency.

The reasoning that led to the identification of HIV as the sole immunosuppressive agent responsible for AIDS is, in fact, based on two questionable assumptions. One of these was pointed out by Dr. Joseph Sonnabend, a physician in New York City who had treated large numbers of gay men with and without AIDS. He and his colleague, Dr. Serge Saadoun, while fully appreciating the need for a stringent and appropriate definition of AIDS, wrote in the journal *AIDS Research* that they were nonetheless troubled by certain aspects of the surveillance definition: "A difficulty with the definition . . . is that it is one of exclusion. All cases of cellular immunodeficiency not accounted for by currently known mechanisms become, by definition, cases of AIDS. This is tantamount to declaring that only one unknown cause of cellular immunodeficiency remains to be discovered—the cause of AIDS. The surveillance definition of AIDS almost demands the existence of a single AIDS agent, as this appears to be the only unifying hypothesis to explain the appearance of an identical disease (by definition) in such disparate groups [as gay men, drug abusers, and hemophiliacs]. In reality, the essential unifying etiologic characteristic is that the immunodeficiency is unexplained in each group. It does not follow that identical mechanisms operate."[3]

The point Sonnabend and Saadoun made is very important. What they were arguing, to return to an analogy introduced previously, is that although all people who drown can be shown to lack oxygen and to have too much carbon dioxide in their blood (the equivalent of HIV and immune suppression in AIDS), the actual mechanisms of drowning can be as diverse as a heart attack or stroke, knocking oneself unconscious while diving, cramps, exhaustion, or murder. The physiological end result is the same, despite the great variety of causes. Thus, although immune suppression is always present in AIDS and HIV is extremely common, it does not follow that one necessarily causes the other any more than lack of oxygen causes too much carbon dioxide. On the contrary, both may be common manifestations of many primary causes. Immune suppression and HIV may therefore be *symptoms* of the actual causes of AIDS.

This phenomenon of similar or identical symptoms resulting from a variety of causes is often found in medicine. T cell immunosuppression in cancer patients, transplant patients, people treated with chronic or high-dose corticosteroids, or children born with genetically inherited causes of immune deficiency is caused by diverse mechanisms that all have the same result: a very high risk of opportunistic infections. Since we know for a fact that T cell deficiencies can have many causes in non-AIDS patients, there is no reason to assume that every AIDS patient must have one and the same cause. As a working hypothesis, it is just as valid to propose that there might be as many different causes of immune suppression at work in people with AIDS as there are in other groups of people who develop similar sorts of immune suppression due to known causes. The implication of this line of reasoning is that AIDS might have many causes and might not be a single disease entity, any more than, say, pneumonia (which can be caused by various bacteria, viruses, protozoa, and chemicals) is a single disease entity. If so, then the search for a single causative agent was, as Sonnabend and Saadoun warned, a mistake from the outset—a mistake we are slowly and unconciously rectifying by introducing cofactors into the picture.

The question of whether a single disease agent is either necessary or sufficient to explain AIDS is only one of the untested assumptions embedded in the AIDS definition. The other is that people who develop AIDS were previously healthy, immunologically normal individuals without any identifiable causes of immune suppression. Not only were there no direct studies of this question until very recently, but this aspect of the definition is again exclusionary. It assumes that the list of identified causes of immune suppression embedded in the definition was complete. But are congenital (or primary) immunodeficiencies, transplant and cancer chemotherapies, steroid or radiation treatments, and diseases of the lymphatic system the only significant causes of immune suppression that can result in T cell deficiencies? The answer to this question, as we will see in the next three chapters, is a resounding "no." People in groups identified as being at high risk for AIDS have a multitude of recognized immunosuppressive factors at work on them long before they encounter HIV, and quite often in its complete absence. These include, in the order they will be addressed, noninfectious agents including exposure to semen, blood, drugs, and malnutrition (this chapter); infectious agents including many of the most prevalent viral infections associated with AIDS (Chapter 4); and the phenomenon of autoimmunity, in which the immune system begins attacking and destroying itself and other parts of the body (Chapter 5). These agents, working in various nefarious combinations, can cause profound immune suppression and susceptibility to opportunistic infections even in the absence of HIV, as I will demonstrate in Chapter 6. Notably, most, if not all, of these agents and their clinically relevant immunosuppressive effects had been identified in the decade prior to the recognition of AIDS.[4] In retrospect, it seems inexcusable that they should have been explicitly ignored as irrelevant (by definition) in the clinical diagnosis and determination of the etiology of AIDS.

The recognition of the role of immunosuppressive agents other than HIV in AIDS will certainly cause much rethinking among AIDS researchers and clinicians. These additional immunosuppressive agents may increase the rate at which AIDS develops in HIV-infected people; they may predispose people exposed to HIV to active infection; they may be necessary cofactors for HIV, without which the virus cannot cause AIDS; HIV may be just another opportunistic infection that takes advantage of and increases the preexisting immune suppression caused by these other agents; or a sufficient combination of such factors may be capable of inducing AIDS on their own, in the absence of HIV. Any, or all, of these possibilities would explain why AIDS has remained almost completely within the originally defined high-risk groups rather than spreading, as other venereal diseases have done, to low-risk groups as well. Any or all of them would explain why HIV has not been able to satisfy Koch's postulates. At the very least, HIV may need cofactors. At most, AIDS has many causes, and no single combination of agents will ever explain the syndrome.

Recognizing the role of immunosuppressive agents other than HIV in AIDS also means that public policies targeted at controlling the spread of HIV may not be sufficient to control the incidence of significant immune suppression. If HIV is not the sole cause of acquired immune suppression among AIDS risk groups, then we may need much more than chastity, safer sex, or clean needle policies. People will need to warned of all the immunosuppressive risks that may predispose them to HIV infections or AIDS.

Moreover, the existence of multiple causes of immune suppression unique to or unusually common among those at high risk for AIDS must also cause us to reconsider the definition of AIDS itself. Recall that AIDS was defined until 1987 as a syndrome characterized by significant immune suppression leading to opportunistic disease in previously healthy people without identified causes of immune suppression. But every AIDS patient for whom complete records exist can be shown to have had multiple causes of immune suppression other than HIV and to have been ill or unhealthy prior to developing either HIV or AIDS. These pre-AIDS health problems were ignored, and generally continue to be ignored, in considering the causes of AIDS and in carrying out epidemiological studies of risk factors for AIDS. Until they are recognized and the definition is altered to reflect their existence, we may never understand AIDS because we will be utilizing an inaccurate and misleading guide.

This is not a novel conclusion. Dr. Sonnabend and his associates have fought for many years to try to bring the relevant evidence, based upon detailed studies of hundreds of AIDS patients, to the attention of the medical community.[5] Their conclusion, and mine as well, is that AIDS did not suddenly appear out of nowhere among otherwise completely healthy young men and women whose only immunosuppressive risk was HIV. This is a myth. The people who developed AIDS during the initial phases of the epidemic, like the people who continue to develop AIDS today, were and are, virtually without exception, unhealthy, immunocompromised people even before they contract HIV and long before they develop AIDS. The evidence for this statement existed in 1980. It is, if anything, even more apparent today.

Immunosuppression due to Semen Components

The noninfectious immunosuppressive agents associated with one or more AIDS risk groups are extensive and diverse. They include a variety of addictive and recreational drugs, pharmacological agents, malnutrition, and exposure to the cells of other human beings (or, in immunological jargon, "alloantigens," or "allogeneic stimuli," meaning molecules that can induce an immune response that derive from a genetically similar, but not identical, individuals).

One of the oldest identified alloantigens known to cause immune suppression is semen. In normal circumstances, semen is neither treated by the body as an antigen (or foreign material to be immunologically eliminated) nor is it immunosuppressive, but if it obtains access to the bloodstream or lymph of an individual, it can become both. The most common way that semen obtains such access is through anal exposure. Receptive anal intercourse, in which a person allows a man to insert an erect penis into his or her rectum, is the most hazardous sex practice associated with the development of AIDS for both men and women. Some AIDS researchers believe that anal exposure to semen is particularly risky not because it is a more efficient means of transmitting HIV than, say, vaginal intercourse or oral sex but because it is an effective means of presenting semen components to the immune system. The immunological processing results in significant immune suppression.

Evidence for the immunomodulatory effects of semen dates back to at least the 1890s, when scientists in Elie Metchnikoff's laboratory at the Institute Pasteur in Paris began studying whether animals could react immunologically to their own or other animals' cells. In 1898, Metchnikoff and his colleague Serge Metalnikoff inoculated a guinea pig with its own sperm. They found that the guinea pig did indeed produce antibody against its own sperm, but that for reasons that could not be explained at the time, this antibody response was not active. Somehow the immune system suppressed its own response to the sperm. It also suppressed its response to foreign sperm from nonrelated guinea pigs.[1]

Experiments over the next several decades demonstrated that the immune suppression observed by Metalnikoff had much broader effects. Dr. F. Duran-Reynals of the Rockefeller Institute for Medical Research demonstrated during the late 1920s that testicle extracts (including semen components) greatly enhanced lesions produced by vaccinia virus and bacteria such as the *Staphylococci* in rabbits and guinea pigs. His work was extended to other viruses and bacteria by other investigators, all of whom showed that immunological response to sperm resulted in a significantly suppressed immune response to infectious agents. Giovanni Favilli, also of the Rockefeller Institute, found in 1931 that such testicle extracts also had the property of greatly increasing red blood cell fragility.[2] This observation may have significance for understanding the thrombocytopenia that is a common concomitant of AIDS, especially among homosexual men.

These early studies were relatively crude compared with modern standards. Little was known about the immune system, and nothing about immune suppression. Thus, ninety years of research allow us to say much more about the effects of semen on the immune system than Metchnikoff and Metalnikoff could even conjecture. By the mid-1970s, investigators had demonstrated that semen has a profound immunosuppressive effect on lymphocyte functions, including responses to foreign stimuli, the ability to proliferate, and activation of antibody response.[3] Adverse effects were particularly pronounced on T-lymphocyte functions, the very functions that are most depressed in AIDS patients.[4] The one limitation of this research was that it was done on lymphocytes cultured in test tubes. No one had demonstrated that the same phenomenon occurred in living animals or in human beings.

The AIDS era opened up other avenues of research. J. M. Richards, J. M. Bedford, and S. S. Witkin demonstrated in 1983 that inoculating semen into the rectums of rabbits resulted in profound and easily demonstrable immune suppression, specifically affecting T cells.[5] Similar experiments involving the inoculation of mice with small amounts of semen into their veins also produced T cell immune suppression.[6] These experiments demonstrate that if semen can gain access to the immune system, then unwanted side effects may result, some of which mimic aspects of AIDS.

Part of this immune suppression appears to be due to sperm itself. Semen, the substance that is ejaculated by a man during sex, has several components, of which the sperm cells are only one. Sperm cells, as distinct from the rest of the seminal material, have been demonstrated to induce animals and people to produce specific antibodies that not only recognize the sperm as a foreign antigen, but also, inappropriately, mistake T lymphocytes as being foreign antigens as well.[7] The result of this immunological confusion is that the antibody-producing B lymphocytes actually attack the T lymphocytes. The resulting civil war within the immune system itself has to be one of the most bizarre forms of immune suppression yet discovered.

Semen also contains a significant proportion of lymphocytes. Not only may these lymphocytes transmit latent viruses such as HIV, cytomegalovirus, and Epstein-Barr virus, they may also act as allogeneic antigens themselves. The most important of these antigens in the context of AIDS are undoubtedly the MHC, major histocompatibility complex (or HLA, human leukocyte antigen) and CD4 proteins that exist on the cell surfaces of T helper cells, monocytes, and B cells. If antibodies are made against these proteins, there is a reasonable chance that they will act as autoantibodies that can also attack the host's own immune system cells. In fact, many AIDS patients have been found to have anti-HLA autoimmunity.[8]

Seminal plasma, the gelatinous liquid in which the sperm is ejaculated, also contains immunosuppressive substances.[9] These semen substances, many of which have yet to be fully characterized, have been shown to inhibit the maturation of T lymphocytes from their precursor cells;[10] to inhibit natural killer cell activity[11] and macrophage ingestion of microbes;[12] and to interfere with lymphocyte activation to specific antigens.[13] It is thought that the usual function of these substances during heterosexual intercourse is to prevent the woman from immunologically interfering with sperm as they travel toward the cervix in their quest to inseminate the egg. Clearly, the same mechanisms would operate in the intestines and bloodstream to inhibit normal immune responses to both sperm and to any bacteria, viruses, or other microbes that might be introduced along with the sperm. Thus, it has been found that it is much easier to transmit viruses such as Epstein-Barr and cytomegalovirus via semen than it is by, for example, kissing.[14]

One other possible role semen-induced immune suppression may play is to increase the probability of intestinal infections, as so often occurs in gay bowel syndrome, and of systemic infections with sexually transmissible disease agents, such as cytomegalovirus, Epstein-Barr virus, hepatitis B virus, syphilis, and gonorrhea. The intestines are studded with a series of Peyer's patches, a variant of lymph nodes filled with lymphocytes. Normally, these Peyer's patches protect the intestines from infection and the gut from being infected by material in the intestines. The introduction of immunosuppressive agents such as sperm directly in the intestines will suppress the activity of the lymphocytes in the Peyer's patches, thereby making infection with any concomitantly transmitted infection much more likely. Clearly, all sexually transmitted diseases, including syphilis, cytomegalovirus, and HIV, can be spread through contact with semen.

In short, semen has multiple immunosuppressive actions, and anyone who engages in activities that promote repeated or chronic immunologic exposure to semen antigens is likely to develop measurable immune suppression. It has been demonstrated that men who engage in unprotected receptive anal intercourse (but not those who are exclusive sperm "donors") develop antibodies to semen and sperm antigens[15] and are at highest risk of developing AIDS.[16] Interestingly, neither oral nor vaginal sex carries significant risks of developing sperm-associated immune suppression. It is worth considering why.

Immunological contact with sperm, or material carried in sperm, is increased in anal, as contrasted with vaginal or oral, intercourse. One reason has to do with the physiological differences of the rectum, vagina, and upper gastrointestinal tract. Vaginal tissue differs markedly from rectal tissue. The vagina has thick, muscular walls covered by a deep layer of epithelial (skin-like) cells that are easily sloughed off and secrete a lubricating mucus to decrease the possibility of abrasion. Even if abrasion does occur, the capillaries that are embedded in vaginal tissue are far from the surface and difficult to breach. There are also very few lymphocytes directly in the vagina, most of them being located higher up, near the cervix. The rectal tissue presents an entirely different picture. The rectum is comprised of an extremely thin layer of tissue, densely entwined with capillaries. It lacks the thick layers of epithelium that protect the vagina and its ability to produce a protective mucus. Moreover, the intestines are studded with Peyer's patches. Located along with the Peyer's patches are concentrations of M cells, which apparently function as portals through which the resident lymphocytes constantly sample the contents of the intestines for foreign material. These M cells have been shown to permit viruses such as HIV to gain access to the immune system from the rectum.[17] Thus, unlike the vagina, the rectum represents a place in the body through which the immune system can easily be reached, even under normal conditions. Since microscopic tears and bleeding can accompany anal intercourse and infections[18] but are rare in vaginal intercourse, anal exposure confers another means for semen components (and viruses) to enter the bloodstream, there to be immunologically processed.[19]

Oral forms of sex appear to confer little or no risk of immunological contact with viruses and semen for rather different reasons. The mouth contains enzymes in its saliva that are part of the first line of immunological defense. These enzymes are capable of inactivating some infectious agents. Those that pass beyond the mouth into the gut are usually digested by the stomach acids. The same mechanisms result in the digestion of semen if it is ingested. It is degraded into nothing more than its molecular components and treated as just another set of nutrients. Thus, oral sex confers little or no risk of immunological contact with viral or semen components, unless, perhaps, oral, esophageal, or stomach ulcerations exist as portals to the immune and circulatory systems. Unfortunately for many people with AIDS and pre-AIDS syndromes, such ulcerations are common concomitants of their various infections.

Immunosuppression Due to Addictive and Recreational Drugs

"It is estimated that every day approximately 5000 people in the United States try cocaine for the first time." Thus begins an article on the immunological effects of cocaine abuse written in 1988 by Dr. O. Bagasra and Dr. L. Forman of the University of Medicine and Dentistry of New Jersey. Current figures estimate that there are about 5 million regular users of cocaine in the United States and that the United States currently consumes half of the world's supply of this drug.[1] It is not surprising, then, that the United States is the AIDS capital of the world, since intravenous drug abusers, particularly those who inject cocaine and heroin, are the most significant, growing group of people developing AIDS.[2] As A. R. Moss wrote in 1987, AIDS stemming from intravenous drug abuse is the real heterosexual epidemic that we must acknowledge and control.[3]

Despite more than a century of worldwide abuse of addictive drugs and certainly a much longer history of limited abuse prior to that, almost nothing is known about the immunologic effects of addictive drugs.[4] What we know about heroin and morphine is extremely limited. What we do not know about cocaine and other stimulant drugs would fill a book and keep hundreds of researchers busy for decades.[5] As Dr. Ronald R. Weston of the University of Arizona recently wrote in the preface to the book *Drugs of Abuse and Immune Function*, "What is the role of drug induced immunomodulation on host defense against pathogens and tumor cells? Is there a role for continued intravenous drug abuse in immunomodulation and progression to AIDS after viral infection? Do the changes in immune functions in man cause a change in disease resistance, or is the immune system so resilient that a measurable change may yet cause no significant change in overall resistance? Only some of these questions can be partially answered and only to a varying degree for different drugs."[6] This is the limited knowledge with which we are trying to fight the dual epidemic of drug abuse and AIDS. As Dr. Everett Ellinwood, Jr., and Dr. Frank Gawin comment, it is like trying to fight a war without weapons.

The limited information that does exist suggests that all addictive drugs probably have significant effects on the immune system. The earliest report that I have been able to find concerning the effects of cocaine on immune function is a 1983 paper by Dr. R. E. Faith demonstrating that cell-mediated (T cell) immunity is enhanced in mice following cocaine exposure. The same effect was subsequently demonstrated in HIV-seropositive drug abusers as well.[7] Other investigators have demonstrated that this immunological enhancement is due to increased suppressor T cells (T8 lymphocytes) and natural killer cell activity. The antibody response is also increased.[8] One might think that cocaine use would actually be an antidote to AIDS, but this is not the case.

One limitation of these studies is that all were performed at low doses of cocaine. The immunological effects of cocaine turn out to be dose dependent. At low doses, cocaine stimulates T helper cells; at high doses (5 milligrams per kilogram of weight or greater), substantial suppression is observed.[9] No direct effects on B cells are observed, and therefore no defects in antibody response. These results suggest that increased levels of antibody are not due to more active B cells but to disruption of the normal helper and suppressor T cell functions. Dr. Thomas W. Klein, Dr. Catherine Newton, and Dr. Herman Friedman of the University of South Florida have also reported that high doses of cocaine seriously impair the ability of T cells to proliferate in response to some antigens.[10] Thus, chronic high doses of cocaine should lead to lowered immunity and increased infection. This prediction, however, has never been tested.

The precise implications of these findings for AIDS are still being worked out. The dose dependence of cocaine's effects suggests that significant immune suppression is likely to be observed only in addicts using large, frequent "hits." The observation that suppressor cell activity is increased at even small doses suggests, however, that although T-helper cell numbers and activity may be normal, functional suppression may be occurring. Moreover, the observed stimulation of T cells (including natural killer, or NK, cells), while at first sight encouraging, may in fact have detrimental effects on patients infected with HIV or other T cell–specific viruses such as cytomegalovirus or herpes simplex. T cell activation has the unfortunate consequence of transforming latent viral infections into active ones. Thus, in stimulating the immune system as it does, cocaine may mimic the effects of massive infection, leading to synergistic effects with these immunosuppressive viruses. In fact, Dr. D. Fuchs has argued that activation of the immune system by agents such as cocaine or cofactor infections may be a necessary precondition for the development of AIDS.[11]

Direct evidence that immune stimulation or a synergistic effect on viral infection is caused by cocaine abuse has recently been reported by Dr. P. K. Peterson, and Dr. Ronald Schut and their associates at the University of Minnesota Medical School.[12] They added extremely small quantities of cocaine (10 picograms per milliter) to the medium in which HIV-infected T cells were being co-cultured and observed that the rate of HIV replication, cell killing, and cell infection increased by a factor of three. In short, cocaine promotes retroviral activity, at least in test tube conditions. We do not know what effects it has on herpes viruses, such as cytomegalovirus or herpes simplex, on hepatitis B virus, or on bacterial growth. No one has yet looked, and people may be dying, not only of AIDS but of other syndromes as well, because of this ignorance.

Perversely, low doses of cocaine, such as are often taken by heroin abusers, seem to have some protective effects against T cell losses in those who inject both drugs. R. M. Donohoe and A. A. Falek of the Emory University School of Medicine have reported that addicts using cocaine and heroin concurrently suffer significantly less T helper cell loss than those who use heroin only. They also found that alcohol eliminates the protective effect of cocaine, so heroin addicts who take cocaine and also drink excessively have significantly lower T cell counts than those who do not drink.[13] Since many people with AIDS abuse more than one drug, including alcohol, these results are relevant.

As reference to the work of Donohoe and Falek suggests, heroin has profound immunosuppressive effects. This fact has been known for nearly a century. In 1898, in Elie Metchnikoff's laboratory at the Institut Pasteur, several investigators, including Metchnikoff and his colleague Dr. J. Cantacuzene, demonstrated that macrophages, whose job is to "eat" up foreign invaders, were completely paralyzed at extremely small doses of heroin, so that bacteria that were normally nonpathogenic could easily overwhelm the immune system. Animals treated with heroin and then inoculated with bacteria died at much higher rates, and with much smaller doses of bacteria, than did animals spared the drug.[14]

A continuous, if astoundingly sparse, literature on the immunosuppressive effects of heroin, and its most active ingredient, morphine, exists from the publication of Cantacuzene's experiments until the present. Among the striking, but apparently long-forgotten, results are a number of great interest in the light of AIDS. Gheorghiewski performed a series of experiments in which he vaccinated animals against specific bacteria. Despite the vaccination, animals that were subsequently addicted to narcotics almost invariably developed active infections from these bacteria when reinoculated, and most died. Apparently the immune system was so paralyzed that, although it had the appropriate macrophages and antibody-producing cells available to fight off infection, these could not be mobilized due to the narcotic-induced immune paralysis.[15] Reynolds[16] and Alan Arkin[17] then demonstrated that the same sort of immune paralysis occurred in human abusers of narcotics, with the same increased risk of infection. Arkin showed that the adverse effects could be found when as little as one one-thousandth of 1 percent of heroin was present in the blood. Three French physicians, Dr. Atchard, Dr. Bernard, and Dr. Gagneux, even applied these results to developing a leukocyte test for heroin addiction in 1909.[18]

Beginning in the 1960s, more sophisticated measures of immune function began to reveal other aspects of heroin and morphine abuse. Generalized lymphadenopathy (unusual lymph node swelling) was reported as a general concomitant of intravenous drug abuse and would later characterize AIDS as well.[19] Elevated absolute numbers of lymphocytes and unusually high amounts of antibody were observed next,[20] and, by the mid-1970s, the failure of T cells to respond appropriately to foreign antigens had been demonstrated.[21] These and other studies confirmed that addicts had an increased incidence of many infectious diseases, including opportunistic infections such as candidiasis, pneumonias, tuberculosis, and cancers, which correlated with their decreased immune function.[22]

Subsequent investigators have further characterized this T cell immune suppression. Robert McDonough and his collaborators at Emory University and the CDC reported in 1980 that T cells taken from addicts were unable to recognize and bind to sheep red blood cells. This test, known as the erythrocyte (red blood cell) rosette formation test, is highly correlated with clinically relevant immune suppression.[23] McDonough also found that heroin abusers had significantly altered T cell numbers and types compared with non-drug abusing controls.[24] Subsequent studies by McDonough's colleagues at Emory, Arthur Falek and R. M. Donohoe, confirmed the depressed E-rosette formation and found it to be increased with longer duration of addiction.[25] They also found in 1982 that 24 percent of the addicts they studied had T-helper/T-suppressor ratios typical of AIDS patients, although subsequent retesting of these individuals in 1985 showed that only 12 percent were HIV positive.[26] This result suggested that HIV was not the main cause of immune suppression in these addicts. Similar results were reported by Joseph Layon and his colleagues.[27]

The plausibility of heroin or morphine's causing the immune suppression reported by Donohoe and his associates has been confirmed by animal studies. Animals addicted to opiates display alterations in immune function similar to those reported in humans, including cellular atrophy of immune system organs such as the thymus.[28] Mice treated with as little as 10 to 40 milligrams of morphine per kilogram of weight for four days had their lymphocyte counts drop 10 to 90 percent, showed 25 to 50 percent decreases in T cell activity, up to 80 percent decreases in lymphocyte-mediated killing of *Candida albicans*, and 40 to 80 percent decreases in the ability of macrophages to recognize and ingest *Candida*.[29] As a result of these decreases in T cell and macrophage functions, mice infected with *Candida* were much more prone to develop overt illness and to die. Rats addicted to morphine are also more susceptible to cancer.[30] These findings clearly corroborate the previous observations of increased susceptibility to infection and cancers in human addicts. Indeed, the typical addict who develops AIDS lives less than a year following diagnosis.

An additional problem for all intravenous drug abusers is that the drugs they take are almost always cut with adulterants, such as sugar, talc, and other compounds. These additional compounds are rarely sterile and often contain materials perceived by the immune system as being foreign. The presence of such adulterants can result in repeated episodes of immunologic stimulation. In some cases, these adulterants have been found to be both physiologically and immunologically toxic.[31] They may stimulate latent infections by much the same mechanisms described for cocaine.

Among the infections that opiates may stimulate are latent viruses. Morphine certainly activates latent HIV infection in laboratory studies.[32] Whether it has a similar effect on latent infections of other viruses, such as cytomegalovirus, Epstein-Barr virus, and herpes simplex viruses, has apparently never been studied. It is possible, however, that drug abuse not only impairs immune defenses but actually facilitates many retroviral and viral infections.

One final insult to the immune system that is also associated with opiate abuse is inhibition of mucus secretion.[33] Mucus secretion is a fundamental aspect of the primary layer of immune defense. Mucus not only contains various enzymes, such as lysozyme, that are capable of destroying some types of bacteria, but the continuous flow of mucus moved along by the cells lining the nasal passages, throat, eyes, bronchial tubes, and vagina serves to wash out many foreign contaminants. By interfering with mucus secretion, opiate abuse may greatly increase the probability of upper respiratory and vaginal infections. Given the very high incidence of pneumonias and tuberculosis among drug addicts as compared with other people, this effect appears to be clinically relevant.[34]

Few drug abusers partake of only a single drug. It is not uncommon for addicts to combine addictive drugs with recreational drugs such as amphetamines, barbiturates, ethyl chloride, marijuana, methaqualone, nitrite inhalants, and excessive amounts of alcohol. Although virtually nothing is known about the immunological effects of most recreational drugs, none should be considered safe, and several, including inhalant nitrites, marijuana, and alcohol, are thought to have significant immunosuppressive effects when taken chronically and at high doses.

Nitrites are a class of simple liquid compounds that vaporize readily. Amyl nitrite was first introduced into medicine as a treatment for angina (heart pain) in 1897 and has had various diagnostic uses since. It acts as a muscle relaxant and vasodilator (increasing blood flow through capillaries). Many gay men have used such nitrites to facilitate anal intercourse, and as an aphrodisiac, since it gives users a rush of blood to their extremities (including the penis) and face and a general sense of well-being. Harry Haverkos, an expert on nitrites at the National Institute on Drug Abuse of the Alcohol, Drug Abuse, and Mental Health Administration in Rockville, Maryland, has described amyl nitrite as follows: "It is a clear, yellowish liquid that is packaged in a cloth-covered ampule and administered by inhalation. When the ampule is broken, it makes a snapping sound; thus the nicknames 'snappers' or 'poppers.' In the 1960s sales of amyl nitrite rose sharply because of its abuse as an inhalant for getting 'high' and as an aphrodisiac. This led to its reinstatement as a prescription drug by the Food and Drug Administration (FDA) in 1968. Since 1968, other volatile nitrites containing isomers of amyl and butyl alcohol and butyl nitrite [which are not regulated by the FDA] have been marketed as 'room deodorizers' and sold in bottles containing 10 to 30 ml of liquid for $5 to $15 each. Since their labels state that they are not to be inhaled, they are sold legally over the counter in bookstores, pornographic and 'head' shops, and by mail order under such names as *Locker Room* and *Rush*. Until recently, male homosexuals indulged in these drugs more than did other groups. More recently, abuse among adolescents has been reported more."[35]

Only two studies of the immunological effects of nitrite use seem to be on record. One, carried out by a U.S. government agency, studied the effects of butyl nitrites on mice exposed for two to four weeks to doses measured in parts per million. This is the sort of dose one might get if one used these nitrites as room deodorizers. They found no detrimental effects on immune function but did note that mice exposed to the nitrites for longer periods of time began developing degeneration of the thymus.[36] Since the thymus is essential for activating T cells, these results suggest that long-term exposure to even very low doses of nitrites might be deterimental to immune function.

The other immunologic study of the effects of nitrites was performed to match actual user exposure more closely. When a nitrite user sniffs an ampule or container of volatile nitrite, he or she receives a dose on the order of parts per hundred or parts per thousand in the inhaled air. Blood concentrations as great as a tenth millimolar (one part in ten thousand) have been measured in cases of overdose. Since one of the effects of nitrites is to cause

methemoglobinemia (the binding up of oxygen-carrying hemoglobin molecules in the blood), hospitals often report having to treat
nitrite abusers for oxygen deficits.[37] When mice are given acute,
high doses of nitrites akin to those that might be experienced by regular abusers, experimenters find that natural killer cell activity was
significantly suppressed, and an inverted T-helper/T-suppressor
ratio typical of AIDS patients is observed.[38] It is probable that
these results, rather than the low-dose results, more nearly mirror
the effects on regular abusers of nitrites. Without further studies,
however, the degree to which nitrites cause immunological defects
must remain unresolved.

Other drugs commonly taken by intravenous drug abusers, homosexual men, and some subsets of heterosexuals are also associated with immune suppression. These include excessive and prolonged use of marijuana and alcohol. The data concerning the
immunological effects of both of these drugs, while substantial,
are controversial. It is doubtful that occasional use has any measurable or long-term effects on immune function. Rather, detrimental effects are probably limited to chronic abusers of these
drugs. Both have been found to depress lymphocyte activity and
to lower the resistance of animals and human beings to infection.
Alcohol, for example, has been associated with immune suppression and increased susceptibility to disease literally since the beginning of the modern age of medicine, when Robert Koch noted in
the 1880s that alcoholics were the most frequent victims of cholera.[39] Various immunologists around the turn of the century followed up his observation by demonstrating that alcohol adversely
affected the ability of macrophages to ingest bacteria and the ability of the immune system to mount a complete antibody response
to them.[40] These effects were correlated with increased mortality
following infections of various bacteria in both animals and human beings.[41] These findings have been confirmed and extended
by more recent research, leading Jacques Descotes, in his influen-
tial book, *Immunotoxicology of Drugs and Chemicals*, to conclude that "ethanol is therefore a proven immunotoxicant and
some of the health complications of long-term ethanol consumption should be considered as consequences of these adverse effects
on the immune response."[42] Whether these effects are additive or
synergistic with other immunomodulating drugs is as yet unknown.

Two things need to be emphasized about drug use in the context of AIDS. First, the various immunosuppressive effects occur independent of the route by which the drugs are administered. It does not matter to the immune system whether the drugs are smoked, injected intravenously, injected by "skin popping" (the technique used in tuberculin testing), or taken by oral or nasal routes. As long as the drug appears in sufficient concentrations in the blood for a long enough period of time, it will lead to both short-term and long-term immune suppression, with specific effects on T cells. A common result, particularly of heroin addiction and high-dose cocaine use is an inversion of the T-helper/ T-suppressor ratio, such as that seen in AIDS. Thus, one important feature of drug abuse that has not been taken into account in defining who is at risk for AIDS is the possibility that nonintravenous drug abusers who are exposed to HIV or other immunosuppressive agents by sexual routes will be at as great a risk of AIDS as are intravenous drug abusers. This fact may help to explain why so many sexual partners of intravenous drug abusers—people who are almost all drug users themselves—are developing AIDS despite the fact that they do not share needles.

The second point is that, despite the large amount that is still unknown about the immunological effects of drug addiction, there is little doubt that continued abuse of drugs contributes significantly to the development of AIDS. If we are to understand or control AIDS, we must learn more about drug abuse. Barbara Bayer and Christopher Flores recently reviewed the literature on this subject: "The finding that opiate addicts almost uniformly exhibit compromised immunity remains the primary impetus for a field that is only slowly delineating the mechanisms which subserve immunomodulation by opiates. . . . The high incidence of AIDS in intravenous drug abusers who may have inherently impaired immunity makes even more urgent the need for a collaborative commitment to elucidate precisely the detrimental immunological effects resulting from the adminstration of exogenous opiate drugs."[43] In short, if we are to understand AIDS, we must understand the drugs that AIDS patients inject, snort, sniff, smoke, skin pop, or otherwise imbibe prior to and during their development of illness. To control AIDS we may have to learn how to get millions of addicts off drugs.

Immunosuppression due to Anesthesia and Surgery

Drugs of abuse and recreation are not the only pharmacological agents that can cause clinically relevant immune suppression. Surgical patients also encounter multiple immunosuppressive risks as a result of their operations.

In 1910, Professor Arthur Dean Bevan of the Memorial Institute for Infectious Diseases and Department of Surgery of Rush Medical College of the University of Chicago suggested to Evarts A. Graham that he investigate the possibility that anesthetics might cause some sort of immune suppression leading to the high incidence of postoperative lung infections observed at the Presbyterian Hospital. Graham took up the challenge, employing ether for his studies. He demonstrated first that ether, added to human immune serum from which the red and white blood cells had been removed, had no effect on antibody activity; humoral immunity was intact. When white blood cells were exposed to ether in test tubes, however, a very different story emerged. Phagocytosis, the process by which antigens are engulfed and prepared for presentation to T cells by macrophages, was greatly diminished. Graham also demonstrated this paralyzing effect on immune response to bacterial antigens mounted by human patients following surgery. Although patients had ample antibody titers and their serum would destroy activity in the test tube, the immune response was almost entirely abrogated in their bodies.[1] It looked as if Professor Bevan's suggestion had been right.

In fact, Professor Bevan seems to have had a bit of help in formulating his hypothesis. A German scientist by the name of Snel had published a report in 1903 that ether adversely affected immunity. Snel, in turn, had been influenced by the work of Cantacuzene, Gheorghiewski, and other scientists on the immune-paralyzing effects of alcohol and opiates. All of this work had been reviewed in 1904 by Bevan's colleague at Rush Memorial College, pathologist George Rubin, who had demonstrated that rabbits anesthetized with ether were much more prone to bacterial infection and death than animals who were not.[2] He had not, however, made the leap to human experimentation that was taken by Graham.

Fifty years of subsequent research validated and extended Graham's work, and in 1974, Dr. Richard J. Howard and Dr. Richard L. Simmons wrote, "Infection, local and systemic, is still a major cause of morbidity and mortality after injury and operation. Microbiologic contamination of wounds is inevitable, and clinical and experimental data lend support to the contention that patients and animals with trauma or surgical operations have altered defense mechanisms. While local and systemic factors, such as wound contamination, collections of fluids and secretions, immobilization in bed, drug therapy, respiratory embarassment, and age are important predisposing factors in the pathogenesis of surgical infection, in the final analysis, it is the body defense mechanisms that are deficient in ridding the body of invading organisms thereby permitting infection to become established."[3]

It has now been established that ether, halothane, nitrous oxide, pentobarbitol, and many other intravenous and local anesthetics cause decreases in absolute numbers of circulating lymphocytes and antibody production lasting a few days to weeks. Anesthesia also appears to potentiate tumor growth and to interfere with allergic responses and phagocytosis. In consequence, cancers and infections are more prevalent and more serious among surgery patients.[4] Antibody responses, even in preimmunized animals, are greatly depressed so that their ability to fight off infection is vastly decreased.[5] T cell activity is particularly affected, so that responses to antigens such as *Candida albicans* and mumps virus are abnormally low for more than fifteen days.[6] Studies of the lymphocyte responses of animals and people after anesthesia but prior to surgery have sorted out which effects are due to pharmaceutical agents and which to surgical trauma.[7] Not surprisingly, both the dose of anesthetic and the number of exposures to anesthetics are factors in the degree of immune suppression experienced by a patient.[8] The alterations in immune function can be so severe that an inverted T-helper/T-suppressor ratio can result, thereby mimicking for a few days, or as long as a week, one of the primary aspects of immunodeficiency typical of AIDS.[9]

Surgery patients are not the only people at risk for this sort of immune suppression. Health care personnel, including anesthesiologists, surgeons, dentists, and nurses, who are frequently exposed to inhalation anesthetics such as ether, halothane, or nitrous oxide, are at increased risk for long-term immune depression, serious infections, and cancers such as lymphomas.[10] It is also relevant to understanding AIDS that nitrous oxide, popularly called laughing gas, besides its anesthetic uses, is employed by some health care personnel and drug abusers as a way to get high.

The physical trauma of having surgery has been implicated as being severely immunosuppressive. Thus, the same loss of circulating lymphocytes, depressed cellular immune functions, and increased susceptibility to infection have been documented for trauma patients as for anesthesia patients. The vast majority of patients who encounter one, of course, are exposed to the other.[11]

In addition to anesthetics and traumatic causes of immune suppression, many surgery patients receive opiate anesthetics and analgesics (for example, morphine and codeine) to dull their postoperative pain. Such treatments have been linked directly to increased occurrence of infections with latent viruses such as herpes simplex.[12] Thus, from an immunological standpoint, there is little difference between the multidrug using addict and the surgery patient save for the fact that the one chronically abuses drugs, while the other is subjected to them for a period of time measured only in days or weeks. The important question in the light of AIDS and the risk of HIV infection is what else is happening to these patients during those days or weeks of acute immune suppression. As we shall see, there is quite a bit.

Immunosuppression due to Pharmaceutical Agents

"A wide variety of pharmacologic agents are capable of altering certain immune functions which are considered an integral part of the normal host defense system. . . . In many cases, immune system dysfunction secondary to various drugs is not the primary reason for their administration. . . . Nonetheless this observed immune system dysfunction can be a striking side effect which may lead to significant potential morbidity."[1] Thus begins a review of "immunosuppression secondary to pharmacologic agents" written by Henry C. Stevenson and Anthony S. Fauci of the National Institute of Allergy and Infectious Disease in Bethesda, Maryland. It is a particularly important review, not so much because of what it says but because Fauci, as director of the National Institute's immunology work, is also its primary spokesman and leader in AIDS research. Thus, what Fauci and Stevenson included or did not include in their review can be viewed as a good indicator of what the NIH considers relevant to understanding AIDS and what is not.

In their review, Stevenson and Fauci discuss a number of drug classes known to be immunosuppressive, including compounds upon which their own laboratories have done significant work, such as the glucocorticoids, cortisone, and other steroids. They also cover in some detail various cytotoxic drugs used in cancer chemotherapies, aspirin, and even gold salts (used for treating rheumatism and arthritis). These agents, they state, are the only ones with significant immunosuppressive activity. To a large extent, this list is similar to that utilized by the CDC in defining the treatments that are exclusionary for diagnosing AIDS. Fauci and Stevenson, like the CDC, write off all other pharmacological agents as being "clinically trivial" in terms of their immunosuppressive potential.[2]

Stevenson and Fauci are demonstrably wrong, as the previous section on anesthetics demonstrates. Most experts agree that anesthetics, at least in conjunction with surgical trauma, are responsible for immune suppression that leads to increased morbidity and mortality in surgery patients. If anything is clinically relevant, illness and death certainly are. Moreover, most textbooks on immunotoxicology (the study of drugs and chemicals that are toxic to the immune system) agree that Stevenson and Fauci's list is dramatically incomplete. The list of known and suspected pharmaceutical agents prescribed by physicians for proper medical reasons that may or do cause immune suppression is frighteningly long and includes some drugs that are so commonly prescribed as to seem beyond suspicion.

It is certainly well established that most antibiotics can cause immune suppression when used in acute, high doses or at moderate doses taken over long periods of time.[3] Many reports were published during the 1950s that high doses of penicillin compounds often resulted in opportunistic infections with various fungi and yeasts, such as *Candida albicans*.[4] Chloramphenicol reduced both humoral and cell-mediated immunity to such an extent that foreign tissue graft survival was demonstrable. Tetracycline, streptomycin, kanamycin, gentamycin, and neomycin are immunosuppressive. Cefotaxime, anikacin, mezlocillin, piperacillin, and clindamycin have been described as "immunomodulatory."[5]

Specific inhibitory effects have been demonstrated on cell-mediated immunity for many of these antibiotics, suggesting T cell–mediated mechanisms.[6] It is thought that some of the antibiotics depress immune function by tying up zinc, selenium, calcium, and other minerals necessary for cell division.

Various antiviral agents, including acyclovir (a common treatment for herpes virus infections), ribavirin, and AZT (zidovudine or Retrovir), have been reported to produce T cell–mediated immune suppression.[7] The effects of AZT are particularly worrisome, since at least half of those patients with full-blown AIDS who take AZT develop drug-associated anemia and/or leukopenia (loss of white blood cells) within a year to eighteen months after beginning therapy. Anemia is, in and of itself, associated with profound immune suppression, and leukopenia is, by definition, a symptom of immune deficiency. These problems require additional corrective measures that include both drugs to force red cell and leukocyte regeneration (the immunological side effects of which are unknown) and, unfortunately, repeated blood transfusions—themselves profoundly immunosuppressive and add to the suppression already being experienced by AIDS patients.

Microbial infections are also treated with antibiotics that interfere with folate metabolism, including trimethoprim, the sulphonamides, and pyrimethamine.[8] Folate, as we will see in the section on malnutrition, is a vitamin essential for proper immune function.

Another set of antibiotics are also thought to have immunosuppressive effects. These are the antiparasitic agents, used to treat various infestations such as helminths (parasitic worms, including those that cause schistosomiasis, a disease endemic to central Africa), protozoa, and amoeba (including many of the organisms associated with AIDS such as Toxoplasmosis, giardiasis, and amebiasis, as well as *Pneumocystis carinii* pneumonia), and *Plasmodium falciparum* (the agent that causes malaria). These antiparasitic agents are of interest in considering the possible immunosuppressive risks of gay men who develop repeated bouts of gay bowel syndrome, a confluence of several of these parasites, and the possible risks of people living in developing nations where such parasites are endemic.

Of particular interest in understanding AIDS risks in Africa is the evidence that various antiparasitic imadazole drugs (Clotrami-

zole, Ketoconazole, and related compounds) and many of the antimalarial drugs, particularly chloroquine, are immunodepressive.
Studies of T lymphocyte proliferation, cell killing, phagocytosis,
and responses to T cell mitogens in laboratory settings uniformly
show 25 to 100 percent decreases in lymphocyte activity. This immunomodulatory effect of chloroquine has recently been applied
to the alleviation of rheumatoid arthritis, an autoimmune disease
of the connective tissue and joints. Apparently no adequate studies
of possible immunological toxicity of most antiparisitic or antimalarial drugs have been carried out under clinical conditions, so it is
difficult to determine the impact of widespread use of these drugs
among homosexual men and central Africans.[9]

One thing seems highly likely: The large number of antimicro-
bials that are thought or known to cause immune suppression
strongly suggests that people who require chronic or frequent
treatments with such pharmaceutical agents are likely to develop
clinically significant immunological problems. Thus, the greater
the disease load—the number of concurrent infections—a person
encounters, the greater the number of pharmacological agents one
is likely to have prescribed, and the greater the number of immunosuppressive agents at work. In the long run, this constant as-
sault on the immune system cannot be good.

In addition to antibiotics, many other medical treatments are
immune suppressive. Some of these are severely so, including high-
dose or long-term treatments employing steroids such as cortisone
and related compounds that are often prescribed for asthma,
chronic inflammatory diseases such as rheumatism and arthritis,
or the sorts of joint injuries many athletes and hemophiliacs develop. Cortisone shots are often given to athletes who have joint
injuries to reduce the swelling and pain so that they can compete,
despite their body's clear-cut signs to the contrary. Corticosteroid
creams are commonly prescribed for treating the itching and in-
flammation of hemorrhoids and a variety of venereal infections
of both the vagina and the anus, such as herpes simplex. These
preparations can be absorbed readily through vaginal, penile, and
rectal tissues, and so they represent an immunosuppressive risk
not only to the regular user of the cream, but to his or her regular
sexual partner(s) as well.[10]

Related to the corticosteroids and glucocorticoids are another
group of widely used drugs, the anabolic steroids. Anabolism is

the process of building things up in the body, or promoting growth. Anabolic steroids increase body mass and the synthesis of essential components such as proteins. No data seem to be available concerning the immunologic effects of anabolic steroids, which have become drugs of abuse among many high school, college, and professional athletes who desire the bigger, stronger, more muscular bodies that such steroids can help to create. If the effects of anabolic steroids are similar to those of other steroids, they represent a great and unappreciated immunosuppressive risk in addition to their already identified, extensive health risks.

Tranquilizers, one of the most overprescribed drugs in the United States, represent another immunosuppressive risk. Opportunistic infections, such as chronic oral candidiasis, and extremely high rates of pneumonias and other severe infections have been associated with tranquilizer therapy among psychiatric patients in private and institutional settings for several decades. The tranquilizers usually involved were chlorpromazine, imipramine, the phenothiazines, and their derivatives.[11] Subsequent studies have implicated nearly all of the newer derivatives of these compounds, with the exception of the benzodiazapines (such as Diazepam) as having significant immunosuppressive effects in cell culture as well. Indeed, several drugs affecting central nervous system function, including the tricyclic antidepressants, have been shown to bind directly to receptors on T cells and apparently modulate immune function by directly interfering with the ability of these T cells to replicate.[12] Once again, few, if any, controlled clinical studies in human patients have apparently been attempted.[13] The one major exception to this generalization is lithium. Lithium treatments have been uniformly shown to increase all measured immunological parameters in human patients.[14] These results suggest that, with some important exceptions, high or chronic doses of many tranquilizers create immune depression sufficient to make some patients susceptible to opportunistic infections and to decreased resistance to all diseases.

In summary, when physicians treat one disease, they may inadvertently promote susceptibility to other diseases through the unexpected or unappreciated immunological side effects of the drugs they prescribe. Such inadvertent side effects are unlikely to be of consequence to otherwise healthy people suffering from occasional infections, but they may be quite important in people with chronic diseases or frequent acute infections that require constant pharmacological management. As we shall see, people in recognized AIDS risk groups, and just about everyone in sub-Saharan Africa where AIDS is endemic, are prone to many of these pharmacological causes of immune suppression in addition to many other immunological affronts.

Immunosuppression Due to Malnutrition

Allogeneic and drug agents are only the beginning of the immuno-suppressive problems that afflict many people who are at highest risk for AIDS. Another risk is malnutrition.

One of the earliest examples of *Pneumocystis* pneumonia reported in an adult was the case of a woman diagnosed as suffering from severe vitamin A deficiency. Her case was complicated by disseminated cytomegalovirus infection.[1] Similarly, the vast majority of *Pneumocystis* cases in children during the period 1930 to 1960 also involved "large numbers of infants, most of whom had been born prematurely or had been severely malnourished in the early months of life."[2] Malnutrition, leading to the specific lack of particular vitamins or minerals, and general starvation continue to be well-recognized risks for the development of opportunistic infections today. This risk has not been investigated as a possible contributory factor in AIDS, despite extensive evidence that it may be relevant, especially in drug addicts and their infants.

The many causes of malnutrition range from inadequate dietary intake of crucial nutrients to outright starvation, and from the side effects of various diseases to the results of certain surgical procedures. Medical reporter Berton Roueché tells the cautionary story of a man who ate every meal for several years at a fast-food restaurant and developed severe malnutrition, despite an abundant intake of calories.[3] Clearly, it is not how much is eaten but what is eaten that counts. In some cases, however, the total intake is simply insufficient. Many drugs, such as the amphetamines, barbiturates, and addictive drugs, demonstrably decrease appetite. Alcoholics often imbibe so many nutritionally empty calories through their drinks that they have no appetite for other, healthier foods. Moreover, alcohol uses up some of the B vitamins during its metabolism and interferes with the absorption and storage of others. Chronic diarrhea due to various infections of the lower gut, such as worms, protozoa and amoeba, cytomegalovirus, or other disease agents, can so inflame the intestines that malabsorption of nutrients results. In such cases, even a nutritionally adequate intake will not result in proper nutrition, because most of the nutrients pass through the gut with the diarrhea. Various surgical procedures that involve taking out part of the gut or bypassing it for a time to promote healing can also result in malabsorption. Finally, many pharmacological agents, including chemotherapeutic agents and some of the antibiotics, selectively interfere with the absorption of crucial vitamins and minerals. Erythromycin, for example, is not supposed to be taken with milk or milk products because the calcium in the milk binds to the erythromycin, resulting in a complex that inactivates both. Other antibiotics apparently bind specifically to minerals such as zinc, and many destroy the bacteria in the gut that produce various B vitamins and vitamin K. Thus, anything from consistently poor eating habits to prolonged chemotherapy can cause significant malnutrition.[4]

The results of malnutrition can be severely debilitating from a purely immunological point of view. Dr. Ranjit Kumar Chandra, director of the Immunology Laboratory at the Memorial University of Newfoundland and one of the world's leading experts on the immunological results of malnutrition, has written that "nutritional deficiency robs the host of many host defenses . . . [producing] multiple rents in the protective umbrella of host defense."[5] These rents can include failure of the integrity of the skin, mucosa, and secretory enzymes (such as lysozyme) that represent the outer barriers against infection; decreased phagocytosis by macrophages, leading to poor response to infection; decreased ability to mount T cell responses and to produce adequate levels of functional antibodies; and failure of the complement and lymphokine systems (the chemicals necessary to activate antibodies and lymphocytes once an antigen is present). Of particular significance with regard to AIDS is the fact that malnourishment can lead to the same profound loss of T cells (fewer than 500 per cubic millimeter of blood) and the inversion of T-helper to T-suppressor populations that are found in the disease.[6] Antibody (B cell) responses are also significantly decreased despite a large increase in the numbers of these cells and the amount of antibody secreted. In short, malnutrition has the potential to impair grossly virtually every aspect of the normal immune response.[7]

Among the best studied causes of immune suppression due to malnutrition are those associated with the lack of essential vitamins and minerals. A severe or prolonged deficiency of several vitamins found mainly in animal foods such as liver, meat, eggs, and fish, including vitamin B6 (pyroxidine), vitamin B12, and folic acid (pteroylglutamic acid, which can also be found in dark green leafy vegetables, beans, and peas), can cause anemia—that is, a failure to produce sufficient red and white blood cells. All three vitamins—B6, B12, and folic acid—are essential for DNA synthesis, and therefore all rapidly dividing cells, including red and white blood cells, require these vitamins. If they are lacking, cell division does not proceed normally, and an absence of functional lymphocytes of all types may result.[8] It is interesting to note that a number of relatively common pharmaceutical agents are associated with folic acid deficiencies, including many cancer chemotherapeutic agents and some of the oral contraceptives.[9] Folate deficiency is also found in pregnancy, chronic alcoholism, drug addiction, and malabsorption syndromes, and it is extremely common in impoverished countries and the inner cities of industrial nations.[10]

Vitamin A deficiency is another common cause of immune deficiency associated with under- or malnutrition. Vitamin A, or its metabolic precursor, carotene, is a common constituent of meats, eggs, fruits, and vegetables, and it is added to fortified milk. Its lack can lead to night blindness, impaired skin and mucous membranes, and immune suppression. Deficiencies of vitamin A in human beings and animals result in increased frequency of infection, as well as increased mortality following infection. T cell numbers and activity are preferentially affected.[11] In the light of these findings, it is not surprising that the woman deficient in vitamin A mentioned above developed several opportunistic infections associated with T cell deficiency.

Several essential minerals are associated with immune deficiency when they are present in inadequate amounts. Perhaps the most common is iron, which is most readily available in red meat, egg, and other protein products. Iron deficiency is extremely familiar in developing countries worldwide, where protein is always in short supply, and it is unusually frequent among menstruating women (who lose significant iron through menstrual bleeding) and among the poor in industrial nations. Iron deficiency results in anemia. Like the other anemias, iron deficiency anemia, which is prevalent worldwide, particularly in developing countries, results in lowered T cell numbers, poor cell-mediated immune responses, failure of lymphocytes to proliferate in response to antigenic stimulus, and increased susceptibility to infection and death.[12]

Zinc, found in essentially the same sources as iron, is essential to immune function. It acts as a cofactor for many enzymes associated with DNA replication. The DNA inside lymphocytes must replicate in order for cell division to occur. Cell division, in turn, is a necessary part of the cloning or proliferative process by which the immune system mounts an effective response against infection. Zinc deficiency also results in failure of the lymph tissues to develop properly, low numbers of T cells, failure to respond to T cell antigens, and impaired wound healing. It can result from an inadequate diet, cirrhosis (sometimes associated with alcoholism), chronic or high-dose antibiotic use, hepatitis, and malabsorption syndromes.[13] In fact, there is a human disease, acrodermatitis enteropathica, that is caused by a genetic defect in the enzymes required to absorb zinc from the gut. Children born with this disease have severe immune deficiency similar to AIDS and often die of the same opportunistic infections. A similar state of malabsorption can be induced by chronic gut infections such as those that sometimes precede or accompany AIDS.[14]

At least one other mineral has been widely described as being essential to immune function: selenium. Selenium is found in high amounts in seafood, whole grain cereals, meats, eggs, and garlic. It acts as a cofactor with vitamin E, and both have antioxidant activity associated with increased immune function. Conversely, selenium (and vitamin E) deficiency reduces antibody responses and superoxide production by lymphocytes, making them less capable of fighting off infectious agents.[15] Experiments in animals have demonstrated that selenium deficiency results in an ineffective response to various opportunistic pathogens such as *Candida*, and increases the severity of viral infections.[16]

Finally, general malnutrition, especially as measured in terms of protein-calorie malnutrition, is almost invariably accompanied by immune deficiency. Protein-calorie malnutrition can take two forms: marasmus and kwashiorkor. Marasmus means "wasting away" and is literally the result of starvation. Kwashiorkor is Ghanian for "second child" and refers to the fact that when a second child is born too soon after the first, the first becomes sick from lack of breast milk. Kwashiorkor results specifically from lack of protein and can develop even in children or adults who otherwise have an adequate caloric intake. Because protein is a necessary component of all cells, conditions that lead to significantly low levels of proteins in the blood result in inadequate production of all types of rapidly dividing cells, including lymphocytes. Moreover, it must be remembered that the crucial receptors (such as the MHC class II and CD4 receptors on lymphocytes) and antibodies involved in immunity are themselves proteins, which cannot be synthesized in sufficient amounts in the absence of adequate nutrition. It is also worth noting that both marasmus and kwashiorkor are associated with anemia and vitamin and mineral deficiencies; the additive effect can be devastating. In fact, it is well known that malnutrition increases susceptibility to many opportunistic infections, including *Pneumocystis carinii* pneumonia.[17] Thus, malnutrition, as we shall see, is very relevant to understanding the risk for AIDS.

Immunosuppression due to Blood Transfusions and Clotting Factors

As if the preceding set of immunosuppressive agents were not enough to cause us to pause for thought, yet another set of agents is known to be immunosuppressive and is highly associated with people at risk for AIDS: Blood and blood-derived products such as the clotting factors used by hemophiliacs to control their susceptibility to bleeding. Recognition of the immunosuppressive risk of blood and blood factors is relatively recent, as the following story indicates.

In 1976, a sixty-year-old man—call him "Mack"—developed incontinence. It was found that Mack had incurred a bladder papilloma (a type of benign tumor) that interfered with the closing of his ureter, so urine dribbled out whenever it began to accumulate in his bladder. The papilloma was treated by burning it off, a process called diathermy. Two years later, three more tumors appeared and were treated in the same way. The treatment sites became infected within twenty-four hours, and Mack developed peritonitis, an infection of the interior cavity of the gut. He was treated with antibiotics, without effect, and collapsed the next day. Exploratory surgery revealed that both his bladder and intestines had been perforated by infection. Corrective surgery was successfully performed. Despite aggressive antibiotic treatment, however (or perhaps in part due to it), Mack developed pneumonia in both lungs within three weeks of his surgery, followed quickly by complete failure of his kidneys. His surgical wounds reopened, and new infections and new perforations in his organs were discovered. Despite further surgery, he bled continuously from several orifices, and his pneumonia overwhelmed him. Mack died ten days later.

The cause of Mack's death was listed as overwhelming cytomegaloviral pneumonia, but the pathological report lists a who's who of opportunistic infections. His lungs were found to contain cytomegalovirus, the fungus *Aspergillus*, and the bacterium *Pseudomonas*. The peritoneum yielded more *Pseudomonas* as well as *Candida albicans*. In other words, Mack died of surgery-associated AIDS. The pathologists explained the obvious evidence of immune suppression in the absence of any underlying factors or immunosuppressive treatments by suggesting that the "very prolonged, complex, and ultimately fatal set of catastrophes after a minor urological operation . . . [along with] the length and severity of the illness created a situation equivalent to immune paralysis." They also mention, but only in passing, that during "the course of this six-week illness, 35 units of blood were transfused." Other than the fact that the blood may have been the source of the cytomegalovirus infection, the pathologists found nothing striking about the blood transfusions.[1]

During the weeks that Mack lay dying, experiments being conducted by another set of physicians may very well have caused the sick man's doctors, had they known about them, to rethink their attitude toward transfusions. Oscar Salvatierra, Jr., a physician at the University of California, San Francisco Medical Center, was leading a team intent on developing a new approach to renal (kidney) transplant procedures. The team members had noticed that a number of previous investigators had reported better survival of both transplanted organs and of animals receiving the transplanted organs when the animals were pretreated with blood transfusions from the donor animal. Dr. Gerhard Opelz and Dr. Paul Terasaki, two members of the Salvatierra team, had themselves unexpectedly observed the same phenomenon in a pair of patients who had been transfused with relatives' blood prior to undergoing transplant procedures.[2] And Salvatierra and his colleagues had noticed that transplant patients who had received unrelated blood transfusions prior to their operations also accepted their transplants with fewer problems.[3] They convinced the Human Use Committee at UCSF to allow them to pretreat patients undergoing renal transplants with blood transfusions from their donors to see if they could increase the probability that the organs would be accepted by the recipients.

The outcome of these experiments was anything but foreordained. Two reports published during the time that Salvatierra was conducting his experiments quite explicitly argued that pretreating organ transplant patients with blood transfusions was not a medically sound concept.[4] If anything, these investigators argued, such pretreatment would only sensitize the recipient to the very antigens that would then be presented to it en masse in the form of the donor kidney. The subsequent immune response would therefore be much greater—and the risk of organ rejection much higher.

The reasoning was logical. Under normal circumstances, the immune system quickly recognizes foreign tissue as nonself, mounts an active immunological response, and eliminates the foreign tissue. To get around this rejection of the foreign tissue during an organ transplant, physicians attempt to match the donor's tissue type as closely as possible to the recipient's, so that few differences are present for the immune system to recognize as nonself. Even so, unless the transplanted tissue has come from an identical twin, all organ recipients must have their immune systems chronically suppressed with special drugs in order to accept the foreign tissue. And in spite of this chemical immunosuppression, most transplant recipients still experience one or more episodes of organ rejection in which the immune system, weak as it is, manages nonetheless to attempt to kill off the foreign tissue. All too often it succeeds. Many medical researchers therefore expected that pretreating people with the blood of the organ donor would serve only to strengthen the already quite potent ability of the immune system to reject the new organ by giving it a taste of what it was soon to experience and the time to prepare a formidable response.

As in so many other scientific discoveries, experience belied logic. An incorrect assumption had been made about the immunologic effects of exposure to blood. Salvatierra and his co-workers clearly demonstrated during 1978 and 1979 that pretreating organ recipients with blood transfusions from their organ donors not only failed to enhance rejection episodes but actually increased survival of transplanted organs. They speculated that one of the primary ways in which these blood transfusions affected the immune system was by suppressing it.[5]

The speculation that blood transfusions cause immune suppression has turned out to be correct. It has been demonstrated that blood transfusions directly suppress T cell–mediated immunity in both animals and humans in a dose-dependent fashion resulting in inverted T4/T8 ratios similar to those seen during the early stages of AIDS.[6] The more blood a person receives, the more immune suppressed he or she becomes. Although this effect can be beneficial for patients receiving organ transplants, it is not at all beneficial for those with other medical problems. Several studies have demonstrated in both animals and humans that patients who receive transfusions during cancer therapy have significantly lower probabilities of survival than those who do not receive transfusions.[7] People whose immune systems are not up to par are less able to fight off any threat to the integrity of their body—whether it is the threat of infection, an invasion of foreign tissue, or cancer.

These results are of great significance not only for understanding why Mack developed such profound immune suppression that he died of a potpourri of opportunistic infections following his multiple transfusions in 1978; they are also important for understanding the occurrence of such opportunistic infections among transfusion patients in the AIDS era. Approximately 2.5 percent of all AIDS cases are associated with transfusions. The typical patient who develops AIDS following blood transfusions has had sixteen to twenty-one units of blood—some five times the average amount received by the average transfusion patient.[8] These are not usual hospital patients or even typical surgery patients. These are people who are literally on death's door either before their operations or as a result of problems encountered during or after their original surgeries. To say that their only risk of immune suppression is an HIV infection is nonsense. There is no doubt that these people become severely immunosuppressed, whether there is HIV present in the blood they receive or not.

The immune suppression caused by the blood transfusion itself is only the beginning of the story. The blood itself all too often contains one or more viral disease agents or reactivates latent infections in the patients who are transfused. These viruses include cytomegalovirus, Epstein-Barr virus, and hepatitis viruses—all of which can cause immune suppression characterized by inverted T-helper/T-suppressor ratios. Until recently, the acquisition of these infections from blood transfusions was not considered significant, since the rate of infection was not very high and rarely caused overt disease in patients with normal immune function.[9] A number of patients infected by transfusions have experienced clinical symptoms, however. For example, one patient exposed to cytomegalovirus developed night sweats and fever for about ten days after being released from the hospital. A second developed prolonged fever and a sudden drop in his white blood cell count.[10] Such symptoms are common enough that physicians have a clinical name for them: posttransfusion syndrome. Not only is this syndrome accompanied by significant depression in immune function, in patients with additional immunosuppressive risks and in infants this syndrome is often fatal.[11]

In the light of these immunosuppressive risks, the fact that Mack died of symptoms that differ in no way from AIDS suggests that at least some HIV-negative patients have always died of various forms of acquired immune suppression and probably continue to do so whether HIV is present. The failure to perform the necessary studies to determine whether this conjecture is true hampers studies of AIDS and understanding of the efficacy and limitations of blood transfusions as one of the most widespread therapies currently employed by physicians.

Surgery patients are not the only ones who receive transfusions. So do patients with sickle cell anemia, a genetically inherited trait in which the hemoglobin that transports oxygen in red blood cells is defective. One result of this defect in the hemoglobin is to cause the red blood cells, which are normally round, to take on sickle-like shapes when oxygen is sparse. Two other effects are poor oxygen transport and anemia, a relative lack of red blood cells. The benefit of the sickle cell trait is that it confers resistance to malaria. Thus, the trait evolved and became widespread in sub-Saharan Africa, where it continues to affect a significant proportion of the population. Many blacks in other parts of the world also carry the trait, as do other people in whom it occurs spontaneously in rare instances. The anemia associated with sickle cell disease often necessitates frequent blood transfusions to raise the functional red cell level back to normal. People in both Africa and other parts of the world who are frequently transfused develop many of the same symptoms as do people with AIDS: an extremely high incidence of overwhelming infections often resulting in death, statistically significant increases in the incidence of active cytomegalovirus, Epstein-Barr virus, and herpes simplex infections, lymphadenopathies, loss of T cells and T cell functions, inverted T-helper/T-suppressor ratios, and a vastly increased probability of HIV infection. Some of these immunological abnormalities, including the susceptibility to opportunistic infection and impaired T cell activity, were observed prior to the recognition of AIDS.[12]

Another group of people who also receive blood transfusions or components of blood and are also at increased risk for AIDS are hemophiliacs. Until about 1970, most hemophiliacs were treated with blood serum, from which the red and white blood cells were filtered out. This serum contained the clotting factors that hemophiliacs do not produce or produce in inadequate concentrations. There are apparently no studies as to the possible immunological effects of serum treatment, but sufficient problems, both of immunological reactions to the serum and its failure to be universally effective, led researchers to investigate ways in which to produce the specific blood clotting factors needed by different types of hemophiliacs. Around 1970 the first techniques were developed for mass producing such concentrated clotting factors. The most common of these, factor VIII and factor IX, are associated with a significant increase in the presence of circulating immune complexes (diagnostic for autoimmunity)[13] and incidence of immune suppression in the absence of HIV contamination.[14] The cause of this acquired immune suppression is not known, although immunological response to the blood factors themselves,[15] multiple infections with viruses contaminating the concentrates, and autoimmunity have been suggested to play a role. What is certain is that hemophiliacs, like people with sickle cell disease and people who receive massive transfusions, are often immunologically deficient in the total absence of HIV infections.

Immunosuppression Associated with Age

One final noninfectious immunosuppressive risk must be added to this extensive list: age. There are two groups of people who incur demonstrable immune deficiencies associated almost completely with their age: premature infants and the elderly. Although infants and the elderly make up only a small proportion of AIDS patients, it is important to recognize their additional risks. Moreover, we will see later in this book that significant evidence exists implicating an aging factor in the increased susceptibility of hemophiliacs, blood transfusion patients, and their sexual partners to HIV infection and the development of AIDS. Most gay men and intravenous drug abusers who contract AIDS do so at such uniformly young ages that no such age-related factor has been implicated for them, although as survival rates increase, such a factor may one day become apparent.

"In general, T cell–dependent cell-mediated functions decline with age," writes Dr. Marguerite Kay in the introduction to a review of the subject. She then goes on to summarize the essential facts as follows. The older a person gets, the more likely he or she is to experience immune deficiency. Normal immune function can begin to decline as soon as sexual maturity is achieved, since the thymus involutes and begins slowly to atrophy at that time in most people. Because the thymus is necessary for activating new T cells, T cell replacement and activity declines along with thymus activity. Thus, the older a person becomes, the poorer his or her T cell–mediated immune functions become. These changes have been implicated in the increased susceptibility to infections (one reason that older people are urged to get flu shots each winter), to cancers, and to a variety of autoimmune complications.[1]

Premature infants also tend to be immune deficient. There are two possible contributory causes. First, they are born with fewer T cells than term babies, even when age and weight are factored in. They therefore display a diminished degree of T cell activity, and they also have unusually low levels of antibody. In consequence, preterm infants experience much higher rates of infection and death than do term infants.[2] A second problem that sometimes increases the immunological problems of the preterm infant is that his or her mother may have some form of disease state involving acquired immune deficiency that is implicated in the premature delivery. Since an infant is normally protected from infection in part by the immune system cells and antibodies it inherited from the mother, immunological problems in the mother can cause additional problems for the infant.[3] Thus, both gestational age, and old age must be considered possible factors increasing susceptibility to all infections, including HIV and other disease agents associated with AIDS.

CHAPTER 4

MULTIPLE, CONCURRENT
INFECTIONS AND AIDS

Antigenic Overload

Noninfectious immunosuppressive agents are only one of three classes of factors that can impair immune function in people at risk for AIDS. The second set, discussed in this chapter, are infectious agents other than HIV that cause immunosuppression themselves or act synergistically with each other and HIV to create increased immune suppression. Unlike most noninfectious immunosuppressive agents, some infectious agents have been considered as possible contributory factors in AIDS. Prior to the discovery of HIV, some, such as cytomegalovirus and human T cell lymphotropic virus type 1, were serious contenders for the leading role in this tragedy.

The importance of multiple, concurrent infections in AIDS was first pointed out shortly after AIDS was recognized by Dr. C. Navarro and Dr. J. Hagstrom in a brief letter to the British medical journal, the *Lancet*. Noting that every AIDS patient was actively infected with half a dozen or more microbes simultaneously, they suggested that the immune system might simply become overloaded and eventually collapse.[1] Alvin Friedman-Kien, one of the first physicians to diagnose AIDS in New York, City put forward a similar idea.[2]

The antigenic overload theory has come in for some hard knocks since 1982. Mirko Grmek, a physician and historian of medicine, comments in his book, *History of AIDS*, that "this explanation was but a metaphor, as unscientific as that of 'mental overload' in certain psychiatric disorders. Often in the past, human populations had sustained a much greater challenge to their immune system than this recent group of American homosexuals."[3] Grmek is wrong. As this chapter will demonstrate, perhaps no other group in history has ever sustained anything like the disease load experienced by highly promiscuous homosexual men and intravenous drug abusers, with the sole exception of peoples who live in Third World nations such as the sub-Saharan regions of Africa where AIDS is also endemic. This fact does not mean that AIDS is a result of antigenic overload, but it does signify that AIDS appears, almost without exception, among groups of people who have an unusual panoply of immunosuppressive risks. Moreover, specific mechanisms have been elucidated to explain how combinations of infections may cause immune defects that individual infections do not.[4]

The primary goal of this chapter is to summarize the evidence that the typical AIDS patient has multiple, concurrent infections made up of some subset of bacteria, viruses, fungi, yeast, and parasites. A substantial number of these agents are known to infect macrophages, T lymphocytes, and B lymphocytes and to disrupt normal immune function. Moreover, several of the viruses are known to induce inverted T-helper/T-suppressor ratios and to result in lymphocyte death similar to those attributed to HIV. Thus, it is impossible to attribute the immune suppression characteristic of AIDS solely to HIV.

A second goal of this chapter is to show that multiple concurrent infections begin to affect people at risk for AIDS prior to and in the absence of HIV infections. Thus, although many AIDS researchers, such as Robert Gallo, consider all infections other than HIV to be opportunists that take advantage of the immune suppression supposedly caused by prior infection with HIV, many, if not most, infections are present prior to HIV infection.[5] They set the stage for AIDS.

Third, combinations of various subsets of these infections can have pathological consequences that are quite different from their individual effects. People who develop multiple, concurrent infections are at risk for unusual complications, among them a number of syndromes associated with AIDS that cannot easily be attributed to HIV (some of which will be discussed in this chapter) and autoimmune complications (discussed in the next chapter).

Finally, HIV infectivity and cell killing depends on the presence of activated, reproducing lymphocytes, such as those produced during infection. If HIV can be transmitted and its effects promoted only in the presence of other immunologically active agents, as Luc Montagnier and other researchers are proposing, then it is impossible to undertand AIDS, and impossible to control it, without understanding and controlling potential cofactors.

The Diseases Associated with AIDS

In 1983, Dr. Martha Rogers and her colleagues at the CDC reported the results of a survey of infections afflicting homosexual men with AIDS and their "healthy" friends. They found that not only were Kaposi's sarcoma and *Pneumocystis* pneumonia unusually common among these men, but so were syphilis, gonorrhea, hepatitis B, hepatitis A, non-A, non-B hepatitis, herpes simplex type 2, cytomegalovirus, Epstein-Barr virus, varicella-zoster virus, and enteric parasitic infections (yeast, amoeba, protozoa, and fungi of the intestines). Every patient examined, regardless of whether he or she had AIDS, had several of these infections concurrently, and a few had all of them and a number of others simultaneously.[1] This study was done prior to the discovery of HIV. We now know that most of these men were also infected with HIV.

As Dr. Rogers's study shows, there are a very large number of diseases, both opportunistic and pathogenic, associated with AIDS: more than two dozen according to the latest count in the CDC's definiton of the syndrome. There is neither room nor reason to describe all of these diseases in a book of this nature. There is, however, both need and reason to describe some of these diseases, since they are known to cause signficant immune suppression individually and because they are under investigation as possible cofactors for HIV. Among the most prominent of the immunosuppressive infectious agents thought to be possible cofactors for HIV are cytomegalovirus, herpes simplex viruses, Epstein-Barr virus, hepatitis B virus, human T cell lymphotropic viruses, *Mycobacteria* species, *Mycoplasma* species, *Candida* species, and various parasitic diseases, including trypanosomiasis and malaria. These nine types of infectious agents and their immunological effects will be described in detail. The remaining AIDS-associated diseases will be characterized only in the briefest summaries within broad categories such as protozoal infections, fungal infections, and other viruses and bacteria. Their description will set the stage for unveiling the unexpected, and largely unappreciated, context of infection in which HIV plays out its actions.

Cytomegalovirus (CMV) is one of the most prevalent viruses associated with AIDS. All AIDS risk groups have significantly increased incidence of active CMV infections compared with healthy heterosexuals, and nearly all patients with full-blown AIDS have active CMV infections.

CMV is one of the class of herpes viruses that include Epstein-Barr virus, herpes simplex viruses, and varicella-zoster virus. All of the viruses are characterized by their ability to become latent. A virus becomes latent when it infects the cells of a host, inserts its genetic material into the genome of the cells, and then lies dormant without replicating or infecting other cells for long periods of time. Scientists believe that once a person is infected with a latent virus, that person will carry the infection for the rest of his or her life. The fact that the virus is present does not mean either that the person will display symptoms of disease or necessarily be infectious to other people. Cytomegalovirus, Epstein-Barr virus, and herpes simplex type 1—the virus responsible for many cold sores—are examples of this principle. Few individuals ever get more than a single, acute case of CMV mononucleosis, and as bothersome as cold sores can be, we all know that they are infrequent problems for the vast majority of healthy people. Usually latent viruses require a trigger, often unidentified, to activate viral replication and infection. Some of the known triggers, such as emotional distress, ultraviolet light, radiation and chemotherapy, corticosteroid therapy, and other infections, are known to be immunosuppressive, suggesting that latency may be broken when the immune system is suppressed.

CMV directly infects B lymphocytes, monocytes, and macrophages. The result of such infections is the same sort of inversion of the T-helper/T-suppressor ratio that HIV is supposed to cause in AIDS. The absolute number of T-helper cells decreases, and the absolute number of T-suppressor cells increases. The inversion of T cell subsets is accompanied by immune suppression, in which T cell responses to antigens are substantially reduced.[2]

The symptoms of CMV infection are similar to the flulike

symptoms often attributed to an initial HIV infection. A healthy individual will develop headache, nausea, diarrhea, muscle aches, a low-grade fever lasting about two weeks or more, and a slight cough. Liver tests may show some abnormalities, and there is evidence of lymphocyte killing and low white cell counts. CMV can be transmitted by saliva, urine, and semen and continues to be shed for a year or more following infection.[3]

Herpes simplex viruses (HSV), of which several types are known, also infect adult human T lymphocytes and macrophages, and some reports of B cell infections exist. Active virus replication appears to require stimulation by other antigens. Herpes viruses can also infect ganglions, a type of nerve cell, where it remains latent, as it also does in lymphocytes. Reactivation of HSV infections often results in painful stimulation of the ganglionic cells, and unlike CMV infections, recurrent HSV infections are common. Like CMV, herpes viruses decrease T-helper cell activity and promote production of increased numbers of T-suppressor cells.

The three most important types of HSV associated with AIDS are HSV 1, HSV 2, and HSV 6. HSV 1 is a common infection in the general population of all countries and can be passed by direct physical contact such as kissing. It is associated with the appearance of cold sores. HSV 2 is a less common infection, usually sexually transmitted, and it can cause cause genital sores. HSV 6 may be of particular importance in AIDS because Robert Gallo's laboratory has demonstrated that it is common among people at risk for AIDS and acts as a cofactor to increase infectivity and cell-killing by HIV under test tube conditions.[4]

Epstein-Barr virus, the virus usually associated with infectious mononucleosis, otherwise known as "mono" or "kissing disease," also causes immune suppression. Epstein-Barr virus was the first virus found to have a specific affinity for infecting lymphocytes. It infects B cells. One of the consequences of infection with Epstein-Barr virus is transient immune suppression involving increased numbers of T-suppressor cells and B cells but a relative decrease in the numbers of T-helper cells. Although the actual numbers of lymphocytes are not drastically decreased even in chronic cases of Epstein-Barr virus infection, as they are in AIDS, it is nonetheless clear that the nearly 100 percent incidence of active or reactivated viral infection among AIDS patients and those at high risk for AIDS produces an increased burden upon the immune system that may make it more susceptible to other opportunistic co-infections.

Epstein-Barr virus plays a substantial role in AIDS. Unusually high antibody titers are present in nearly every AIDS patient, and direct evidence of reactivation of latent virus is present in the vast majority of these cases.[5] The presence of active infection is correlated with an unusually high incidence of Burkitt's lymphoma (an Epstein-Barr virus—associated cancer) among AIDS patients.[6]

Hepatitis B virus (HBV) and its near relations, non-A, non-B hepatitis viruses (sometimes known as hepatitis C and hepatitis D), are another set of viruses that are nearly ubiquitous among AIDS patients and groups at high risk for AIDS. Hepatitis B is the first of what has become a growing class of viruses known as the hepadnaviruses, or viruses that infect the liver. It infects liver cells, apparently without killing them, and can remain latent in these cells. The immunological response triggered by active infection or, perhaps, by other infections or liver toxins can cause the immune system to attack liver cells containing HBV antigens. It is this immunological attack, rather than actual viral activity, that results in the liver damage associated with HBV infections.

HBV, unlike its more easily transmissible cousin hepatitis A virus (HAV), is usually transmitted by direct contact with contaminated blood (as might occur through blood transfusion or sharing of needles by drug addicts) or by contact with fecal material. Most homosexual men, drug addicts, and hemophiliacs have antibodies to HBV, and many remain chronically infected for years. HBV infections are extremely rare, on the other hand, among nonrisk groups.

Both acute and chronic infections with HBV can have serious effects on the ability of the liver to cleanse the blood of accumulating toxins, drugs, and alcohol. Thus, HBV infections can potentiate the destructive effects of such drugs and toxins by preventing their timely elimination from the body.

The immune system effects of HBV have been likened to CMV. HBV directly infects T-helper and T-suppressor lymphocytes. Acute and chronic infections cause a depletion of T-helper lymphocytes, resulting in reduced or low helper/suppressor ratios and depressed cell-mediated immune functions.[7]

The human T cell lymphotropic viruses, types I and II are

among the newest viruses known to medicine. Like HIV, they are retroviruses; their genetic information is encoded in ribonucleic acid (RNA) instead of deoxyribonucleic acid (DNA), which is more common in viruses. Also, like HIV, these viruses infect lymphocytes and macrophages and can remain latent in them for long periods of time. HTLV I and II cause no known disease, but HTLV I is associated with one form of leukemia, and both are considered possible cofactors for stimulating HIV replication and increasing the rate at which AIDS develops in HIV-infected people.[8]

Mycobacteria are responsible for some of the oldest recorded human diseases, including tuberculosis and leprosy. These classical manifestations of mycobacterial disease were increasingly rare in developed nations (although not in Third World Nations) until the AIDS era. Now both classical tuberculosis and less common types of *Mycobacteria*, usually subsumed under the general name atypical *Mycobacteria*, are becoming increasingly prevalent, perhaps because people are no longer immunized against them by prior exposure to common tuberculosis species. Some forty species of *Mycobacteria* are now generally recognized.

Both typical and atypical *Mycobacteria* play various roles in AIDS. *Mycobacteria tuberculosis*, the causative agent of tuberculosis, an infection of the lungs, affects many AIDS patients and a substantial percentage of those at risk for AIDS. In immunocompromised individuals, tuberculosis can become disseminated to the joints of bones, the brain, and other parts of the body and in this form is diagnostic for AIDS. Both drugs and a vaccine, Bacillus Calmette Guerin (BCG), exist for preventing or treating tuberculosis, although a new drug-resistant strain has complicated the AIDS picture.

Atypical mycobacteria, such as the *Mycobacterium avium complex* (MAI), affect the majority of AIDS patients, infecting bone marrow, spleen, liver, lungs, stool, intestines, and brain. Because of these sites of infection, atypical *Mycobacteria* may result in or complicate immune suppression of hepatitis, pneumonia, diarrhea, and dementia.

Notably, all *Mycobacteria* infect macrophages and, during active infection, maintain their activity from within the immune system itself. As a result of this infection, *Mycobacteria* cause nonspecific immune suppression that interferes with the phagocytosis and antigenic presentation of foreign material. Only a very small proportion of those infected by *Mycobacteria* ever develop symptomatic infections, however.

Mycoplasma is a genus name for about fifty different species of bacteria. *Mycoplasmas* differ from most other bacteria in being relatively small and lacking an outer cell wall. They are often among the most difficult bacteria to isolate. They can cause a range of disease manifestations, including pneumonia, when present in the lungs, and proctitis, when they infect the rectum. Animals infected with *Mycoplasmas* often become immune suppressed.

The most common types associated with AIDS are *Mycoplasma hominis*, *Mycoplasma genitalium*, *Mycoplasma pneumoniae*, *Ureaplasma urealyticum*, and *Mycoplasma fermentans*. M. *hominis*, M. *genitalium*, and U. *urealyticum* are apparently transmitted by means of vaginal and anal intercourse and are highly associated with premature deliveries and sterility in women. M. *pneumoniae*, as its name suggests, infects the lungs, and can cause one of a variety of what used to be known as atypical pneumonias. One form of M. *fermentans*, given the subtype name "incognitus," is a new species of *Mycoplasma* first isolated by Dr. Shyh-ching Lo from AIDS patients and believed by him to be a major cause of disease in AIDS patients. It may also be a potent cofactor for HIV. M. *fermentans incognitus* is common in AIDS patients but not in healthy heterosexuals.[9]

Esophageal candidiasis affects about 10 percent of all AIDS patients, and candidiasis of other orifices, such as the mouth (thrush), anus, vulva, and even internal organs, including the heart and stomach, affect more than half of all AIDS patients. Thus, *Candida* infections are one of the most prevalent infectious agents associated with AIDS.

Candida is a group of yeasts that are nearly symbiotic in human beings. That is, they usually exist in people without causing disease.[10] *Candida albicans* is found so often in the stools of healthy people that it is not considered to be of clinical signficance in most cases. However, once *Candida* becomes established in the gut, there is no evidence that it ever leaves the host. Thus, an infected person becomes a potential source of infection both to others who engage in anal forms of sex (oral or genital) and even to themselves, since the infection may spread from the intestines through the gut to the stomach and esophagus. Moreover, fecal *Candida* infection is 100 percent correlated with the presence of recurrent vaginal candidiasis, which is not only a chronic and troublesome problem in its own right but in some cases presages or accompanies AIDS. *Candida* infections also can complicate antibiotic treatment. Apparently many antibiotics kill off so many bacteria, both useful and pathogenic, that microbes such as *Candida* that are not affected by the antibiotics are allowed to flourish. People with iron and calcium deficiencies, general malnutrition, and diabetes are also at increased risk for candidiasis.[11]

Chronic infections with *Candida* apparently can cause, as well as take advantage of, immune suppression. The yeast has several components that are capable of interfering with normal T cell functions in both animals[12] and human beings.[13]

Malaria, trypanosomes, and helminths also affect many people at risk for AIDS, especially in tropical countries. Malaria is a mosquito-borne disease prevalent in tropical areas. Its causative agent is the parasite *Plasmodium*. A variety of species of *Plasmodia* exist, and many cause malaria in human beings. Trypanasomes are a type of parasite that causes sleeping sickness in humans and related diseases in animals. They are found in central Africa and in Central America. The term *helminth* refers to a broad set of parasites also commonly found in tropical countries, and sometimes in poverty-stricken inner-cities of developed nations where sanitation is poor. *Helminths* means "parastic worms" and generally includes nematodes, flatworms, tapeworms, and roundworms.

Malaria, trypanosomiasis, and helminth infections are known to cause immune suppression in human beings and in animal models of these human diseases. Malaria seems to predispose people with Epstein-Barr virus infections to developing Burkitt's lymphoma, although this connection is not universally accepted. Certainly infection with *Plasmodium* species results in dramatic reductions of antibody production, and has less pronounced but significant effects on T cell–mediated immune functions. The more severe the infection is, the greater is the immune impairment. It is believed that malaria may increase susceptibility to other infections.[14] Notably, both AIDS and HIV infection have been associated with underlying malaria in Africa.

Sleeping sickness in people due to trypanosome infection causes more severe immune suppression than does malaria, and it affects both antibody and T cell functions. The resulting immune deficiencies may last for many months after clinical symptoms of infection have disappeared, and they may increase susceptibility to other infections.[15]

Finally, although little research has been done on the immunological effects of helminthic parasitism, many human patients with such parasites are immune suppressed, and some worms, such as the human blood fluke that causes schistosomiasis, can cause both antibody and T cell deficiencies in human patients.[16] Thus, both poverty-stricken people and people living in tropical, developing nations where public health facilities are poor have greatly increased risks of immune suppression due to such parasitic diseases.

The herpes viruses, hepatitis viruses, retroviruses, *Mycobacteria*, *Mycoplasmas*, and *Candida* yeasts are only the beginning of the infectious agents that take advantage of or contribute to the immune suppression that characterizes AIDS patients and those at risk for AIDS. A number of other viruses are also associated with AIDS, including adenoviruses of various types, herpes zoster virus, and papovaviruses. Adenoviruses, like the herpes viruses, infect lymphocytes and are associated with immune suppression, although it is not known whether they cause immune suppression. They affect at least 5 percent of AIDS patients, and perhaps more. Herpes zoster is a virus with a special ability to infect nerves. It is extremely rare among the general population (perhaps one in a thousand people are infected, and most of these are asymptomatic), but 5 percent of AIDS patients develop active infections that produce clinical disease. The most frequently encountered papovavirus is the JC virus responsible for causing progressive multifocal leukoencephalopathy, an infectious disease of the myelinated nerves that results in dementia, loss of coordination, visual loss, and related complications. It affects about 4 percent of AIDS patients.

Most AIDS patients have several bacterial infections besides those already mentioned. The most common of these are syphilis (*Treponema pallidum*), gonorrhea (*Neisseria gonorrhoeae*), chancroid (*Hemophilus ducreyi*), and Clamydial infections. All of these are sexually transmitted diseases that are unusually prevalent in AIDS risk groups, and each infection has been associated with increased probability of transmission of HIV. Although the usual sites of infections by these diseases are the genitals, infections of the mouth, trachea, esophagus, and anus are not uncommon among homosexual men, and tertiary syphilis may play a role in causing dementia in a small number of patients. Notably, syphilis infections, particularly when they reach the secondary and tertiary stages, are directly immunosuppressive.[17]

All of these venereal diseases are highly associated with increased risk of contracting both HIV infections and full-blown AIDS. It is thought that the presence of open sores caused by these bacteria in the genital regions may provide portals that allow HIV (and other infectious agents) entry into the body. Lymphocytes drawn to the diseased tissue then provide an immediate source of infectable cells.

Another notable group of bacteria associated with AIDS cause pyogenic (pus-producing) and septicemic (blood) infections. These include *Pseudomonas aeruginosa* and various *Streptococci*, both implicated as having immunosuppressive actions.[18] Since pyogenic infections and septicemia are common in intravenous drug abusers, such infections are of great importance in evaluating the immunosuppressive risks of AIDS patients.

AIDS patients and those at risk for AIDS contract many funguses and yeasts besides *Candida*. These include *Cryptococcus neoformans* and various species of *Trichosporon*, *Histoplasmosis*, *Blastomyces*, *Coccidioides*, and *Aspergillus*. Most of these fungal infections are associated with chronic skin infections or unusual forms of pneumonia, although dissemination to other organs, including the brain, has been documented.

Without a doubt the most important group of nonviral and nonbacterial agents infecting AIDS patients are the parasitic infections they may have other than malaria, trypanosomiasis, schistosomiasis, and helminths. *Pneumocystis carinii* is undoubtedly the most common of these. It is a protozoan that causes pneumonia in about two-thirds of all AIDS patients. Like cytomegalovirus and so many other AIDS-associated diseases, *Pneumocystis* is an opportunistic infection that is commonly encountered by healthy individuals without causing overt disease. Indeed, there is a high correlation between CMV and PCP infections in both AIDS patients and other immunosuppressed populations. PCP can be treated prophylactically with great success.

AIDS patients are prone to other parasitic infections as well. These include various *Cryptosporidium* species, most frequently found as causal agents of severe and prolonged diarrhea. *Toxoplasmosis*, another protozoan, is frequently encountered without causing complications in healthy individuals, but can migrate to the brain in AIDS patients causing one of the many forms of AIDS dementia. It is noteworthy in that it hides from the immune system by infecting and multiplying within macrophages. An active attack on *Toxoplasma*-infected cells requires an attack on part of the immune system itself. Approximately 30 percent of AIDS patients eventually develop toxoplasmosis, for which there is no effective treatment. Homosexual men are unusually prone to developing amebiasis and giardiasis as part of gay bowel syndrome. *Entamoeba* and *Giardia lamblia* are two more protozoa that can be transmitted by oral-anal and anal-genital contact, causing severe forms of diarrhea that can persist for months in people with impaired immunity.

This incomplete list of the principal infectious agents associated with AIDS demonstrates the complexity of AIDS. Of utmost importance is the fact that several of these infections—including adenoviruses; CMV; EBV; HBV; HIV; HSV 1, 2, and 6; HTLV I and II; *Mycobacteria*; *Toxoplasma*; and *Varicella zoster virus*—infect macrophages, T lymphocytes, or B lymphocytes, causing immune disregulation. CMV, EBV, and HBV are known to cause the same inversion of the T-helper/T-suppressor ratio that is associated with HIV infection (Table 1). There is also some evidence that severe bacterial infections producing septicemia (a persistent infection of the bloodstream, as is common in intravenous drug abusers) and pneumonias can also cause significant drops in total numbers of T cells, and specifically in T-helper cell numbers.[19] Thus, it is impossible to attribute the immune suppression typifying AIDS solely to HIV. More important, it can be demonstrated that almost everyone at risk for AIDS has been infected repeatedly or chronically with several of these disease agents prior to developing AIDS.

TABLE 1. SUMMARY OF AIDS-ASSOCIATED INFECTIOUS AGENTS KNOWN TO REPLICATE WITHIN MACROPHAGES AND LYMPHOCYTES

Virus	Monocytes/ Macrophages	T Cells	B Cells	Low or Inverted T4/T8	Transactivates HIV	HIV Pseudovirons
Adenoviruses	−	+	+	+	?	?
Cytomegalovirus	+	−	+	+	+	+
Epstein-Barr virus	−	−	+	+	−	+
Hepatitis B virus	−	+	+	+	+	?
Herpes simplex viruses	+	+	+	−	+	+
HIV	+	+	−	+	+	
HTLV I, HTLV II	+	+	+	−	+	+
Mycobacteria	+	−	−	−	−	−
Toxoplasma gondii	+	−	−	−	−	−
Varicella-zoster virus	+	?	?	−	−	?

Source: Adapted from Bendinelli M, Specter S, Friedman H. 1989. Conclusions and prospects. In: Specter S, Bendinelli M, Friedman H, eds. 1989. *Virus-Induced Immunosuppression*. New York: Plenum Press, 444–445; Grange JM. 1988. *Mycobacteria and Human Disease*. London: Edward Arnold, 66. See also Root-Bernstein RS. 1992. HIV and immunosuppressive cofactors. *EOS. Revista di immunologia ed immunofarmacologia* 12 (4). Table reprinted with permission of the publisher.

The Incidence of Multiple, Concurrent Infections in Various Risk Groups Compared with American Heterosexuals

Those at risk for AIDS have much higher rates of infections than do those in nonrisk groups. Consider, for purposes of comparison, the disease profile of a typical twenty- to forty-year-old, heterosexual North American or European male or female who does not abuse drugs, is not undergoing cancer chemotherapy, and is not a hemophiliac. Such a person will, about a quarter to half the time, have been exposed to CMV, EBV, HSV 1, and HSV 2 as can be documented by the presence of antibody to these viruses in their bloodstream.[1] In the majority of cases, these individuals will not even know that they have had these infections, and only a very small fraction of 1 percent of them will be carrying active infections, as measured by the presence of viral antigens or actual isolation of the virus.[2] Exposure to other immunosuppressive viruses such as HBV and HTLV will be similarly rare (less than 1 in 10,000 people per year), with active infections almost unheard of except in frequent travelers to Third World Countries.[3] Classical sexually transmitted diseases, such as *Clamydia* species and gonorrhea, will also be infrequent, syphilis rare (less than one in 1000 people), and enteric pathogens, such as amebiasis, giardiasis, and shigellosis present only in 1 in every 1000 to 10,000 people.[4] In short, the typical young, North American heterosexual is extremely healthy and is highly likely to be free of active infection by all of the disease agents associated with AIDS.

There is one point on which clarification about the "typical" heterosexual is necessary. Speaking in terms of populations (not individuals, each of whom differs widely), blacks and Latinos tend to have higher rates of all infections than whites. For example, the seroprevalence of CMV antibody in young Caucasian adults is about 25 percent, but over 40 percent among nonwhites.[5] Similarly, among sexually promiscuous blacks, CMV seropositivity is present in about 80 percent of women as compared with 60 percent in whites, reflecting more partners and an earlier age of first coitus.[6] The rates of HTLV I and II infections in whites in the United States are 8 per 10,000 people, but for blacks the rate is 74 per 10,000 people, and for Latinos, 66 per 10,000 people.[7] Whereas 9 in 100,000 whites contract tuberculosis (*Mycobacterium tuberculosis*), 50 in every 100,000 nonwhites will contract it.[8] Similarly, black Americans develop HSV 1 and HSV 2 anti-bodies at much younger ages than do white Americans, and twice as many blacks as whites eventually become infected with HSV 2.[9] This increased rate of infection is thought to be due to earlier sexual experimentation, greater promiscuity and poverty-related malnutrition. These figures may be relevant to understanding AIDS, since AIDS is three times as prevalent among people of color than among whites in the United States.

Our "typical" healthy heterosexual, whatever his or her skin color, stands in stark contrast to "typical" individuals in all identi-fied AIDS risk groups. Hemophiliacs are already under immuno-logical pressure from their more or less constant contact with for-eign blood and blood products. In addition, the typical young to middle-aged adult hemophiliac has about a one in two probability of having an active EBV infection, a one in five probability of having an active CMV infection, a one in ten probability of having an active HBV infection, and an equal probability of having an active non-A, non-B hepatitis infection.[10] Liver damage and mea-surable problems with liver function are quite common in older hemophiliacs. HTLV infections are apparently very uncommon.[11] One major exception are Brazilian hemophiliacs, in whom 13 per-cent of HIV seropositives had a concurrent HTLV 1 infection.[12] In addition, at least 75 percent of all hemophiliacs have been ex-posed to HIV through the use of HIV-contaminated factor made prior to 1985. Of these HIV-exposed hemophiliacs, it is estimated that about 20 percent have an active infection and the rest prob-ably carry a latent infection. These rates of active infection are, for each separate infection other EBV, hundreds of times higher than are found among comparable, healthy heterosexuals. Hemo-philiacs are the "healthiest" of the identified risk groups.

Intravenous drug users (IVDUs) are, as a group, next on the list of increasingly unhealthy adults. Added to the immune suppres-sion caused by the drugs they inject are a slew of infectious agents. These users have elevated levels of active CMV, EBV and HSV infections compared with nondrug abusers, although the rates of exposure, as measured by antibody seroprevalence, are not enor-mously elevated. HTLV infections are much more common among drug addicts than among the general population. Rates of infection among addicts have ranged between 5 and 48 per hun-dred depending on the country or city studied,[13] as compared with less than 1 in 10,000 among unselected blood donors.[14] All of these viruses can be transmitted by needle use.

Among the other diseases spread by needle use, the most prevalent are hepatitis A and B. The rate of hepatitis A antibody positivity ranges between 30 and 60 percent in addicts compared with less than 4 percent in the general population.[15] Drug abusers have unusually high frequencies of infections with Toxoplasmosis and *Mycobacteria*, particularly tuberculosis. Thirty-six percent of intravenous drug abusers in New York City had antibody against *Toxoplasmosis*, whereas none of the non-drug-using controls was seropositive.[16] Similarly, although the incidence of tuberculosis in the average American is about five in every 100,000 people, the incidence in intravenous drug abusers is about 5 in every 100 people, and exposure to tuberculosis is about five times this figure. Eight percent of HIV-seropositive drug abusers develop tuberculosis each year in the United States.[17]

Drug abusers (including non-needle-using addicts) have unusual incidences of all of the classical sexually transmitted diseases, including syphilis and gonorrhea. The increase in sexually transmitted diseases is apparently linked to the fact that many addicts—men and women—trade sex for drugs or prostitute themselves to earn the money necessary to support their habits.[18]

Most signicant, intravenous drug users have extremely high rates of pyogenic and septicemic bacterial infections, including *Staphylococcus*, *Streptococcus*, *Pseudomonas*, and *Escherichia coli*. Their very high incidence of bacterial infections is undoubtedly due to the generally poor habits of sterilization employed by most drug addicts. Unlike a physician or nurse, who swabs the patient's arm with alcohol or iodine before giving an injection and who uses a properly sterilized or new needle and syringe, the typical drug abuser never swabs the injection site and is as likely to reuse an old, unsterilized needle as he or she is to obtain a new or properly cleaned one. Thus, it is not just HIV that is transmitted from individual to individual when needles are shared but many other infectious agents, including CMV, EBV, HBV, HTLV, and bacteria, as well. The result is that active or long-term intravenous drug abusers are likely to have one or more active bacterial infections and at least one active viral infection at any given time. Dr. E. Fernandez-Cruz and his collaborators in Madrid, Spain, reported in 1988 that among the HIV-seronegative heroin abusers they had studied, "over 56 percent had pyogenic [pus-forming] infections [presumably bacterial], 17 percent had hepatitis-B-virus infections, 4 percent had *Mycobacteriuum tuberculosis* infection and 4 percent had *Treponema pallidum* [syphilis] or *Candida* infections."[19]

Homosexual men have the highest disease load of any other North American or European risk group. Virtually every homosexual man has been repeatedly exposed to CMV, EBV, and HSV 1. Their rates of active infection, particularly of CMV, are extremely high compared with heterosexuals. During the first half of the 1980s, as many as 98 percent of homosexual men (and in some studies, 100 percent[20]) had antibody to both CMV and EBV, and 15 to 40 percent of these men had active infections.[21] This compares with active infection in less than 4 percent of unselected male blood donors (some of whom were doubtless gay, bisexual, or drug addicts[22]), and no viral isolations in several heterosexual control groups screened for risk factors.[23] In addition, rates of exposure to HSV 2, HBV, and HTLV are also much higher than in heterosexuals, with active infections more likely.[24] The annual incidence of HBV infection, for example, was 4 to 6 percent in homosexual men (with as many as 30 percent developing infection each year during the late 1970s) as compared with an 0.3 percent rate of infection in heterosexual men.[25] Thus, over 60 percent of 3,816 homosexual men examined in five U.S. cities had immunological or other evidence of hepatitis B infection in 1979[26] as compared with only 5 percent of heterosexual men. Over 40 percent of homosexual men had antibodies to hepatitis A virus as well by 1983.[27] About 70 percent of homosexual men who are HIV seropositive are also HSV 2 seropositive, and it has been found that seroconversion to HSV 2 from a seronegative state is highly predictive of both HIV seroconversion and an increased rate of AIDS development.[28] The most dramatic disease differences between heterosexual and homosexual men relate to infections associated with anal forms of sex. Amebiasis and giardiasis, which are almost never seen among heterosexuals (except those traveling to Third World nations), were present in about one in every ten gay men during the early 1980s. Mycobacterial infections of the anus, also extremely rare in heterosexuals, could be found in nearly one in every five gay men. The rates of syphilis and gonorrhea were several hundred times those found in the general population.

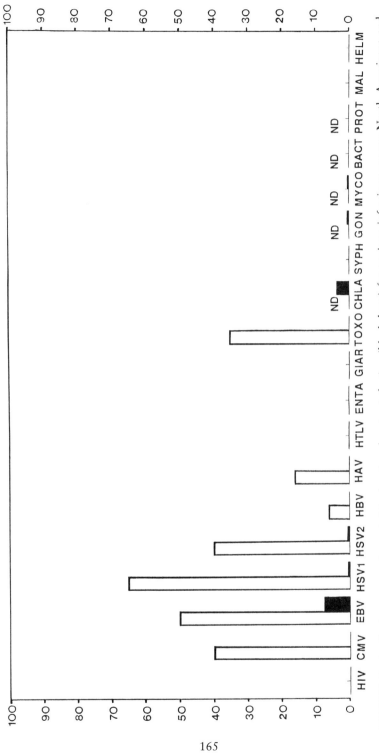

FIGURE 3. Antibody seroprevalence (white bars) and active infection (black bars) for various infections among North Americans and Europeans at risk for AIDS.

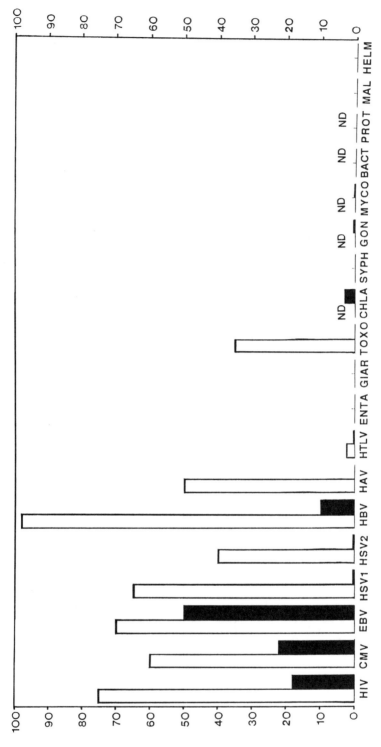

FIGURE 4. Antibody seroprevalence (white bars) and active infection (black bars) for various infections among North American and European hemophiliacs.

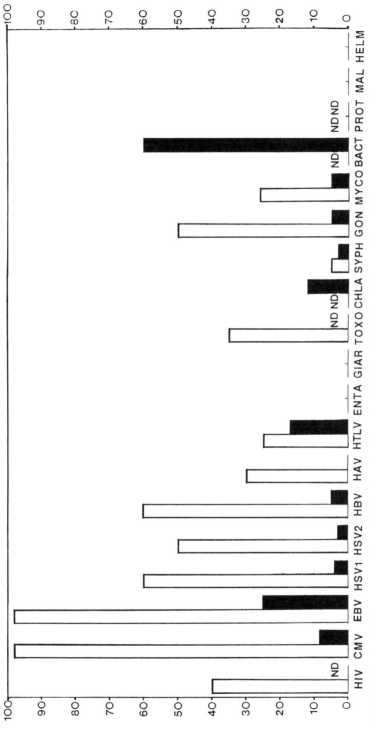

FIGURE 5. Antibody seroprevalence (white bars) and active infection (black bars) for various infections among North American and European intravenous drug abusers.

167

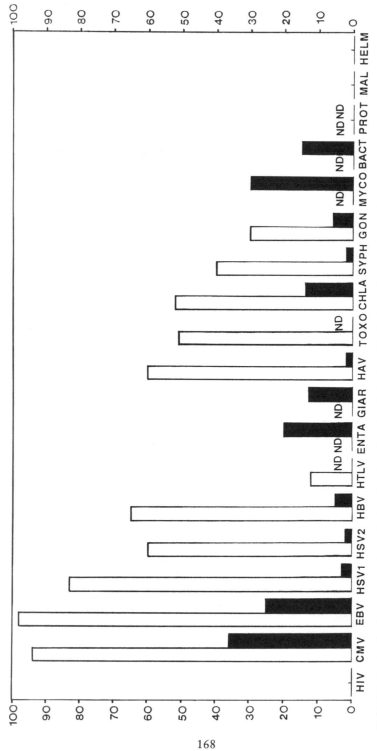

FIGURE 6. Antibody seroprevalence (white bars) and active infection (black bars) for various infections among North American and European HIV-seronegative homosexual men.

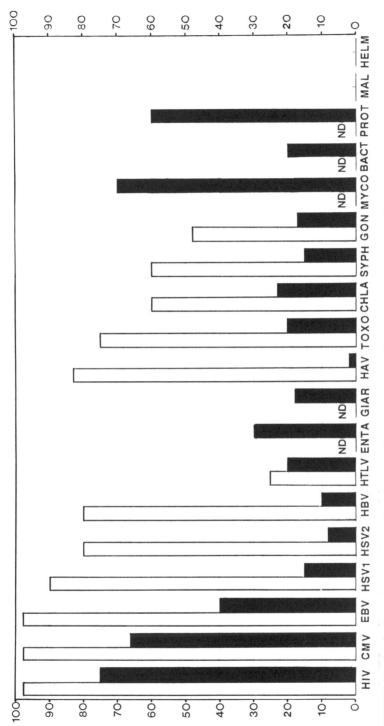

FIGURE 7. Antibody seroprevalence (white bars) and active infection (black bars) for various infections among North American and European people with AIDS.

169

Notes Concerning Prevalence of Infection Charts (Figures 3–7)

These charts should be considered to represent relative incidences of antibody positivity to particular infections (white bars) and relative incidences of active infection (black bars) as measured by microbial or antigen isolation. The infections listed are, from left to right, HIV, CMV, EBV, herpes simplex 1, herpes simplex 2, hepatitis B virus, hepatitis A virus, HTLV I and II (most studies do not distinguish between the two, unfortunately), *Entamoeba*, *Giardia lamblia*, *Toxoplasmosis*, *Chlamydia* (all species), syphilis, gonorrhea, *Mycobacteria* (both tuberculosis and atypical forms), pyogenic and septicemic bacteria (*Pseudomonas*, *Escherichie coli* in the bloodstream, *Staphyloccocus*, etc.), protozoa (including *Pneumocystis carinii*), malaria, and helminths.

The data are, by necessity, derived from a large variety of sources.[29] Different studies often disagree on the exact percentage of patients with antibody or active infection. Thus, the numbers presented here should be interpreted as representative numbers rather than exact ones. In addition, when several different tests exist to measure any particular infection or antibody, it was sometimes necessary to make educated decisions as to what measure of infection to use. For example, in Epstein-Barr virus infections, the data plotted in the charts are IgG antibody to EBV early antigen because this yields the most clear-cut differences among risk groups.[30] In other cases, no direct data were available as to prevalence of antibody or active infection. In these cases, the letters ND ("no data") have been placed on the charts. The fact that I did not find the relevant data does not preclude such data existing somewhere in the medical literature. I can only say that many weeks of intensive searching did not yield the desired numbers.

To summarize, a comparison between the disease profiles of the "typical" heterosexual, the "typical" intravenous drug abuser, the "typical" HIV-negative homosexual male, and the "typical" AIDS patient reveals that intravenous drug abusers and HIV-negative homosexual men are about half way to becoming AIDS patients (in terms of total active infections) even in the absence of an HIV infection. Particularly striking is the incidence of active infections with viruses such as CMV, EBV, and chronic hepatitis infections, which are both immunosuppressive and opportunistic. HIV simply adds to a disease load that is already very high in risk groups.

The importance of the high incidence of multiple, concurrent infections preceding or working in tandem with HIV has been explored by a number of investigators. Fernandez-Cruz and his co-investigators, for example, noted that HIV-infected intravenous drug users had the highest rates of infections among addicts and speculate that "in HIV-infected drug addicts, heroin, as a chemical agent, together with other infectious micro-organisms, acting through a common final pathway may produce immunoregulatory changes that, resulting in further breakdown of the host's immune status, could contibute to the development of AIDS."[31] Similarly, Dr. Steven Specter, Dr. Mauro Bendinelli, and Dr. Herman Friedman note that co-infections may explain why some people exposed to HIV develop AIDS quickly and others not at all:

> Immunocompromised individuals have no greater likelihood of being infected by a virus than do normal individuals but, once infected, they have a higher morbidity and mortality. This may be the key factor as to why only certain HIV-infected people develop AIDS. More than 95 percent of AIDS victims are either homosexual, drug addicts, hemophiliacs, transfusion recipients, or a combination of these. None of these persons is truly in a normal state of health. For example, each has a much higher than average incidence of other viral infections, most notably the hepatitis and herpes group viruses. These agents are capable of severely altering the functioning of the immune system, which may leave these persons susceptible to the most severe consequences of HIV infection. Alternatively, these viral infections may stimulate immunocompetent cells so that they become better host cells for HIV. It has been reported that HIV replicates well in activated T helper cells, but not in resting cells. Thus, any immunologic stimulus may increase susceptibility to infection. . . . This seems to be the story of HIV, a vicious catch-22. Immunosuppressive superinfections, viral or otherwise, may be cofactors that enhance susceptibility to AIDS in HIV-infected individuals; HIV in turn further suppresses the system, allowing even more severe consequences attributable to opportunistic infections, which would normally not be pathogenic.[32]

For Specter, Bendinelli, and Friedman, HIV is not the whole story of AIDS any more than it is for Fernandez-Cruz's group. Both believe that multiple immunosuppressive agents act synergistically with each other and HIV to create what they call a catch 22—although "vicious cycle" is a more appropriate term—in which each new agent produces further immune suppression, allowing yet other infections to catch hold. Their point, which is crucial for understanding the epidemiology of AIDS: no one at risk for AIDS has a normal immune system to begin with. The question is not what begins the process of immune impairment but how HIV and other agents take advantage of this impairment.

Synergism and Viral Transactivation as Causes of Immune Suppression

There is more to the problem of multiple, concurrent infections than simply the immunosuppression caused by each disease agent by itself. Co-infections can be synergistic. A well-known model of synergism is the effect of combining alcohol with other drugs such as barbiturates. Neither alcohol nor barbiturates are usually fatal, save in extremely large doses. Combine them, however, and very small amounts of each can result in death. Infections can also be synergistic. Neither CMV nor EBV, for example, is serious by itself. Combine them with some other infections, however, and the immune system may be overwhelmed.

Immune suppression due to the synergism between unrelated infectious agents has been studied since the 1970s to explain how various combinations of antigens exert unexpected effects on T cell function.[1] Perhaps the most persistent investigators of this phenomenon were Dr. Shih-wen Huang and Dr. Richard Hong. They tracked down numerous examples of children and young adults who developed combined infections in which the lymphocyte count dropped to a third or less of normal, clinically evident immune suppression resulted, and death sometimes followed. Many of these cases involved combinations of infectious agents seen in AIDS patients: *Herpes simplex* and other viruses with *Mycoplasma* co-infections, or *Varicella-zoster*, measles, or adenoviruses combined with other viral infections. These combined infections resulted in such severe and prolonged immune suppression that in two cases disseminated *Varicella-zoster* and disseminated candidiasis complicated by *Staphylococcus* resulted in death. Both cases qualify as presumptive AIDS cases retrospectively.[2]

Many other combined infections also result in death much more frequently than do single infections. For example, toxic shock syndrome is usually associated with improper use of high-absorbency tampons by women who consequently develop uncontrolled vaginal infections of the bacterium *Staphylococcus aureus*. The bacterium produces very large quantities of toxins that lead to an illness usually lasting several weeks that can be fatal. In 1987, Dr. Kristine MacDonald of the CDC and her colleagues in the Minnesota Department of Health reported that toxic shock syndrome could also complicate influenza outbreaks. Notably, the combination of flu with a *Staphylococcus* infection was so deadly that many of the patients they described died within days of the onset of their symptoms, often before they could even be hospitalized.[3] Similarly, Eunice Carlson of the Michigan Technological University has found that a combination of *Staphylococcus* with *Candida albicans*, a yeast commonly found in AIDS patients as well as in the vaginal and anal orifices of healthy women, results in up to a 70,000-fold decrease in the number of bacteria necessary to cause death in a mouse model of toxic shock syndrome.[4] In other words, a bacterial infection that might easily be controlled by a healthy person could result in the death of a person who also contracts a *Candida* infection simultaneously.

The importance of the concept of disease synergy for AIDS is twofold. First, as Montagnier has suggested, AIDS may be caused by one or more combinations of immunosuppressive agents acting synergistically, rather than by a single infectious agent. In this case, the search for a single cause of AIDS would be misdirected. The application of Koch's postulates and other standard criteria for evaluating disease causation would then be impossible to apply to AIDS. Second, many of the particular manifestations of AIDS, and perhaps even the initiation of immune system breakdown, may have their origins in the synergism of the various infectious agents present in people at risk for AIDS. Cytomegalovirus and Epstein-Barr virus might be of little consequence to healthy individuals but potentially lethal to people unlucky or unwise enough to develop concurrent infections.

Recall Mack, the man described in the previous chapter who died in 1978 of disseminated cytomegalovirus infection complicated by *Pseudomonas* and *Candida* infections contracted from abortive surgery and massive blood transfusions. Just two years prior to his death, Dr. John R. Hamilton, Dr. James C. Overall, Jr., and Dr. Lowell A. Glasgow of the University of Utah College of Medicine had demonstrated that precisely this combination of infections dramatically increased mortality in mice. Whereas no deaths occurred from individual infections with cytomegalovirus, *Pseudomonas*, *Candida*, or *Staphylococcus* at very small doses, combined infections of CMV with any of the other agents at the same low doses resulted in death 90 to 100 percent of the time. Moreover, the mice, like Mack, often died within four or five days of the onset of their combined infections.[5] A similar, deadly synergistic effect was subsequently demonstrated between CMV and *Escherichia coli*, a ubiquitous bacterium that sometimes invades the bloodstream of gay men and intravenous drug abusers to cause severe infection.[6] Combined EBV, HBV, and CMV infections create unusually severe illness in human patients as well.[7] The incidence of these sets of combined infections in AIDS patients makes these laboratory and clinical studies intriguing.

Other intriguing observations of combined infections are also relevant to AIDS. For example, the majority of hemophiliacs contract hepatitis B, non-A, non-B hepatitis, CMV, and EBV infections. In some patients, these infections become chronic and coexistent. Dr. R. E. Enck and his coinvestigators found that one symptom of these chronic infections was progressive liver malfunction. Surprisingly, however, no single one of these infections, including HBV, was correlated with liver disease in their patients. Their patients with liver disease had either two or all three infections simultaneously. "It is intriguing to speculate," they commented in their paper, "that past history of infection with EBV might interfere with the host's ability to eliminate another transfused infectious agent [such as CMV or HBV]. When this [second] agent is introduced and interference occurs, evidence for persistent infection with that agent (abnormal liver function) develops."[8] Similarly, combined infections of HBV with tuberculosis (*Mycobacteria*) have been found to produce unusual liver complications that are extremely difficult to treat compared with the individual infections.[9] Thus, it is almost beyond doubt that at least some of symptoms subsumed under the rubric of AIDS are due to synergistic effects of the combined infections that characterize those at risk for the syndrome.

Cofactors can also influence HIV activity in a very different way. Recall that some of the viruses associated with AIDS infect either macrophages or T-helper cells—the same cells infected by HIV. When viruses infect cells, they insert their genetic material directly into the genome of the host cell, so that it becomes integrated into the normal genes of these cells. If two viruses infect the same cell, then both sets of viral genes will be present. In that case, the phenomenon of transactivation may occur. *Transactivation* is a technical term from genetics meaning that a gene in one chromosome can turn on the activity of a gene on another chromosome. Since viral genes insert themselves into human chromosomes, one may transactivate another. Thus, when one virus begins to replicate, its activity will turn on the replication of the other virus as well. HIV has been found to be transactivated by CMV, HTLV 1, *Herpes simplex* viruses, and several other viruses in laboratory studies.[10] In addition, HIV has been found to transactivate CMV and HTLV 1.[11] Transactivation works both ways. Triggering one virus to replicate inevitably causes the others to replicate, and a vicious cycle of immunosuppressive viral infections is initiated. For example, the more CMV that is present in a person, the more cells are infected, and the greater is the chance that HIV and CMV will co-infect the same cells. Having once infected the same cells, transactivation mechanisms ensure that a deadly symbiotic dance will be initiated in which each virus seduces more and more of the other viruses into actively participating in the destruction of the immune system.

Ample evidence exists to support the concept that HIV interacts synergistically with CMV, HTLV 1, HBV, and other infectious agents, and several mathematical models now exist to describe the interactions.[12] Combined infections of HBV and HIV, for example, result in increased levels of HBV antigens and DNA in the blood and an increased rate of immune decay.[13] Studies of hemophiliacs have demonstrated that the presence of either Epstein-Barr virus or cytomegalovirus in HIV-seropositive individuals significantly increases the probability that the individual will develop an abnormal T-helper / T-suppressor ratio and results in increased immune suppression as measured by loss of T-helper lymphocytes.[14] Moreover, it is clear that such combined infections can lead to consequences not seen as a result of single infections. People who contract Epstein-Barr virus infections in Western countries rarely experience anything more serious than mononucleosis. These infections in co-infected AIDS patients, however, may result in a range of diseases, including non-Hodgkin's lymphomas,[15] hairy cell leukemia,[16] and oral hairy leukoplakia.[17]

Active cytomegalovirus infection can also act synergistically with HIV. A statistically significant correlation between HIV and unusually high CMV antibody titers has been reported among intravenous drug abusers, suggesting synergistic coactivation.[18] Similarly, a study of thirty-eight HIV-infected infants found that those with clinically evident AIDS were much more likely to have a CMV infection (eleven of twenty-four) than were symptomless infants (one of thirteen). CMV infection also conferred a statistically significant increase in the probability of these infants' dying within two years of birth.[19]

Similarly controlled studies have found that HTLV 1 is highly prevalent in AIDS patients but not in asymptomatic carriers of HIV, once again suggesting a cofactor role for HTLV.[20] Investigators have also found that drug abusers multiply infected by HIV and HTLV I and II "had more clinical symptoms related to immune deficiency" than abusers infected with either retrovirus alone.[21] Gay men co-infected with HIV and HTLV 1 developed more severe immune suppression and more quickly proceeded to clinically evident AIDS than those infected only with HIV.[22]

Co-infections may vastly increase the transmissibility and infectivity of HIV. Since most HIV-infected people are multiply infected with other sexually transmissible diseases, a person contracting an HIV infection is extremely likely to contract more than just HIV. We have already seen that the probability of transmitting most sexually transmitted diseases from one person to another is hundreds or thousands of times higher than it is for HIV. Since many of these additional infections are themselves immune suppressive and can act as cofactors of HIV, the result of such co-infection should be disastrous. Indeed, Van Griensven and his colleagues at the Municipal Health Service in Amsterdam report that having unprotected sex with a person with full-blown AIDS, as opposed to an asymptomatic HIV-seropositive individual, often leads to much more rapid development of AIDS in the partner following seroconversion. "A history of having sexual intercourse with a person who had AIDS points to the more virulent properties of HIV in these persons."[23] The European Study Group on AIDS found that having sex with a hemophiliac who has full-blown AIDS, particularly unprotected anal intercourse, results in transmission of HIV at up to fifteen times the rate that it does in unprotected sex with an asymptomatic HIV-seropositive hemophiliac. Once again, co-infections may be responsible.[24]

How do co-infections produce these sinister results? A number of possibilities exist. One for which there is direct evidence is that co-infections expand the range of cells that HIV can infect. Normally HIV infects only cells that have the CD4 protein on their cell surface—mainly macrophages and T-helper cells. Briefly, HIV carries a "key" (the MHCII-like protein, GP120) that "recognizes" the CD4 protein as its corresponding "lock." The "key" opens the "lock," allowing HIV to infect cells carrying the CD4 protein. HIV cannot enter cells having other "locks" because it does not have appropriate keys for them. Thus, if HIV is on its own, it can only open certain "doors," and these doors are found only on certain monocytes and lymphocytes.

Under certain circumstances, the range of "doors" through which HIV passes expands considerably. One of the first observations that Luc Montagnier made concerning the activity of HIV after isolating it in 1983 was that it was unable to infect B cells. This observation has been amply confirmed by subsequent research and makes sense in the light of the notion that HIV can infect only cells bearing the CD4 protein "lock" on their cell surfaces. B cells normally lack CD4 molecules, and therefore the HIV "key" is useless. On the other hand, Montagnier was surprised to discover that in the presence of Epstein-Barr virus, which does have a "key" for B cells, HIV would also infect B cells.[25] His observation has been confirmed by many subsequent investigators.[26] HTLV I,[27] herpes simplex viruses,[28] and cytomegalovirus[29] each expands the range of cells that can become infected by HIV and thereby allow HIV to create much more havoc than it could otherwise do.

The precise mechanisms by which co-infections allow HIV to open the "doors" of cells lacking the CD4 "lock" is an area of considerable speculation and research. Various investigators have suggested that perhaps damage to cells caused by previous viral infections allows HIV to "leak" into these cells; others speculate that HIV may co-infect susceptible cells such as T-helper cells and get its molecular components mixed up with those of other viruses to form odd, hybrid viruses called "pseudovirons"; and yet others think that perhaps various viral infections can induce cells that do not normally have the proper CD4 protein "lock" for HIV's key to produce it. There is no direct evidence for any of these mechanisms in human beings, and very little even in test tube experiments.[30]

Some interesting clues have emerged from such studies, however. One is that the presence of a second virus expands the range of HIV infection only to the type of cell infected by the concomitant infection. For example, Epstein-Barr virus has a molecular "key" that permits it to open the "locks" on the "doors" to B cells, the antibody producing white blood cells. In the presence of EBV, HIV can co-infect only these B cells and not any other type of cell. Why this should be the case is not clear. It is known that infecting B cells with EBV does not result in the B cells' suddenly producing a CD4-like molecule or becoming permissive to any infectious agent whatever, so these factors cannot account for how HIV follows Epstein-Barr virus into B cells. No HIV-EBV pseudovirons have been reported so far. There is apparently something specific about the way that Epstein-Barr virus "opens" the B cell door that allows HIV to sneak in with it. So how do such productive co-infections work?

Dr. Sheila Hobbs of the Parke-Davis Pharmaceutical Research Division of Warner-Lambert and I have proposed a controversial, speculative, but simple mechanism that may explain how many viruses expand the range of HIV infectivity. A simple analogy will make the mechanism clear. Imagine that you possess a special key or pass to a single door—say your office—but that you need to get into a series of locked rooms for which you have no pass. Imagine that the passes to each door are the property of separate individuals and that they are nontransferable. For example, each door might be guarded with locks that can be opened only when a person with the correct fingerprint places his or her finger on the scanning mechanism. Imagine further that all of the doors are designed so that only a single person can enter at a time. This scenario is not too different from the viral situation. The "key" possessed by each type of virus is as unique and integral to it as your fingerprint is to you. In the case of a virus, its fingerprint is one of the proteins that comprise its outer surface. Each viral fingerprint opens only a limited range of doors, and only viruses having the right fingerprint normally infect any particular cell, regardless of the other types of infectious agents that may be present. So how do you get through these apparently impassable doors?

The answer is quite simple, as anyone who has ever ridden piggyback knows. Two people can easily pass as one through a door if the one who has the appropriate "key" carries the other one, or vice versa. All that is necessary for this mechanism to work at the viral level is some means for one virus to attach itself specifically to the other, so that the one can carry the other into cell types from which the passenger virus would normally be excluded.

As far-fetched as this piggyback theory of expanded HIV infectivity may sound, it is based on some intriguing evidence. Viruses have simple structures, usually involving a core of genetic material surrounded by a layer of proteins, often called generically "core proteins." Surrounding these core proteins is often another layer of proteins, the "membrane" or "surface" proteins. These membrane proteins contain, among other things, the "keys" that permit the virus to open the "locks" to the cells they infect. Dr. Hobbs and I have found that these same membrane proteins often contain regions that mimic cellular "locks." In particular, the cell surface proteins of Epstein-Barr virus, cytomegalovirus, the herpes simplex viruses, and HTLV I and II contain large segments that very closely mimic various regions of the CD4 protein of T lymphocytes. In other words, these viruses contain proteins that look like the "lock" for which HIV has the "key." HIV may be able to attach itself to these other viruses as it would during the first steps of infecting a CD4-bearing T4 cell. In this fashion, EBV, CMV, and other viruses can give HIV a piggyback ride into cells for which it does not have the "key." When these infect their preferred cell-types (B cells in the case of EBV; intestines or lung in the case of CMV), then the HIV gets carried along. (See Figures 8, 9, and 10.)

Direct evidence of this mechanism exists for sperm. Human sperm cells, like the viruses just mentioned, have surface proteins with significant regions that mimic the CD4 protein of T4 lymphocytes. Dr. B. Gobert and his coworkers have demonstrated that antibodies made against CD4 protein will recognize and bind to these CD4-mimicks on sperm.[31] Dr. O. Bagasra and his colleagues have demonstrated that HIV will bind directly to the surface proteins of sperm.[32] Thus, sperm cells may act as carriers for

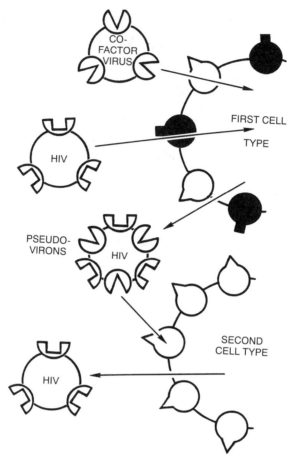

FIGURE 8. A simplified model of pseudoviron production. Pseudovirons are human immunodeficiency viruses that have the proteins of another virus mixed into their outer membranes. These other proteins allow HIV pseudovirons to infect types of cells that they cannot normally infect.

HIV among actively infected men. People infected with HIV by sexual means would therefore be immunologically exposed to sperm as well.

There are several implications for AIDS that are implicit in the piggyback theory of expanded HIV infectivity. One is that people who are co-infected with HIV and other infectious agents, such as Epstein-Barr virus, HTLV 1, cytomegalovirus, or other piggyback agents, may transmit their HIV infections to other people much more efficiently than people lacking these coinfections, since all of

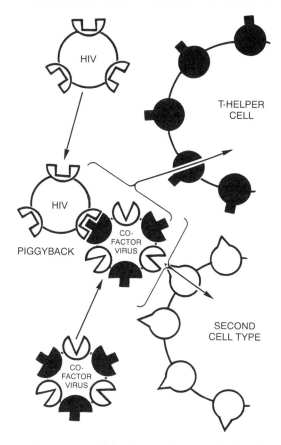

FIGURE 9. An alternative model of pseudoviron production, Part 1: Many viruses associated with DIS have proteins on their outer membranes that resemble the CD4 protein that HIV uses to infect T-helper cells. Thus, HIV may bind to these other viruses. Then each virus may carry the other into a type of cell it could not normally infect by a process of piggybacking.

these viruses are much more readily transmitted than is HIV. This mechanism would help to explain why HIV and AIDS follow hard upon unprotected sexual contact with people with AIDS but less frequently with unsymptomatic HIV-infected individuals. The piggyback theory also explains why HIV is limited almost solely to those at risk for AIDS. They are already being immunosuppressed by concomitant exposure to CMV, EBV, sperm, and so forth. Piggybacking would facilitate transactivation of HIV by cofactor viruses. And piggybacking, or a similar mechanism, would

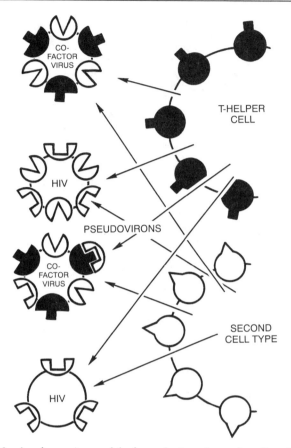

FIGURE 10. An alternative model of pseudoviron formation, Part 2: If a cell is infected by both HIV and another virus, then pseudovirons like those illustrated in Figure 8 may be produced, as well as cofactor virus pseudovirons.

explain why multiple, concurrent infections are associated with an increased rate of development from HIV infection to AIDS among some people.

The piggyback mechanism of increased viral infectivity might be utilized, *mutatis mutandis*, to spread other pairs of infectious agents in both AIDS and other diseases. Dr. Hobbs and I have found that a number of viruses and bacteria other than HIV have class II MHC-like proteins on their cell surfaces that might be able to act as the "locks" for viruses such as EBV and CMV that have CD4-like protein "keys." These class II MHC mimics include proteins on *Mycoplasmas*, *Escherichia coli*, and *Pseudomonas* bacteria, as well as various blood coagulation factors. Thus, the presence of multiple, concurrent infections in AIDS patients likely promotes a large series of unusual events that serve to increase vastly the pathological and synergistic effects of the individual infectious agents.

In summary, the implication of expanded HIV transmissibility, infectivity, and morbidity in the presence of other infections, whatever the mechanisms turn out to be, is that many aspects of cell destruction usually attributed solely to HIV could not occur except for multiple co-infections. More simply stated, people who have HIV infections in the absence of co-infections should be much healthier for much longer periods of time and have much better prognoses than people who have HIV infections combined with many other co-infections. That is in fact the case. Conversely, extensive evidence now exists that co-infections greatly enhance HIV activity and are predictive of much quicker progression to AIDS following HIV infection. Thus, one of the most important ways to control AIDS is to control exposure to the cofactors that may act synergistically with HIV and with each other to promote infectivity and to increase immune suppression. AIDS is more than just HIV.

But the observation that AIDS is more than just HIV reveals a glaring lacuna in our knowledge about co-infections. Despite intensive efforts to determine the possible range of synergistic interactions that may occur between HIV and other viruses in AIDS, I can find only a handful of studies of the possibility of similar interactions between other viruses. It is not known, apparently, whether CMV and EBV, or CMV and HTLV I, for example, can transactivate each other as they do HIV or whether their combined infections are significantly more immune suppressive then individual infections. Clearly, we need to know whether such synergistic interactions are possible if we are to evaluate correctly the risks of individual AIDS patients who may be co-infected with these or other microbes. One of the costs of focusing so exclusively on HIV has been the possibility that we have overlooked equally relevant discoveries concerning the other infectious agents that characterize the syndrome. These discoveries may be necessary parts of understanding, preventing, and curing AIDS and other currently mysterious diseases such as chronic fatigue syndrome.

CHAPTER 5

AUTOIMMUNE PROCESSES AND IMMUNE SUPPRESSION IN AIDS

AIDS Progresses Independently of HIV

People with AIDS and most of those at risk for the syndrome have multiple concurrent infections and many noninfectious causes of immune suppression. As if these problems were not enough, there is yet a third cause of immune suppression that affects all people with AIDS: autoimmunity. Autoimmunity results when the immune system turns against the body it is supposed to protect and begins attacking it instead. In AIDS patients, this autoimmunity can take many forms, including an attack on T cells.

Autoimmune theories of AIDS were first proposed to explain many of the anomalous observations about AIDS that have been described in previous chapters of this book: the paucity of HIV-infected T cells contrasted with the immense amount of immune destruction; the fact that antibody to HIV signals progression of AIDS rather than protection against it; and the long latency period between infection and disease manifestation.[1] All of these characteristics are typical of autoimmune diseases rather than infectious diseases.

So is one other aspect of autoimmune diseases: they progress independently of the initiating infections. Again, Rock Hudson's story is both cautionary and enlightening. During the course of his illness, he traveled to Paris to seek treatment with an experimental anti-HIV drug, HPA 23. The treatment was administered by Dr. Dominique Dormont of the Percy Hospital. After four weeks of treatment, Dr. Dormont reported that there was no longer HIV detectable in Hudson's blood. He warned Hudson not to celebrate, however. "The culture can be negative and the person still has AIDS," cautioned Dormont. "The virus creates disorders in the immune system, after which the disease progresses independent of the virus."[2] Certainly it did in Hudson's case, for he died despite the apparent short-term success of the treatment.

As time has shown all too well, Hudson's demise is all too typical for people with AIDS. No antiviral treatment for AIDS—AZT, DDI, Carrisyn, ribavirin, dextran sulfate, soluble CD4, Compound Q, AL-721, Peptide T, Kemron, isoprinosine, imuthiol, Imreg -1, polio vaccine, suramin, nonnucleoside reverse transcriptase inhibitors, or anything else for that matter—has succeeded in controlling AIDS for more than a few months.[3] In other words, treating HIV has so far proved to be universally ineffective. Dr. Dormont seems to have been right: AIDS does seem to progress independently of HIV. From the standpoint of autoimmune theories of AIDS, this result is not surprising. From the standpoint of treatment, it means that we may have to take a very different tack to cure AIDS.

The Range of Autoimmune Processes in AIDS

Autoimmunity has a wide range of manifestations in AIDS patients and in people at risk for AIDS. Consider the case of a thirty-eight-year-old male homosexual—call him Mike—who was admitted to the National Institutes of Health in Bethesda, Maryland in March 1982. Mike's first intimations that he might have AIDS were a progressive weakness in the limbs on his left side and an inability to walk properly. He was diagnosed as having Kaposi's sarcoma of the mouth and lymph nodes. Despite (or possibly in part due to) aggressive treatment of his Kaposi's sarcoma with (immunosuppressive) chemotherapies, including adriamycin, bleomycin, vinblastine, actinomycin D, and vincristine, as well as repeated radiation treatments, his Kaposi's sarcoma continued to progress, and Mike became increasingly worse. By June, he had lost control of his bladder and bowels, indicating severe central nervous system deterioriation. By September, he had developed global dementia, meaning he had lost most higher mental functions, including the abilities to speak coherently, to understand conversations or simple instructions, and to remember names and faces. His progressive neurologic deterioration was exacerbated by persistent oral herpes simplex, disseminated cytomegaliovirus, *Pneumocystis pneumonia*, chronic nausea and abdominal pain, vomiting, diarrhea, and intestinal bleeding. He died in February 1983 after a long, grim illness.

Because AIDS was still a new and novel disease, an unusually thorough autopsy was performed on Mike. It, along with a previous biopsy, revealed that he had substantial demyelinization of his nerves in his limbs, spinal cord, and brain. Myelin is a protein in the cells that surround many of the major nerves in the human body. This myelin sheath acts much like the insulation on an electrical wire. When the insulation on the wire is stripped off, short circuits often result. A similar phenomenon occurs when the myelin sheath is destroyed around a nerve. The messages that are sent down the nerve never arrive or arrive in a garbled form, similar to the short-circuited message that might arrive on a stripped wire. Closely associated with the foci of demyelinization were prominent areas of perivascular infiltrates, or "cuffs" surrounding nearby blood vessels. These cuffs are made up of huge conglomerations of lymphocytes, all aggregated around the blood vessels in the regions of demyelinization, so that they resemble the white cuff of a man's dress shirt surrounding his arm. These are sure signs of an autoimmune process and are found in other demyelinating diseases such as multiple sclerosis. In fact, many of Mike's symptoms were identical to multiple sclerosis but a form of multiple sclerosis that was greatly speeded up. Thus, one might say that his perivascular cuffs were signs that he was being "dressed" for death.

As if these autoimmune symptoms were not more than enough, Mike had several other associated conditions as well. His thymus and lymph nodes were infiltrated with unusual lymphocytes, characterized by fibrotic growths, and significantly degenerated. His bone marrow, which should have been full of burgeoning new lymphocytes and red blood cells, was relatively depleted. So were his Peyer's patches in his intestines. And, perhaps most odd of all, his testes were significantly atrophied, his sperm count was very low, and the few sperm he had failed to develop completely. Once again, significant perivascular cuffing was present along with lymphocyte infiltration in the testes. He evidently had an autoimmune reaction against his thymus, lymphocytes, and testes in addition to the one against his nerves.[1]

A rather different set of autoimmune phenomena were observed a few years later in a group of forty-three hemophiliacs studied at the Institute of Microbiology in Bologna, Italy. Roberto Conte and his colleagues examined each of eleven HIV-seropositive and thirty-two HIV-seronegative patients for the presence of lymphocytotoxic antibodies. *Lymphocytotoxic* is a German-style word that is pieced together from several others. The *lympho* part refers to lymphocytes. The *cyto* fragment refers to cells, so that *lymphocyte* means literally "cells of the lymph system," or white blood cells. And *toxic* means just what toxic means in everyday life: deadly. Thus, a lymphocytotoxic antibody is an antibody that attacks and kills white blood cells. This rare phenomenon can occur only when the body's B cells—the antibody producers—turn on its T cells in the immunological equivalent of civil war. The results are just as damaging to the cellular participants as is a human civil war to the people who take part in it.

In the case of the hemophiliacs, the results of this civil war were quite evident. Conte and his associates found that both HIV-seropositive (ten of eleven) and HIV-seronegative (ten of thirty-two) patients had lymphocytotoxic antibodies present in their blood and that the presence of these antibodies was correlated with diminished numbers of T cells. HIV-negative hemophiliacs with lymphocytotoxic antibodies were found to have half the number of T-helper and T-suppressor cells that those without such antibodies had. These patients also had unusually high amounts of circulating antibody, suggesting that the B cells were actively carrying on their sorties against the T cells. Their immunological civil war was producing demonstrable destruction.[2]

Both types of autoimmunity described—autoimmunity directed against tissues and organs such as Mike experienced, and autoimmunity directed against parts of the immune system itself, such as many hemophiliacs experience—are typical of people with AIDS or people in high-risk groups for AIDS. In fact, lymphocytotoxic antibodies, or LCTAs for short, that are directed against both T cells and B cells are frequently, if not ubiquitously, found in these groups.[3] For example, Kiprov and Anderson found LCTAs in both HIV-positive and HIV-negative homosexuals.[4] Pruzanski, Jacobs, and Laing found that all of their homosexual patients with AIDS or AIDS-related complex had LCTAs against both T and B cells, but so did thirteen of seventeen symptom-free homosexual men.[5] Kloster, Tomar, and Spira reported similar results.[6] Nine of ten AIDS patients had LCTAs, ten healthy heterosexuals did not, while three of five healthy homosexual men also had evidence of LCTAs. Further tests were done on two of these three men, and both showed significantly depressed T cell activity and counts. In controlled studies, LCTAs specifically directed at T cells have been demonstrated in up to 93 percent of homosexual AIDS patients, up to 60 percent of AIDS-related complex patients, but only rarely in healthy homosexuals and never in heterosexuals.[7] Moreover, the presence (though not the titre or absolute amount) of these T cell–reactive LCTAs correlates with disease progression in AIDS.[8] Lack of LCTAs is conversely associated with continued health even among HIV-positive homosexuals and drug addicts.[9]

LCTAs directed against MHC class II proteins on T and B cells have been demonstrated in over 82 percent of 210 HIV-seronegative and HIV-seropositive intravenous drug abusers. Most of these drug abusers also had significant levels of antibody directed against MHC class I proteins as well.[10] Only 1 out of 100 healthy blood donors was positive for anti-MHC antibody. The presence of these LCTAs was not associated with infection with hepatitis B virus or HIV, but previous studies have demonstrated that *Candida albicans* infections can result in antibodies that cross-react with MHC class II proteins.[11] Other infectious agents causing pyogenic and septicemic infections may also be related to such autoantibody responses.

LCTAs are prevalent in HIV-positive hemophiliacs, with 46 percent having autoantibody to T cells and 84 percent to B cells.[12] Conte and his colleagues, for example, describe eleven HIV-seropositive, but AIDS symptomless, patients with types A and B hemophilia whom they compared with thirty-two type A and B hemophiliacs who were HIV seronegative.[13] Ten of the eleven HIV-positive patients had LCTAs. Five of the thirty-two HIV-negative patients also had LCTAs correlated with decreased numbers of T helper lymphocytes. These findings suggest that HIV is not necessary to induce such autoimmunity. Similar results were reported by Daniel, Schimpf, and Opelz, who found that LCTAs directed at both B cells and T cells were prevalent among HIV-positive hemophiliacs and also present among some HIV-negative patients.[14] They concluded that LCTAs must play a role in the pathogenesis of acquired immunodeficiency.

Some hints exist as to possible inducers of lymphocytotoxic autoimmunity. Researchers found from the outset that semen components may play a pathogenic role in AIDS. Receptive anal intercourse is the most significant of the risks identified for homosexual and bisexual men[15] and for heterosexual women.[16] The same mechanisms proposed for promoting HIV infection—abrogation of the anal or intestinal tissue due to anal forms of sex or untreated sexually transmitted or enteric diseases—would also permit sperm cells to enter the bloodstream or lymph system. These cells could then induce the formation of antibodies. Although women who practice anal intercourse and subsequently develop AIDS have not specifically been tested for antisemen antibodies, 30 percent of unselected female AIDS patients in one New York study had significantly high antibody to human sperm extract, protamine, and fertilization antigen as compared with healthy controls.[17] The prevalence of such antibodies in male homosexual AIDS patients is much higher: 40 percent have significant antibody against human sperm extract, 70 percent against protamine, and 70 percent against fertilization antigen.[18] These antisperm antibodies have been demonstrated to cross-react with host T cells (that is to act as LCTAs) in AIDS risk groups and to cause immunosuppression.[19] Indeed, lymphocytotoxic antibodies have been found not only in HIV-positive but in many HIV-negative homosexuals as well.[20] These antisperm antibodies may be the cause of the antitesticular autoimmunity that is present in virtually all gay men who develop AIDS. Aspermatogenesis, sperm maturation arrest, and testicular fibrosis are the rule rather than the exception in such patients.[21]

Another possible cause of antilymphocyte autoimmunity in AIDS may be foreign lymphocytes. Recall that semen contains a high proportion of lymphocytes, as do whole blood transfusions, granulocyte transfusions, and some other blood products. Recall also that Stott and his colleagues found that human lymphocytes were able to protect monkeys against simian HIV, indicating that the lymphocytes were acting as antigens to induce an immune response. In fact, most studies of AIDS patients report that the LCTAs they isolate are directed at T-helper (T4) cells and monocytes or macrophages—the two types of cells that become most seriously depleted during AIDS—but are less commonly reactive against T-suppressor (T8) cells, B cells, or most HLA-bearing cells—cell types that are generally less affected during AIDS.[22] Particularly noteworthy are reports of anti-CD4 autoantibodies in HIV-infected individuals, since the CD4 protein is the badge that specifically identifies T-helper lymphocytes and macrophages.[23] Several investigators have demonstrated that such anti-CD4 antibodies inhibit lymphocyte activity in both the test tube and the body and must therefore account in some measure for the immune suppression characterizing AIDS.[24] This makes sense in the light of the way in which the immune system works. Just as HIV could theoretically disrupt proper immune function by destroying T-helper cells and macrophages, so could LCTAs directed at the CD4 protein on their cell surfaces.

The obvious question at this point is what role HIV plays in triggering autoimmunity. Specifically, does it induce LCTAs directed against T cells? If it does, is it the only antigen that is capable of doing so, or may other antigens, such as semen, play some role in the development of autoimmuninty in AIDS? Since some people in high-risk groups develop antilymphocyte autoimmunity in the absence of HIV infections, it seems unlikely that HIV is necessary to induce this form of autoimmunity. On the other hand, the much higher incidence of LCTAs among HIV-infected individuals than among those not infected argues either that such people have higher risks not only for HIV but for one or more other autoimmunity-inducing agents or that HIV can act synergistically to promote such autoimmunity. HIV probably does not induce such autoimmunity on its own. Zarling's and Stricker's groups could find no evidence of LCTAs among HIV-infected chimpanzees.[25] This observation is extremely important, since no HIV-infected chimpanzee has yet developed AIDS. Perhaps the presence or absence of LCTAs, not HIV seropositivity, is the factor that determines the true risk of AIDS.

The importance of determining the cause or causes of autoimmunity in AIDS cannot be understimated. As the case of Mike so clearly demonstrates, autoimmunity directed at lymphocytes is only one of the many forms of autoimmunity that manifest themselves during the process of AIDS. Peripheral neuropathies and multiple sclerosis–like demyelinization occur in over half of AIDS patients. Autoimmune orchitis, in which antibodies and T cells invade and slowly destroy the testes and their sperm-producing cells, is extremely common among both AIDS patients and promiscuous gay men in general. Many AIDS patients develop autoimmune forms of arthritis; autoantibodies directed at muscle proteins; and symptoms similar to both Sjögren's syndrome and systemic lupus erythematosus, including skin rashes, kidney damage, and antibodies against DNA, thryoglobulin, and adrenocorticosteroids. Autoimmune forms of thrombocytopenia and coagulopathies (in which blood fails to clot properly) caused by antibodies against either platelets or blood coagulation factors are also extremely common among AIDS patients and those at risk for AIDS. So is autoimmune anemia. And, as was mentioned above, large numbers of homosexual men and some women with AIDS or in AIDS risk groups have antisperm antibodies in their bloodstream that have been shown to recognize T cells as targets for destruction.[26] Clearly, it will be impossible to treat AIDS if some allogeneic or infectious agent, or set of such agents, can provoke these various manifestations of immunological civil war and then, like provocateurs, leave the scene before the danger to themselves becomes too great. What are the provocateurs that are responsible for these diverse autoimmune conditions in AIDS patients and in those at risk for AIDS? To answer this question, it is first necessary to know more about how autoimmunity may be initiated.

An Introduction to Autoimmunity

Ten years ago, I sat in an office in the Salk Institute for Biological Studies puzzling over the problem of autoimmunity with Fred Westall, another investigator in Jonas Salk's Autoimmune and Neoplastic Diseases Laboratory. I knew nothing whatever about autoimmunity or its causes at that time, but I was eager to learn. Dr. Westall was equally eager to teach me since he thought my ideas concerning the ways in which peptides (small proteins) interact might throw some light on an outstanding anomaly concerning autoimmunity. The story that Dr. Westall told me over several months is actually quite simple—so simple, in fact, that it turned out to be a great help that I knew nothing about the field, since it gave me an excuse to ask the kind of basic questions that more sophisticated investigators are too proud or too well trained to ask.

Dr. Westall had established his reputation by characterizing the fragments of a protein called myelin basic protein. Myelin basic protein is the protein in Mike's nerve sheaths that was being destroyed in his brain and central nervous system. (Recall that these sheaths act like the insulating plastic on electrical wires.) One of the ways in which myelin sheaths can be stripped off nerves is by an autoimmune attack on the myelin basic protein in the sheaths. The result in human beings may be any of several diseases, including Rasmussen's encephalitis, Landry-Guillain-Barré syndrome, or multiple sclerosis, a disease characterized by progressive paralysis beginning with the extremities and slowly advancing toward the spinal cord and brain.

It is possible to create a disease that mimics multiple sclerosis in animals such as rats by inoculating them with a mixture of material including myelin basic protein. The resulting disease is called experimental allergic encephalomyelitis, or, more simply, EAE. EAE is one of the best-studied models of autoimmunity we have; it was discovered during the 1930s and has been investigated continuously ever since. Indeed, much of what we know about multiple sclerosis and all of the attempts to develop cures for it have begun with studies of EAE.

Fred Westall was one of the key figures in determining which parts of the very long myelin basic protein molecule were necessary for inducing EAE in various animal models. He and several other people in Jonas Salk's laboratory had found that only a very short portion of the protein, a peptide, was required.[1] His next step had been to determine whether this peptide was also sufficient to cause EAE. It was not. And that was odd, since all of the textbooks and articles on EAE quite clearly stated that EAE was caused by an autoimmune reaction directed at myelin basic protein and induced by myelin basic protein.

Recall from Chapter 2 that a normal immune response proceeds in the following fashion. Some foreign material, an antigen, manages to invade the body by evading digestion or bypassing the skin and mucosal membranes that act as the first lines of defense. Inside the body, the material is "examined" by various lymphocytes that pass by. These white blood cells "test" the material to see if it is "self" or "nonself." If it is self, they generally leave it alone.

If the material is nonself, then either a B cell or T cell response, or both, is mounted. B cells secrete antibodies into the blood and lymph that specifically recognizes and inactivates the antigen. T cells also produce proteins similar to antibodies, but these remain attached to their cell surfaces. The T cells then scavenge the antigen, like huge armored personnel carriers that transport whole squadrons of soldiers right to the place they are needed. The response tends to be quite specific. A single antigen will induce the production of only a handful of related antibodies or activate only a few T cell clones.

Standard accounts of autoimmunity maintain that somehow this normal immunological response becomes subverted. A mistake is made at some point, and the immune system can no longer distinguish self from nonself. Standard accounts of EAE maintain that myelin basic protein from one animal is "recognized" as being "foreign" by the immune system of the animal into which it was inoculated. In immunological jargon, the foreign myelin basic protein is considered to be an antigen. Unfortunately, in EAE and other autoimmune diseases, the antigen—in this case, myelin basic protein—looks so similar to the body's own myelin basic protein that when an antibody or T cell response is mounted, some of the "soldiers" trained to destroy the foreign myelin basic protein mistake the body's myelin basic protein for foreign material, and they attack it. In this view of autoimmunity, the damage that is done to the body during an autoimmune reaction is of the accidental sort inflicted by so-called friendly fire during wartime. In the chaos that ensues, however, more and more "soldiers" become "confused," resulting in ever more severe attacks on their "own men" until the body finally capitulates by dying. This theory has been espoused by many prominent investigators, including M. B. A. Oldstone, Robert Fujinami, E. C. Alvord, Jr., and P. R. Carnegie. They and others have demonstrated that several viruses, including measles, influenza type A, and Epstein-Barr virus, have proteins that mimic myelin basic protein. Thus, they suggest that the immune response to these viruses may trigger autoimmunity.[2]

Dr. Westall revealed problems with this view of autoimmunity that convinced me it was untenable. To begin with, almost everyone becomes infected with measles, influenza, or Epstein-Barr virus sometime, and yet no more than 1 in 100,000 people develops a resulting autoimmune demyelinization. Surely this fact argues that something more is required to induce autoimmunity than mimickry between the antigen and some self protein. Moreover, Dr. Westall pointed out to me at the start of our conversations that the standard accounts of EAE were not quite accurate. Even in the animal models of autoimmunity, one needs more than just the foreign myelin basic protein to trigger disease. One needs, in addition to myelin basic protein or some active peptide fragment of it, something called an "adjuvant." Adjuvants, in immunological jargon, are defined as "nonspecific immune potentiators." In other words, they are compounds that supposedly cause the immune system to react more strongly to almost any foreign material. They hype-up the immune response to make it more effective—just the sort of thing that one might think would help AIDS patients. Because these adjuvants are supposedly nonspecific—somewhat like giving every soldier amphetamines to make them all more alert regardless of their mission—many accounts of EAE ignore the adjuvant. Soldiers, after all, can fight whether they are given amphetamines or not, and by analogy, one would expect that the adjuvant in EAE was helpful but not necessary. No one, as far as we could tell, had ever considered the possibility that the actions of adjuvants in inducing multiple sclerosis or other human forms of autoimmunity might actually be specific and necessary. Fred Westall thought this was an important oversight.

The reasons for his opinion were carefully based in experiment. Having identified the key peptide necessary to induce EAE, Dr. Westall and his colleagues had inoculated animals with larger and larger doses of the peptide in hopes of producing EAE without using an adjuvant. If the adjuvant did nothing more than increase the immune response to the myelin basic protein, then presumably a large enough dose of myelin basic protein should have the same immunologically stimulating effect. The more antigen was injected, the greater the immune response to it should become, went the reasoning. Surprisingly, this was not the case. In fact, the more myelin basic protein Dr. Westall and his colleagues injected into an animal, the less of an immune response they observed. Some large or repeated doses actually caused immune suppression.[3] These observations led Dr. Westall to suspect that the adjuvant was more than just a nonspecific immune potentiator in EAE. Perhaps it was necessary.

Two groups of immunologists independently confirmed Dr.

Westall's hunch. While he had been busy characterizing the immunological properties of the various peptide fragments of myelin basic protein, Dr. E. Lederer in France[4] and Dr. Y. Nagai in Japan[5] had been investigating the precise molecular structure of the adjuvant. This adjuvant turned out to be a small fragment of the cell wall of various bacteria such as *Mycobacterium tuberculosis*, the bacterium that causes tuberculosis in humans. Dr. Nagai had demonstrated that just as there was a minimal peptide fragment of myelin basic protein necessary to induce EAE, there was also a minimal adjuvant fragment, called the "adjuvant peptide," that was necessary to induce EAE. Altering the adjuvant peptide chemically or substituting other similar peptides for it did not result in inducing EAE in animals. In other words, EAE required not just a specific immunological reaction directed against some part of the myelin basic protein; it also required a second specific immunological reaction directed against the adjuvant. There were two branches of the armed forces involved in this immunological warfare, and not just one. The question that Fred Westall and I wrestled with for several months was: Why? Why must one have two peptides—myelin basic protein and adjuvant—and two sets of antibodies (or, in the case of EAE, two sets of T cells) rather than just one?

Our question was made even more interesting by some other well-established observations about EAE. For one thing, it turns out that if you inoculate an animal with myelin basic protein and then wait two weeks and inoculate the same animal with a combination of myelin basic protein plus adjuvant, it does not develop EAE. The opposite experiment yields the same results. If you inoculate an animal with adjuvant, wait two weeks, and then inoculate it with a combination of myelin basic protein plus adjuvant, it does not develop EAE either. In other words, both of the antigens that are necessary to induce EAE can act as vaccines to protect an animal from developing EAE if injected by themselves and sufficiently in advance of the encephalitogenic dose. Even more perplexing, it is possible to *cure* an animal that has EAE by injecting it with even more of either the adjuvant or the myelin basic protein alone. Conventional immunological theory has no explanation for these results at all. They are among the great unsolved paradoxes of the field.

Or they were. In 1983, Dr. Westall and I finally came up with an explanation of these and many other anomalous findings concerning EAE and other animal models of autoimmunity that made sense of them all without resort to "accidents," "mistakes," or other non-scientific concepts. And as with all other good scientific theories, it is both simple and commonsensical. It is also controversial.[6]

Dr. Westall and I believe that autoimmunity is a natural and conventional immune response to an unusual set of circumstances. Begin by considering a normal immune response. An antigen elicits its chemically complementary antibody. In lay terms, "chemically complementary" means that the two chemicals fit into one another, the same way that matching puzzle pieces fit. Just as matching puzzle pieces must share corresponding shapes, colors, and patterns, so must the antigen and antibody share corresponding shapes and chemical properties. When they do, they combine. The resulting combination suffices to begin an immune reaction that eliminates the antigen from the body and leads, in the case of infection, to its control. It is believed that a couple of weeks after the initial antibody is produced, a second antibody that is complementary to the first is elicited, and the second antibody acts as a brake upon the first, both controlling the amount present and perhaps ensuring that some antibody continues to be produced as a sort of immunological memory to prevent recurrent infections. That is the normal function of the immune system.

Now imagine that the immune system is faced not with one, but with two antigens and that these two antigens have a special relationship to one another: they are themselves chemically complementary. In other words, they can fit together like two pieces in a jigsaw puzzle. Both antigens will elicit antibodies in the normal manner. Antigen 1 elicits antibody 1 that is chemically complementary to it. Antigen 2 elicits antibody 2 that is chemically complementary to it. But now something very odd becomes apparent: Because antigen 1 is complementary to antigen 2, the part of antibody 1 that recognized antigen 1 looks very much like antigen 2. Conversely, antibody 2, because it is chemically complementary to antigen 2, looks very much like antigen 1. The result is mayhem.

The immune system is now faced with an unresolvable conundrum: It can no longer distinguish self from nonself. Antigen 1 looks like antibody 2, and antigen 2 looks like antibody 1. Which is self and which is nonself? The body cannot tell. If antibody 1 begins to search for nonself antigen 1, it may end up finding and trying to destroy self-antibody 2. Conversely, if antibody 2 searches for nonself antigen 2, it may find and destroy self-antibody 1. The result will be a civil war within the immune system itself. The same thing will happen if chemically complementary T cell clones are activated, as occurs in EAE (and as may occur in AIDS).

Several consequences follow. One is that the results of this civil war should be obvious in the blood of people with autoimmune conditions. Large clumps of antibodies or T cells should be apparent—and they are: perivascular cuffs and circulating immune complexes. In fact, both perivascular cuffs, such as those characterizing Mike's AIDS demyelinization, and circulating immune complexes are used to diagnose autoimmune conditions. Another prediction is that the pair of chemically complementary antigens should produce immune disregulation, since the complementary antibody that is normally produced a couple of weeks after the initial antibody to regulate it is being produced at the same time as the original antibody. There is now no complementary antibody to produce immune regulation. Both antibodies are primary responses to different antigens. Thus, the immune system no longer knows what is self or when to turn itself off. It indiscriminately "shoots" at everything that "looks" foreign.

If one or both of the antigens are similar to self proteins, even more dire consequences may follow. Imagine one or more of our chemically complementary antigens—in this case virus or bacterial antigens—are similar to a self protein such as myelin basic protein. Not only do all of the interactions already discussed occur; another reaction does too. T cells or antibodies activated against antigen 1 will not only attempt to eliminate antigen 1, they will also seek out and attempt to destroy its look-alike, myelin basic protein. In short, not only does this scenario of complementary antigens lead to a mechanism for immunological civil war, it also predicts just the sort of attack on the body that is observed in autoimmune conditions.

This, then, is the theory that Fred Westall and I elaborated. We call it the theory of multiple-antigen-mediated autoimmunity, or MAMA for short (Figure 11). It is, like Montagnier's cofactor theory of AIDS, a theory that requires the presence of two or more antigens simultaneously. We have demonstrated that the two antigens involved in producing EAE and the pairs involved in produc-

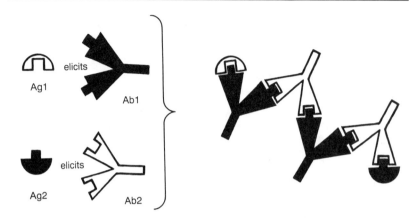

FIGURE 11. Basis of the multiple-antigen mediated autoimmunity (MAMA) theory. Two chemically complementary antigens (Ag) elicit antibodies (Ab) that are idiotype–anti-idiotype pairs. These not only attack the antigens but each other as well. Autoimmunity therefore begins with the immune system's attacking itself, resulting in disregulation of its normal functions.

ing several other animal autoimmune diseases are, in fact, chemically complementary as predicted.[7] We have evidence that the T cells elicited by the two antigens necessary to produce EAE do, in fact, recognize each other as foreign and do kill each other off.[8] And we have tested various other predictions made by the theory. It is, however, still a theory and still very much outside the mainstream of immunological research. With that caveat, however, it has nonetheless turned out to be extremely useful for exploring the possible causes and varieties of autoimmunity in AIDS.

Interestingly, at almost exactly the same time Dr. Westall and I proposed our theory of autoimmunity, Dr. Paul H. Plotz of the Connective Tissue Diseases Section of the National Institutes of Health proposed a somewhat different theory that leads to exactly the same outcome. Plotz proposed that a *single* antigen can induce the entire sequence of events we had envisioned. Plotz's mechanism works as follows. An antigen invades a person. The person naturally responds by creating a series of antibodies against this antigen. This antibody is chemically complementary to the antigen. The specific, unique part of the antibody that recognizes the antigen is called the idiotype. (*Idiot,* in the original Greek does not mean "stupid" but rather "unique." The idiotypic part of the antibody is therefore the unique part that recognizes one, specific antigen.) After the person has manufactured enough of the idiotypic antibody to control the antigen, the immune system responds with a second antibody response. In this second response, an antibody is formed against the idiotypic, or unique, portion of the first antibody. This second antibody is therefore called an anti-idiotypic antibody. It regulates the amount of the first antibody that is present and active in the person and may also be involved in ensuring that a continuous, low level of antibody is always stimulated to protect against future infections.

Since the antigen elicits a chemically complementary antibody and the first antibody elicits a chemically complementary antibody in turn (the second antibody), the second antibody will mimic the antigen. Think of it in this way: If you place your hand on some wet plaster or cement and then let it harden, you will obtain a cast of your hand. If the hand is the antigen, the cast is the idiotypic, or first, antibody. If the first cast is used to make a second—that is, if it is filled with plaster or rubber—then one will obtain a second image, which is very similar to your hand. Thus, the images are maintained in a complementary fashion. Antibodies can, in fact, transfer their images in this manner. Plotz suggested that this idiotype–anti-idiotype image transfer mechanism may lead to autoimmunity under certain circumstances.

Plotz recognized that one outcome of such a system of image–anti-image or idiotype–anti-idiotype antibodies is that the first, or idiotypic antibody, will be able to bind to both the original antigen (or hand) and the second, idiotypic antibody (or cast of the hand). Thus, these antibodies will bind to one another. If, in addition, the first antibody has a shape similar to that of some protein within a person's cells, then the second, or anti-idiotypic antibody, will attack the first antibody and this cellular protein as well. In this way, a virus or bacterium could potentially initiate an immune response against a very different protein within the body of the person it has infected.[9] The result will be exactly the one that Westall and I had foreseen. The only major differences are that Plotz assumes only one viral initiator instead of our two, and he cannot explain why the normal idiotype–anti-idiotype system that regulates immune function becomes dysfunctional in autoimmunity.

In summary, there are now at least three theories of how autoimmunity may be induced. The Oldstone-Alvord type of model says that the similarity of a virus or bacterium with some protein in the body may directly induce an antibody that will create autoimmunity. Plotz maintains that this autoimmunity is indirectly produced not by the original antibody but by the anti-antibody (or anti-idiotype antibody). And Westall and I believe that autoimmunity can be induced only when two chemically complementary antigens, at least one of which is similar to a self protein, elicit two sets of antibodies that have an idiotype–anti-idiotype relationship. The long and the short of it is that none of the theories of autoimmunity has yet been proved to the satisfaction of any but a very small minority of researchers. Notably, however, both Plotz's idiotype–anti-idiotype model and the Root-Bernstein–Westall model lead to the same outcome, despite their initial differences, and both have been used to explain AIDS. These two theories are of some interest in the present context.

Idiotype–Anti-idiotype Theories of Autoimmunity in AIDS

The idiotype–anti-idiotype theory of autoimmunity was the first to be applied to AIDS by Ziegler and Stites and by Andrieu and his colleagues in 1986, and in less complete forms by other investigators subsequently.[1] They observed that several HIV proteins have sequences similar to the sequence of the MHC class II (HLA-D) protein of various T lymphocytes and B cells. In other words, parts of HIV mimic the immune system itself. Recall from Chapter 2 that MHC class II (the security system that distinguishes self from nonself) interacts with the CD4 protein (identification badge) of T-helper cells (the security guards) in the presence of a foreign antigen (nonself). This interaction is a necessary part of several steps in the activation of both T cell and B cell responses to infection. Anything that interferes with this step will interfere with immune competence. Thus, if antibody is made against the HIV protein sequences that mimic MHC class II and if these antibodies mistake MHC class II proteins for foreign antigens, then the immunological security system will be destroyed.

The problem with this scenario is that one would expect that the major thrust of autoimmunity induced by HIV would be against the MHC-bearing cells. This is not, in fact, the case. Instead, the major targets of LCTAs are T-helper cells and monocytes, with T-suppressor cells, natural killer cells, and other lymphocyte subsets less conspicuously depleted. Ziegler and Stites therefore proposed a clever adaptation of Plotz's basic theory of autoimmunity. Just as Plotz suggested that the original, or idiotypic, antibody gives rise to an anti-image, or anti-idiotypic antibody that is the true cause of autoimmunity, so Ziegler and Stites proposed that HIV induces an idiotypic antibody, which in turn induces an anti-idiotypic antibody. It is this anti-idiotypic antibody that they think causes T cell destruction in AIDS (Figure 12).

The Ziegler-Stites hypothesis is reasonable on the face of it. Some HIV proteins mimic MHC class II protein sequences. HIV and MHC class II proteins are chemically complementary to CD4 proteins and bind directly to them. That is, they fit into one another just as the hand (HIV–MHC class II) fits into its mold (CD4). The antibody that binds to HIV–MHC class II proteins will therefore resemble the mold that fits the MHC hand. If this original CD4-like antibody were to induce the formation of an

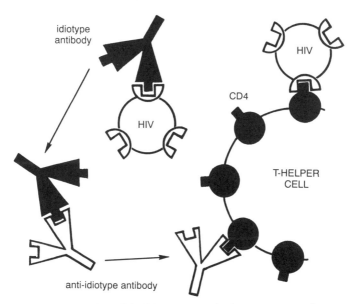

FIGURE 12. A general model of the way in which HIV may product anti-idiotype antibodies that attack T-helper lymphocytes, initiating autoimmunity.

anti-idiotypic antibody, then this second antibody would resemble the original hand that made the mold. In other words, this second, anti-idiotypic, antibody would resemble HIV itself. The immune system would be mass producing proteins in the image of its own enemy. It follows that anti-idiotypic antibody would fit into (and attack) both the original antibody (a mold shape) and cells, such as T-helper cells and macrophages, that have the CD4 protein (a similar mold) on their surfaces. Thus, HIV might be able to induce autoimmunity against the very cells that are selectively destroyed as AIDS progresses.

An almost identical theory, also based on this idiotype–anti-idiotype theory of autoimmunity, has been suggested as well by Geoffrey Hoffmann, a well-respected immunologist at the University of British Columbia. In 1986 he and some of his colleagues found evidence for a process similar to that predicted by Plotz in 1983. An antigen (unrelated to AIDS) could induce an antibody. This antibody could induce another antibody, and this latter antibody mimicked the binding activities of the original antigen.[2] In 1988, they applied their observations to AIDS. In essence, Hoffmann's theory is identical to that proposed a few years before by Ziegler and Stites, but his terminology differs considerably. Instead of idiotype–anti-idiotype, Hoffmann prefers the terms MHC-image and anti-MHC-image, or MI-AMI.[3]

Several major problems exist with the type of theories proposed by both Ziegler and Stites and by Hoffmann and his colleagues. First, although the existence of the sort of idiotype–anti-idiotype antibody pairs proposed by both groups are known to exist, the standard interpretation of these antibody pairs within immunology is that they exist to *control* the immune response, not as a consequence of a pathological condition. Thus, the expected result of idiotype–anti-idiotype or MI-AMI processes according to most textbooks on the subject is to maintain the distinction between self and nonself and thereby to *prevent* autoimmunity. Why HIV should be able to disrupt the distinction between self and nonself by subverting the normal network of regulatory immune processes is not addressed by Ziegler and Stites, Hoffmann, or anyone else utilizing similar theories.

A second major problem with these theories is that they address only a very limited subset of the autoimmune diseases that afflict AIDS patients. As important as it is to explain how LCTAs are induced to kill off T cells, it is equally important that the causes of autoimmune orchitis, autoimmune demyelinization and dementias, thrombocytopenia, the systemic lupus erythematosus–like symptoms, and so forth be elucidated. It is clearly impossible, at least in my opinion, that HIV can cause all of these different disease manifestations on its own. If it can, it is a unique virus.

Third, no one has addressed the question of what makes HIV uniquely capable of such destruction, when other microbes, such as cytomegalovirus, Epstein-Barr virus, and *Mycobacteria*, also have proteins that mimic both MHC class II and CD4 proteins.[4] If HIV can induce T cell destruction and AIDS by itself, then these other microbes should be able to do so as well. Clearly they do not.

Fourth, investigators such as Ziegler and Stites and Hoffmann have not addressed the manifest difficulties to their theory that are raised by evidence that many HIV-seronegative people in AIDS risk groups nonetheless have LCTAs correlated with diminished T cell counts. Since HIV is not present to induce autoimmunity in these people, what has done so?

Finally, there is no direct evidence that HIV or HIV proteins can, in fact, induce autoantibodies in either human beings or our anthropoid cousins. Recall that one of the differences between human beings with AIDS and healthy chimpanzees infected with HIV is the lack of LCTAs. HIV, in short, may be a player in the autoimmune drama that characterizes so much of AIDS pathology, but it is not likely to be the only actor.

Multiple-Antigen-Mediated Autoimmunity and AIDS Dementia

My belief is that HIV is only one of a multitudinous cast that cooperate to produce autoimmunity. What has been overlooked by Hoffmann, Ziegler and Stites, and many of their colleagues is the huge number of other infectious agents that are also present in AIDS patients, often concurrently. These other infections are, according to my multiple-antigen-mediated theory (MAMA) of autoimmunity, not mere bystanders or even a supporting cast but costars in the tragedy that produces LCTAs and other manifestations of autoimmunity in AIDS. Moreover, the MAMA theory predicts that the specific type or types of autoimmunity that result from these multiple infections will differ from patient to patient (as is actually observed) depending on their specific subset of infections. What makes AIDS patients unique from the perspective of the MAMA theory is not their HIV infections but their extremely unusual set of multiple, concurrent infections.

The application of the MAMA concept to AIDS has come in two stages. One of the earliest predictions that Dr. Westall and I made was that the autoimmune complications that follow various infections such as measles, or rabies, and those that sometimes accompany their vaccinations, would be found to occur only when specific pairs of infections were present in the same patient simultaneously.[1] Such a combination might consist, for example, of a virus that has proteins that mimic some set of human proteins, accompanied by a bacterial infection that can act as its adjuvant. There are, in fact, some limited data from hospitals that patients suffering postvaccinal complications often have multiple infections,[2] and all physicians are warned in the standard textbooks to avoid vaccinations when a patient is ill or feverish since these conditions have been found by experience to increase greatly the potential for postvaccinal complications. The problem is that such complications occur so rarely—perhaps 1 in every 100,000 or 1 million vaccinations—that data are not collected in a coherent fashion, nor are generalizations easily made. It is difficult to determine what specific sets of viruses and bacteria may cause any particular autoimmune manifestation.

Every AIDS patient, however, develops multiple, concurrent infections, and many of these are very well documented. One would therefore expect that if the MAMA theory of autoimmunity were correct, not only would AIDS patients be extremely likely to develop various forms of autoimmunity—a fact already well established—but one could also hope to elucidate the precise combinations of infections that cause these different forms. In short, AIDS patients may present us not only with the ultimate test of the MAMA theory but the clues as to what causes thrombocytopenia, autoimmune demyelinization, and so forth, in their much rarer forms among the general population.

When this realization finally dawned upon me late in 1989, I began with the disease I knew best, EAE, and worked from there. It took only a matter of a few hours, to find what I was looking for in the library: a series of case histories of dementias in AIDS patients—Mike's case was one—that could hardly have resembled EAE more closely. For example, during February 1984, Dr. Surl L. Nielsen, Dr. Carol K. Petito, Dr. Carlos D. Urmacher, and Dr. Jerome B. Posner of the Department of Pathology at Sutter General Hospital in Sacramento, California, reported on their autopsy studies of forty AIDS patients. Thirty-one of these patients (78 percent) had marked neurological abnormalities due to a variety of causes, including direct infections of the brain by toxoplasmosis, progressive multifocal leukoencephalopathy (a virus-induced degeneration of the brain), or lymphomas. More interesting to me, nineteen of the patients (48 percent) suffered from encephalitis associated with severe dementia. This encephalitis was often the first symptom these patients displayed of AIDS, and it resembled EAE in being characterized by demyelinization, microglial nodules in the brain, and perivascular cuffing. It was Mike's clinical description repeated nineteen times. Even more interesting, all of these patients suffered from systemic cytomegalovirus infections, but CMV could not be isolated from the brains of any of them. In other words, although CMV was associated with the presence of the encephalitis, it could not be the direct cause of it.[3] Nor could HIV be the direct cause of their encephalitis, since nearly all AIDS patients are infected with HIV, but only a fraction of them experience demyelinization. In fact, sometimes the same symptoms are found in HIV-negative homosexual men, cancer patients, and renal transplant patients, and many such cases were reported long before AIDS or HIV were recognized.[4] Could the cause of this AIDS-associated demyelinization be a combination of CMV with some other infectious agents?

Further autopsy studies of the AIDS dementia complex by Dr. Bradford A. Navia, Dr. Barry D. Jordan, and Dr. Richard W. Price of the Memorial Sloan-Kettering Cancer Center, the New York Hospital, and Cornell University Medical College confirmed the cytomegalovirus link to demyelinating autoimmunity and pinpointed a probable cofactor.[5] Navia, Jordan, and Price studied differences between AIDS patients who developed various types of dementia and those who did not. Encephalitis, like that described by Neilsen and his colleagues, was only one of the more common dementias found, and the results were not published in a way allowing encephalitis to be considered separately from the other dementias. Nonetheless, the data were fairly clear. No significant differences were found by age, sex, or risk factor for AIDS. HIV was not considered, since all AIDS patients were infected with HIV but only some developed dementias. Kaposi's sarcoma, *Pneumocystis carinil* pneumonia, opportunistic infections either alone or in combination with Kaposi's sarcoma, and lymphomas were equally prevalent in both demented and nondemented groups. Demented patients, however, had *Mycobacterium avium-intracellulare* (MAI) infections 2.5 times as frequently as nondemented patients (50 percent versus 21 percent; $p < 0.05$ by chi-square analysis), and systemic CMV infections 1.3 times as frequently (67 percent vs. 50 percent). These figures suggested that a combination of CMV with *Mycobacteria* might be a cause of demyelinization. The specific combination of CMV with MAI infection as a risk factor of dementia was not considered in Navia's analysis, however. The figures were merely suggestive.

The CMV-*Mycobacteria* lead was supported by a subsequent study by Petito and her colleagues.[6] They found demyelinization accompanied by microglial nodule encephalitis in twenty of eighty-nine consecutive autopsies of AIDS patients. Fifteen of the twenty had systemic CMV infections, six had systemic HSV infections, and seventeen had systemic MAI infections. The other three had other bacterial infections. These data confirm the unusually high correlation between herpes viral infections, *Mycobacterial* or other bacterial infection, and the incidence of demyelinating neuropathies in AIDS patients.

Armed with these leads, I performed a comprehensive review of published autopsy reports on AIDS patients that specifically detailed the type of dementia, presence or absence of demyelinization, and a complete catalog of the infections suffered by the patient. I found that almost every AIDS patient who develops an encephalitic or demyelinating dementia has an active, often disseminated cytomegalovirus, Epstein-Barr virus, or herpes simplex virus, in conjunction with a *Mycobacterium* infection (most often an atypical *Mycobacterium* but sometimes *M. tuberculosis*). Patients with only one of these infections, or neither, never (with only a couple of exceptions) developed demyelinating or encephalitic dementias. Thus, the epidemiologic evidence argues strongly for autoimmune demyelinization being due to a herpes-virus-*Mycobacterium* combination.[7]

The plausibility of the herpes-virus-*Mycobacterium* combination as a cause of autoimmune demyelinization is supported by several additional facts. First, *Mycobacteria* are present in both the AIDS dementia patients and in the animal model EAE. Thus, one of the two components necessary to induce demyelinization is the same. The other necessary component is myelin basic protein or something that resembles it closely. In fact, computer-based homology searches carried out not only by Dr. Sheila Hobbs and me, but Dr. Fujinami and Dr. Oldstone, and Jahnke and his collegues, and P. R. Carnegie have demonstrated that cytomegalovirus, Epstein-Barr virus, hepatitis B virus, various influenza A subtypes, measles, adenoviruses, and a variety of other human pathogens have proteins that mimic significant portions of myelin basic protein.[8] Thus, it is certainly plausible that cytomegalovirus or other herpes viruses might be able to induce the formation of antibodies or lymphocytes capable of also recognizing myelin basic protein as a target. If that does occur, then the conditions necessary to induce demyelinization in animals would be mimicked exactly in human beings.

In fact, the story has now moved beyond the purely theoretical. I have now demonstrated that antibodies to cytomegalovirus, Epstein-Barr virus, and hepatitis B virus, and, unexpectedly, sperm proteins as well, will all precipitate myelin basic protein.[9] Thus, the theoretically predicted mimickry between cytomegalovirus and other herpes virus proteins and myelin basic protein is real, although the range of viruses and other antigens that might be involved in creating demyelinization is broader than I had assumed on theoretical grounds. In fact, the gp120 protein of HIV induces antibodies that will also recognize myelin basic protein as a target, so HIV may, after all, play a role in the induction of autoimmune demyelinization in AIDS.

Equally important, I have found that a few, specific antibodies to *Mycobacteria* and to *Mycoplasmas* will interact to form the sorts of idiotype–anti-idiotype (or complementary) antibody pairs with cytomegalovirus, hepatitis B virus, and HIV antibodies that MAMA predicts. In other words, at least two types of bacteria have the potential of playing the role of adjuvant to various viruses in the induction of demyelinization, and one of these bacteria, the *Mycobacteria*, are none other than the ones already used in experimental animal models of these syndromes. Whether this adjuvant effect is limited to inducing autoimmune complications or also helps to explain the results that Montagnier and Lo have published concerning the necessary role of *Mycoplasmas* in AIDS remains to be elucidated.

The criteria laid down by the MAMA theory are satisfied in laboratory conditions. Two different antigens, a virus and a bacterium, each produces separate antibodies (Figure 13). These antibodies have the characteristics of being an idiotype–anti-idiotype pair (they are chemically complementary and therefore bind to one another). Thus, antibody to the virus not only acts against the virus but against the bacterial antibody as well, and vice versa. The distinction between self and nonself is abrogated. The immune system can no longer tell friend from foe, and the confusion is increased by the fact that one of the original antigens, in this case the virus, mimics myelin basic protein, a self protein, and may therefore induce an autoimmune reaction against it. Whether

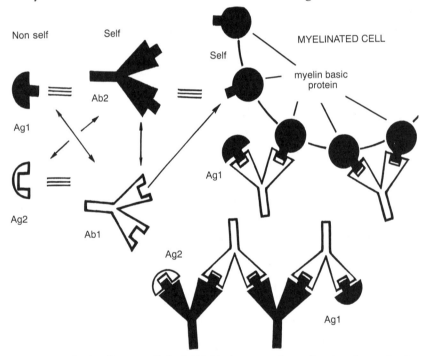

FIGURE 13. Application of the MAMA theory to demyelinating dementias in AIDS. One antigen (Ag) resembles myelin basic protein. The other antigen is chemically complementary to it. Not only do the resulting antibodies (Ab) attack the antigens and each other, but also myelinated cells.

the relevant antibodies will now be found in human patients and whether these leads will result in the development of an animal model on which possible preventative measures and treatments can be tested will require extensive and difficult work. The EAE model does offer hope, since proper vaccination or suppressive measures do exist that might be applicable to AIDS.

The notion that at least some aspects of AIDS dementia are not due primarily to HIV infection is highly controversial. The current dogma is quite clear: "The AIDS dementia complex is a direct consequence of HIV infection of the brain, not a complicating infection," pontificate Jonathan Gold and Donald Armstrong of the Cornell University Medical College.[10] "Central nervous system involvement and AIDS dementia . . . I also regard as due to HIV alone. I think both would occur if humans were infected [with HIV] in a germ-free environment," argues Alfred Evans of Yale.[11] Yet recall that since these very forceful statements were made, just three years in 1989, ago, the *Journal of National Institutes of Health Research* has published a review article concluding that the dogma is foundering: "From the outset, the study of the neurological manifestations of infection by the human immunodeficiency virus-type 1 (HIV-1) has been beset with practical problems, experimental complexities, and theoretical improbabilities. As the years wear on, the problems seem to be growing more complicated."[12] It is no longer clear that getting HIV into the brain is sufficient to cause dementia, nor is it clear that dementias are initiated only after HIV is present. Sometimes HIV seems to follow preliminary manifestation of dementia in people at risk for AIDS.[13]

We must beware that the presence of HIV in the brains of AIDS patients is not an epiphenomenon. One of the consequences of autoimmune demyelinization is to break down the blood-brain barrier that protects the brain from infection. Once the barrier begins to break down, many infectious agents are allowed access to the brain that normally are kept out. HIV is undoubtedly one of these. It may even be brought into the brain by infected lymphocytes that are drawn there to participate in the formation of those deadly perivascular cuffs that accompany demyelinization. My opinion is that we have asked HIV to be responsible for too much of AIDS.

Multiple-Antigen-Mediated Autoimmunity and Immunosuppression

The application of the MAMA theory to AIDS dementia was only the beginning of a story that is still unfolding. The manifestations of autoimmunity in AIDS are diverse: autoimmune orchitis, thrombocytopenia, systemic lupus erythematosus–like symptoms, and lymphocytotoxic antibodies directed against T cells. I believe that many of these autoimmune processes have origins in multiple, concurrent infections just like demyelinating dementias. Continued research may allow us to identify the specific triggering infections, and either vaccinate against them or intervene to prevent autoimmune complications following infection. The MAMA approach is proving useful in this search.

Perhaps the most widespread and certainly one of the most important forms of unexplained autoimmunity in AIDS is the presence of antibodies that attack and kill T cells. The theory of multiple-antigen mediated autoimmunity has provided some clues as to the possible origins of these antibodies. Recall that one of the crucial steps in the immunological processing of most antigens is the interaction between the MHC class II antigen-presenting cells and T-helper cells. The MHC protein binds directly to the CD4 protein of these T-helper cells in the presence of antigen. CD4 protein and MHC class II proteins are chemically complementary; they fit together in the same way that an antibody recognizes its antigen, or the way a hand fits its mold. The key to unlocking the puzzle of LCTAs may exist in this MHC-CD4 complementarity.

According to the theory of multiple-antigen-mediated autoimmunity, all that is necessary to induce autoimmunity is a pair of chemically complementary antigens, one of which mimics CD4 and the other of which mimics MHC class II. Such antigens would themselves be chemically complementary, they would elicit a set of chemically complementary antibodies. And each of these antibodies would react with either MHC class II or CD4 proteins. The result would be not only immunological mayhem but mass destruction of T cells bearing both CD4 and class II MHC proteins, as is, in fact, observed in AIDS (Figure 14).

The prediction that pairs of appropriate antigens are necessary to trigger LCTAs in AIDS is testable. To begin, Dr. Sheila Hobbs and I undertook a series of computer searches designed to identify

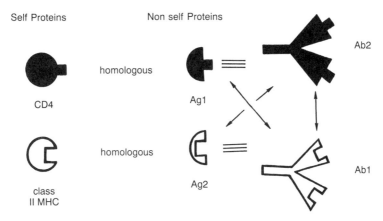

FIGURE 14. Application of the MAMA theory (Fig. 11) to T cell destruction in AIDS. The protein antigens (Ag) of some infectious and allogeneic agents associated with AIDS mimic both the CD4 protein of T-helper lymphocytes and the MHC class II protein to which CD4 protein binds. Since CD4 and MHC class II proteins are known to be chemically complementary, some of their mimicking antigens will also be complementary, and so will the resulting antibodies (Ab). These antibodies will therefore represent an idiotype–anti-idiotype pair with all of the concomitant disastrous results described in Fig. 11.

proteins that mimic some portion of the CD4 protein of T helper cells or the complementary protein, MHC class II. Thousands of researchers have collaborated to make available the results of their studies of the protein sequences of tens of thousands of infectious organisms, enzymes, cellular receptors, and so forth, and all are available on computer databases. Other investigators have developed sophisticated ways of searching these computer databases in order to compare the sequences of one protein with another. Thus, it is now fairly easy to ask a question like, How many proteins in the database have sequences similar to that of the CD4 protein or human T-helper lymphocytes according to the following criteria . . . ? How many proteins from what microbes look like human MHC class II proteins? These are the sorts of questions that we asked.

The results that Dr. Hobbs and I obtained were surprising.[1] Besides the obvious results that mouse CD4 protein looks like rat CD4 protein, which looks like pig CD4 protein, which looks like the human version, and that MHC class II proteins from most mammals look very similar, too, we also found a who's who of the infections associated with AIDS. HIV, I hasten to say, is on the list. Recall that HIV recognizes and infects T-helper cells by binding to the CD4 protein. It therefore has proteins that mimic MHC class II proteins. (Indeed, this is the basis of the Ziegler-Stites-Hoffmann theory summarized above.) What other investigators have not reported is that many proteins from other infectious agents associated with AIDS look as much like MHC class II proteins as does the gp120 protein of HIV. These include, among others, *Mycoplasmas* such as those advocated by Montagnier and Lo as key players in AIDS; various *Mycobacteria*; adenoviruses; herpes simplex viruses; and bacteria such as *Pseudomonas aeruginosa* and *Escherichia coli* that cause septicemia in intravenous drug abusers.

The MHC mimics are only half the story. If the multiple antigen theory of autoimmunity is correct, then a second set of chemically complementary antigens must also exist. Since CD4 protein is chemically complementary to MHC class II proteins, what is needed are agents that are homologous to CD4 protein. There is no shortage of these. Dr. Hobbs and I have found significant similarities between CD4 and proteins of various infectious agents, many of which are suggested cofactors for HIV: cytomegalovirus, Epstein-Barr virus, hepatitis B virus, HTLV-I, *Mycoplasmas*, and *Staphylococcus aereus,* among others. Not only may these CD4 mimics act as cofactors in AIDS, and as possible co-inducers of lymphocyte autoimmunity, they also suggest that the sort of Ziegler-Stites-Hoffmann idiotype–anti-idiotype model of autoimmunity is too roundabout. The question these investigators have been addressing is how HIV, a MHC class II mimic, can induce autoimmunity against CD4 protein on T-helper cells. The answer may be as simple as the fact that it does not: that autoimmunity may be initiated by the presence of other viruses and bacteria that directly mimic CD4 protein itself. HIV—or other MHC class II mimics—may play the role of adjuvant to these other infections. In fact, several of the viruses Hobbs and I identified, including cytomegalovirus and Epstein-Barr virus, are associated with the presence of lymphocytotoxic antibodies independent of either HIV or AIDS.[2]

It is important to emphasize that the similarities between CD4 proteins and viruses are unusual—so unusual that Dr. Hobbs and I entertain the possibility that the reason opportunistic infections are opportunistic is that they hide from the immune system by mimicking it. They are, in a sense, like Romeo, dressed in his mask to visit Juliet at the ball of the Capulets. He is safe as long as his mask remains in place, for he looks no different from his hosts. They see him not as he is but as reflections of themselves. But let the mask slip or let all men with masks become suspect, and then friend and foe alike suddenly spring to awful warfare. So it seems to be with these opportunistic infections. The fact that they so resemble their hosts ensures that when they are unmasked, a goodly portion of the hosts themselves will be killed in the ensuing internecine feud.

What we think happens is the following. An unfortunate person becomes infected with some appropriate combination of infectious agents, one of which is a CD4 mimic and the other a MHC class II mimic. Many possible combinations exist: HIV with cytomegalovirus, HIV with *Mycoplasmas*, hepatitis B virus with *Mycobacteria*, and so forth. Each infectious agent induces an appropriate antibody response. These antibodies not only recognize their respective antigens as targets, but each other and CD4 and class II MHC proteins as well. The result is immunologic mayhem. As in the case of autoimmune demyelinization, the distinction between self and nonself is abolished; the usual idiotype–anti-idiotype regulation network is destroyed; and, most important, the antibody-producing B cells begin attacking the T cells. (Figure 15) This is exactly the immunologic civil war that one observes in AIDS patients who have lymphocytotoxic antibodies (LCTAs).

I have modeled the formation of immune complexes of potential LCTAs in my laboratory.[3] Dr. Shyh-ching Lo supplied me with antibody to the *Mycoplasma incognitus* that he has isolated from AIDS patients. This antibody has various components that specifically bind to cytomegalovirus antibodies, suggesting that a cytomegalovirus-*Mycoplasma* combination would act synergistically. Other types of *Mycoplasmas* also associated with AIDS patients produce antibodies that interact specifically with HIV antibodies. Moreover, all of these active *Mycoplasma* antibodies will bind to antibodies raised against human lymphocyte proteins such as CD4 and MHC class II proteins. Indeed, Dr. Lo informs me that he and his associates have demonstrated that these antibodies will bind directly to the lymphocytes themselves.

I have also obtained other intriguing results. Antibodies against *Staphylococcus aureus* will precipitate antibodies against both the

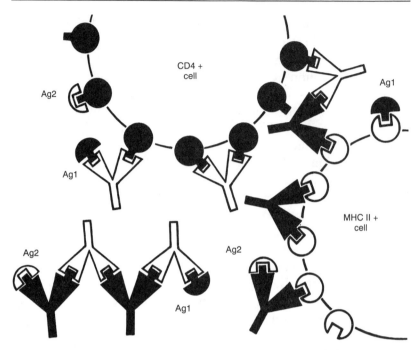

FIGURE 15. The results of the production of idiotype–anti-idiotype antibody pairs from CD4 and MHC class II mimics described in Fig. 14. Circulating immune complexes will be fored. Both T-helper cells and MHC class II cells will be attacked. The immune system will destroy itself in a cellular civil war.

Nef protein of HIV and antibody against MHC class II proteins. This virus-bacterium combination might also cause immune suppression, particularly among intravenous drug abusers among whom *Staphyolococtus* infections are very common. A rather different combination—hepatitis B virus with *Mycobacteria*—may account for the peliosis hepatitis of many AIDS patients, in which the perivascular region of the liver is attacked. Antibody against hepatitis B virus proteins bind to mycobacterial antibodies and antibodies against the CD13 protein that characterizes monocytes.[4] Thus, various combinations of infectious agents may provoke quite different autoimmune responses in AIDS patients and result in a variety of immunosuppressive effects.

In short, laboratory evidence strongly supports Dr. Lo's and Dr. Montagnier's contention that *Mycoplasmas* may play a critical and necessary role in AIDS pathogenesis,[5] in this case by means of the co-induction of lymphocytotoxic autoimmunity in the presence of HIV or cytomegalovirus. But the same experiments also implicate other combinations of infectious agents and other HIV cofactors as well. Research in this area is continuing.

The plausibility of the multiple-antigen-based scenario is confirmed by a number of observations. First and foremost, the opportunistic infectious agents that Dr. Hobbs and I have identified as mimicking the immune system proteins are very common in AIDS patients. Recall that unusually high levels of cytomegalovirus antibody are found in over 95 percent of AIDS patients and active infection in over 50 percent; both hepatitis B and herpes simplex infections are present in about 70 percent of AIDS patients; mycoplasmal infections in no less than 60 percent and perhaps as high as 100 percent; and so forth. If we now calculate the probability that an individual in a high-risk group for AIDS will have any pair of infections, the probability becomes extremely high. HIV plus *Mycoplasma*, for example, would be expected to occur in no less than 60 percent of AIDS patients; HIV plus cytomegalovirus or Epstein-Barr virus in no less than 90 percent of all AIDS patients; and so on. Since dozens of possible combinations of CD4 mimics and MHC class II mimics are possible, it becomes almost necessary that the criteria for inducing multiple-antigen mediated autoimmunity will be satisfied. Indeed, an AIDS patient, by definition, has multiple infections (including HIV), and most of these infections are hundreds of times more easily transmitted than is HIV; the liklihood that a person will contract a combination of HIV plus other infectious agents simultaneously is high. It is irrelevant for current purposes whether this combination of infections precedes, accompanies, or occurs after HIV infection. All three cases have been reported in the medical literature, and the outcome is the same in each: significant immune suppression leading to increased likelihood of further infections, multiple forms of autoimmunity and postinfectious complications, and, usually, death.

The MAMA theory also explains two more important features of AIDS. First, it explains why Koch's postulates have not been satisfied: AIDS may require specific pairs of antigens (of which HIV may be one) to be triggered; no single agent will be identifiable as the cause. Second, MAMA explains why such autoimmunity is so rare among the general population. Compare the probability of appropriate combinations of infections being acquired nearly simultaneously by a person without risk for AIDS. Clearly it is extremely low. Although over half of the general population has antibody to cytomegalovirus and Epstein-Barr virus, many fewer than 1 in 1,000 have an active infection at any given time. The same is true of adenoviruses, hepatitis B virus, *Mycoplasmas,* and *Mycobacteria.* In each instance, antibody seropositivity is present in less than a third of the population, and active infections are almost unheard of among healthy heterosexuals. HIV seropositivity is present in only 3 in 1,000 people in the general population. So the probablity of finding a person in the general population, without known risks for AIDS, who becomes seropositive to HIV at about the same time he or she contracts hepatitis or cytomegalovirus is on the order of one in tens of thousands or less. In fact, the incidence of cytomegalovirus-associated neuropathies (abnormal nerve or brain function) is on the order of 1 in 100,000 cases. If the theory presented above—that cytomegalovirus needs a cofactor such as a concurrent mycobacterial infection to trigger autoimmunity—is correct, then the incidence in the general population is about right. On the other hand, the incidence of such neuropathies among AIDS patients approaches 20 percent, which is again in the correct ballpark according to the probability that such patients will acquire concurrent infections. An incidence of 50 percent active cytomegalovirus infection times an incidence of 60 percent active *Mycobacterial* infection yields a sum of 30 percent for the probability of developing a demyelinating AIDS dementia.

If this were all that Dr. Hobbs and I had found, we would have been quite pleased. It is actually **only** half the story. In addition to infectious agents associated with AIDS, two further sets of proteins also turned out to be highly homologous to either CD4 protein or MHC class II: sperm proteins and the blood factor proteins used to treat hemophiliacs.[6] Interestingly, we did not actively search for these mimics; they emerged in the data we obtained from our previous questions as unexpected, but quite illuminating, results.

As we saw in Chapter 3, semen antibodies have been found in the majority of AIDS patients, and these antibodies have been shown to attack T-helper cells. The similarity between some sperm components and the CD4 protein of T-helper cells and macrophages is so complete that antibody made against CD4 protein also reacts with sperm proteins.[7] In fact, HIV will actually bind to these CD4-like proteins on sperm, just as it does to the CD4 protein of T lymphocytes and macrophages,[8] and various species of *Mycoplasmas* (which share protein similarities with HIV) will also bind to sperm.[9] Sperm, in turn, will bind to MHC class II domains on T lymphocytes.[10] Thus, there is no doubt of the chemical complementarity between the CD4-like and MHC class II–like domains in these cases. It follows that if such sperm-*Mycoplasma*, sperm-lymphocyte, or sperm–HIV complexes were to be immunologically processed together, then the immune system would be faced with precisely the sort of chemically complementary antigens required by the MAMA theory. The target would be both CD4 and MHC class II proteins on lymphocytes—precisely what is observed in AIDS patients.

Our other big surprise was the appearance of the blood factors in the lists of protein mimics printed out by the computer. Factor VIII, used to treat hemophilia A, factor IX, used to treat hemophilia B, and factor VIII–related protein, used to treat von Willebrand's disease, all displayed multiple homologies with MHC class II proteins. We saw in Chapter 3 that autoimmunity to blood factors has often been reported in hemophiliacs and is associated with the most severe cases of the disease. Its cause is not known, but our research suggests that blood factor contaminated with any CD4 protein mimic, such as cytomegalovirus, Epstein-Barr virus, or hepatitis B virus, might be sufficient to induce the observed autoimmunity if HIV or an appropriate bacterial infection were present simultaneously. Not surprisingly, because blood factors mimic the proteins on immune system cells, many reports of autoantibodies that recognize and kill the T and B cells of hemophiliacs exist in the medical literature.[11]

In short, an increasing body of evidence strongly suggests that autoimmune conditions associated with AIDS may result from a wide range of concomitant infectious and allogeneic (foreign cell) stimuli. HIV may be one of the primary players in the induction of autoimmunity directed against T cells due to its mimicry of MHC class II proteins, but the results that Dr. Hobbs and I have obtained strongly argue for an equal role for many other viruses and bacteria, including cytomegalovirus and *Mycoplasmas*, and for allogeneic stimuli such as semen and blood factors, that display equal similarity to T cell proteins. If so, then the specific forms of autoimmunity that appear in individual AIDS patients will vary with the specific combinations of antigens to which they are exposed, and their specific symptoms will alter accordingly. Most important, there will be no single way to prevent, treat, or cure such diverse autoimmune conditions.

Autoimmunity and AIDS: Conclusions and Questions

Several things must be emphasized about the role of autoimmunity in AIDS in concluding this discussion of the topic. First, there is substantial evidence that autoimmunity directed against lymphocytes plays at least some role in causing the immune suppression typifying AIDS. Whether this role is the major factor that causes AIDS or is one of many small insults that eventually add up to a disaster is not yet clear. One major characteristic of AIDS suggests that lymphocytotoxic autoimmunity may be the primary cause of immune suppression and that it explains the very long period over which the disease develops. Very few infectious diseases have subclinical courses that spread over a decade or more before leading to death. Autoimmune diseases such as multiple sclerosis, amyotrophic lateral sclerosis (Lou Gehrig's disease), and systemic lupus erythematosus, on the other hand, are very often characterized by prolonged courses of deterioration, marked by extensive periods of remission and normal or near-normal health. Thus, if HIV, or HIV in conjunction with other infectious agents or allogeneic stimuli such as sperm or blood proteins, were involved in triggering AIDS, all of the current data concerning HIV seropositivity, the need for cofactors, and the variable period between infection and disease manifestations might make sense.

Second, autoimmunity has many manifestations in AIDS besides that directed at lymphocytes. The causes of lymphocyte depletion may be entirely unrelated to the causes of the specific autoimmune symptoms, such as demyelinization and thrombocytopenia, that are frequent concomitants of AIDS. It is possible that HIV may play the major role in one form of autoimmunity, and none in others. A concerted effort is needed to disentangle the many different forms of autoimmunity in AIDS. As these various manifestations become distinct, they will inevitably call for new treatments unrelated to retroviruses.

Third, there is no agreement as to the cause of any human autoimmune disease, let alone the cause of any form of autoimmunity in AIDS. The account I have presented is biased by my own research and preconceptions. I do not present them as being the "truth" but as being one of many possible hypotheses. The purpose of theorizing is to cause us to rethink things we thought we understood in order to go out and ask new questions. Recognizing that autoimmunity may play an important role in causing AIDS and that vast areas of ignorance attend our explorations into this field must cause us great excitement. There may be major discoveries still left to be made not only concerning AIDS but the entire field of immunology—discoveries that may illuminate many diseases besides AIDS. With these discoveries will come new possibilities for treatment.

Only one thing is certain about the role of autoimmunity in AIDS: Every recognized risk group for AIDS has an unusually high probability of developing autoimmunity against their own lymphocytes. This extraordinary risk must be factored into the equation that predicts AIDS susceptibility and the rate at which the syndrome develops. Indeed, there may be no cures for AIDS until we understand its autoimmune component.

CHAPTER 6

WHO IS AT RISK FOR AIDS AND WHY

AIDS Risk Groups and the Variable Latency of Disease

The long lists of infectious, noninfectious, and autoimmune causes of immune suppression just outlined clearly raise an important question: "Is everyone equally susceptible to infection with the retrovirus [HIV] and its consequences?" This question was posed by the Working Group on AIDS of the Clinical Immunology Committee of the International Union of Immunologic Societies of the World Health Organization in 1984.[1] If exposure to HIV is sufficient to cause AIDS, then everyone should be at equal risk, and AIDS should develop at an equal rate among different risk groups once infection has become established. Clearly that is not the case.

Researchers recognized by 1987 that the threat of AIDS to non-risk groups was very small. Some calculations place the figure of contracting AIDS from a heterosexual without risk behaviors as low as 1 in 1 million—or the same risk as being struck by lightning. Even in areas of epidemic AIDS, such as New York City and San Francisco, "very little cross-over to the mainstream heterosexual population has occurred," Dr. Harold Jaffe of the CDC told the *Los Angeles Times* during that year. "We don't need to panic about heterosexual transmission," agreed Robert Gallo.[2]

On the other hand, the high risk groups are still the high-risk groups. The cumulative incidence of AIDS seven years after HIV infection in drug abusers is over 40 percent[3] and about the same in homosexual men.[4] These figures are the basis for the statistical generalization that the average latency period for the development of AIDS is about ten years. In other words, one would expect that about half of all HIV-positive individuals should develop AIDS within ten years, if intravenous drug abusers and homosexual men are typical of the entire population. They are not, of course, as we have seen in detail in the previous three chapters. Hemophiliacs, for example, have very different immunosuppressive and disease risks and have a correspondingly different risk of AIDS. About 15,000 of the 17,000 or 18,000 hemophiliacs in the United States are HIV seropositive.[5] All of them were infected with HIV prior to 1985, when HIV testing of the blood supply essentially eliminated HIV-contaminated blood and blood factor concentrates from general use. This means that all of these 15,000 people were infected between eight and thirteen years ago—that is, an average of about ten years ago. Since the predicted latency period between infection and AIDS is ten years, one would expect that half of all HIV-seropositive hemophiliacs in the United States (about 7,000 people) would either have AIDS or have died of AIDS by now. In fact, by June 1991 only 1,535 cases of AIDS in U.S. hemophiliacs had been reported to the CDC—10 percent of HIV-seropositive hemophiliacs.[6] In other words, it appears that 90 percent of the HIV-infected hemophiliacs who have been carrying the infection for eight or more years are apparently still free of AIDS.

The fact that so few hemophiliacs have developed AIDS is good news for them but a problem for researchers claiming that HIV is the sole necessary and sufficient cause of AIDS. Despite various claims from American and British groups that 20 to 25 percent of HIV-positive hemophiliacs develop AIDS within seven years of infection,[7] which would at least put them in the same ballpark as intravenous drug abusers and homosexual men, several European studies have confirmed that only about 10 percent develop AIDS during this time.[8] It appears that various forms of statistical finagling account for the higher rates of AIDS reported by the American and British studies. But even these studies find large discrepancies between the risk of older and younger hemophiliacs. Over 30 percent of hemophiliacs over the age of thirty-five progress to AIDS within seven years, while less than 10 percent of hemophiliacs under the age of fifteen do so.[9] Clearly, age-related factors—be they accumulated infections, repeated blood transfusions, slow erosion of the immune system by use of painkillers and other drugs, or other unidentified agents—greatly alter the progression to AIDS. HIV is definitely not the sole controlling factor.

Other studies confirm that such disparities in time of infection to development of AIDS are the rule rather than the exception and support the notion that cofactors control the development of AIDS. The average latency period from HIV to AIDS calculated for Americans is ten or more years. For equatorial Africa, this figure is a mere three and a half years, and for Thailand, it is six. Moreover, although the rate of transmission from mother to infant is about 12 percent in Europe, it is 40 to 50 percent in Africa and Haiti.[10] These data strongly suggest that HIV seropositivity varies greatly in its predictive value for the transmission and development of AIDS by region of the world, by risk group, by age, and by degree of continued exposure to immunosuppressive agents.

If we return to the question posed by the Working Group on AIDS in 1984—"Is everyone equally susceptible to infection with the retrovirus [HIV] and its consequences?"—the answer is "no." The consequences of this answer are vital. As the working group itself wrote in response to its own rhetorical question: "If [it] is not the case [that everyone is equally susceptible to HIV and its consequences], predisposing or contributing factors must apply. These could include immunogenetic variation, intercurrent infections, alloantigenic stimulation, or drugs. Any of these potentiating factors could act either to predispose the host to infection or to determine the expression and clinical outcome of the infection. Enhanced susceptibility to the retrovirus may result from a relative immunodeficiency or from prolonged or repeated activation of target cells, which could facilitate penetration or spread of the retrovirus in the immune system."[11] Until we understand exactly what these predisposing factors are for each separate risk group, we will not be able to identify, treat, control, or eliminate the risk of AIDS. The purpose of this chapter is therefore to explain why people at risk for AIDS are at risk by examining in detail their specific, and sometimes surprising, mix of immunosuppressive factors.

Immunosuppressive Risks of Some Homosexual Men

Alex Witchell was captivated by a man she writes about simply as "John" from the first time she set eyes on him at the Yale School of Drama in 1979. His handsome, bronzed, southern California skin, huge gold watch, and white suit stood in stark contrast to the sallow, black-clad artistes who made up her classes. She fell in love. "You can never know everything about someone you love," she wrote recently. "When we were at school, John would drop me at home at 11 at night, and often disappear for most of the next day. He went to the bars, he went to the baths, he danced, he drank, he took drugs. Some days, when I looked at his bad color and bloodshot eyes, I thought his night life excessive, but I figured it was just John, needing to grab up everything at once, as fast as he needed to throw it all away. Isn't that how everyone grows up?" John grew up too fast, or perhaps never grew up enough. After a brief career as a Hollywood agent, he died in 1989 of AIDS.[1]

Hilary Waddles tells a similar story about her "dear friend and 'blood brother' Kenny." Kenny, she writes, "died in 1988 at 27 after losing his third battle with *Pneumocystis carinii* pneumonia. Kenny spent most of his life from 17 to 25 as a prostitute and IV drug user. The former was a matter of survival; the latter—well, that's much more complex. But why does any of that mean that Kenny was somehow unworthy of concern, compassion, and support? Whatever else he was, Kenny was a beautiful, talented person who never tried to do harm to anyone. He had a family who loved him even though they had trouble accepting his lifestyle. He had friends who accepted him as a person and respected him as an artist. . . . [But] Kenny couldn't even afford the food, medicines, and the secure place to live that might have given him a longer and healthier life."[2]

Witchell and Waddles provide us brief, poignant glimpses of the lives of two very different homosexual men. One sold his body; one gave it. One was rich, talented, successful; the other destitute, rejected, a failure. Both were human: they needed to be loved, to have sex, to eat, drink, and sleep. Both needed reassurance, sometimes in the form of drugs, sometimes in the form of friendship. Both had artistic inclinations. Both were unusually promiscuous, at least by heterosexual standards. Both used drugs. Both died of the same dread scourge. The question we must ask is whether their stories, and the stories of other gay men who have contracted AIDS, as varied as they are in their details, have enough in common to tell us anything useful about who is at risk for AIDS and why.

It is, of course, always dangerous to generalize about any group of people, and people with AIDS are no exception. And yet certain generalizations about who is most likely to contract AIDS have proved to be useful from a medical perspective. We recognize that the vast majority of people with AIDS are gay men and/or intravenous drug abusers. These generalizations provide clues about what may cause AIDS, what may predispose people to contract the syndrome, and how the disease may be spread. Equally important is a knowledge of who does not contract AIDS and why they do not. Healthy heterosexuals who avoid drugs and drug users, female prostitutes who do not abuse drugs, and lesbians almost never contract AIDS. These facts also provide clues as to the mechanisms that underlie this syndrome. Yet even more important may be the clues provided by comparing those who develop AIDS within high-risk groups with the people in the same risk group who do not.

Homosexual men are not all at equal risk of becoming infected with HIV or of developing AIDS. Elena Buimovici-Klein and her colleagues in New York have noted that susceptibility to HIV infection and the subsequent development of AIDS are highly associated with multiple concurrent infections and the practice of unprotected anal intercourse.[3] Other investigators have confirmed that the presence of diverse antiviral antibodies, active infections, a history of syphilis, sexual promiscuity, unprotected anal intercourse, and use of multiple street drugs are all risks associated with the development of AIDS in gay men.[4] Conversely, the same studies suggest that gay men who do not engage in high-risk sexual activities, who abstain from drugs, and whose infectious disease load remains low are at low risk for both immune suppression and AIDS.

The most pronounced risk for AIDS appears to be immunologic exposure to semen and fisting (insertion of fingers or hand into the rectum). A number of studies indicate that over 50 percent of gay men engage in both receptive and insertive anal intercourse, another 15 percent in receptive anal intercourse only, and another 15 percent in insertive anal intercourse only.[5] A 1980 study documented significantly higher rates of anal intercourse, fisting, and oral-genital contact, as well as greater promiscuity, among homosexual men in cities such as San Francisco and Los Angeles that developed high rates of AIDS than in cities in which AIDS remained relatively rare, such as Denver and Saint Louis.[6]

Anal forms of intercourse ·are associated with vastly increased disease transmission and AIDS risk. One of the most clear-cut studies demonstrating this risk was performed by Roger Detels, Patricia English, and their colleagues in the Multicenter AIDS Cohort Study conducted by the National Institute of Allergy and Infectious Diseases and the National Cancer Institute in Bethesda, Maryland. They followed a group of 2,915 homosexual men who were HIV seronegative upon enrolling in their study in 1984 or 1985. During the study period, which lasted through 1988, 232 men (8 percent of the group) seroconverted. Of these, all had participated in receptive anal intercourse. Most of the men who seroconverted also participated in insertive anal intercourse during the six months prior to seroconversion. They had fifteen times as many partners as the men who stayed healthy. Condom use cut down the risk of HIV infection but did not eliminate it for these men. "Receptive anal-genital intercourse has been shown by us and others to be the major risk factor for transmission of the human immunodeficiency virus (HIV) among homosexual men. The risk of infection associated with other sexual activities has not been established. . . . The results of this study (a) reaffirm that receptive anal-genital intercourse is the major route of infection among homosexual men of HIV-1, (b) suggest that there is a low risk of HIV-1 infection to the insertive partner in anal-genital intercourse, and (c) suggest that infection may *rarely* occur through sexual activities other than anal-genital intercourse."[7] In short, it is not being gay that puts a man at risk for AIDS but specific activities associated with (but not confined to) the life-styles of some gay men that puts them at risk.

Why does unprotected receptive anal intercourse confer an increased risk of AIDS? One possibility is that it is much easier to transmit HIV to a receptive partner than from a receptive partner. No other sexually transmitted disease behaves this way, however. Although it is usually easier for a woman to contract, say, syphilis or gonorrhea from a man than it is for the man to contract these diseases from a woman, the difference in the rate of transmission is only a matter of 20 or 30 percent. From a germ's-eye view, there is little difference in size between the vagina and the urethra: both openings are huge. HIV would be the first disease agent to be able to make the discrimination, unless some other factor is involved.

Another possibility is that AIDS is not due to exposure to HIV, per se, but to the semen that contains this HIV. Semen is, in fact, transmitted in only one direction—from the active partner to his recipient. If HIV can be transmitted only within semen lymphocytes or if these lymphocytes or the sperm themselves are responsible for inducing immune disregulation, then the extraordinary risk of receptive anal intercourse becomes comprehensible. Unprotected receptive anal intercourse certainly represents an unusual risk for exposure to allogeneic antigens such as the lymphocytes and sperm found in semen. As we have seen in Chapters 2, 3, and 5, lymphocytotoxic antibodies (LCTAs) directed at sperm and foreign lymphocytes attack T cells and can mimic HIV antibodies. These LCTAs are almost ubiquitous among homosexual people with AIDS and are prevalent among men and women who have engaged in repeated anal intercourse.[8] Thus, at best, HIV is only one of the immunologically damaging factors associated with anal intercourse.

One other factor may also play a role in this risk. Recent studies have demonstrated that infection by both Epstein-Barr virus and cytomegalovirus is greatly enhanced when it is present in semen, as opposed to transmission by oral contact, for example. The semen interferes with normal lymphocyte activity and thereby promotes infection.[9] The same enhancement of transmissibility may occur with other viruses, including hepatitis B and HIV when they are present in semen. Anal intercourse is also associated with unusually high rates of anal infections with AIDS-associated diseases such syphilis, gonorrhea, *Chlamydia*, *Mycoplasmas*, and papilloma virus.[10] Thus, the disease load of promiscuous gay men exposed to semen anally is extremely high. Many of these diseases can cause direct immune suppression, and many are candidates for HIV cofactor status.

In some cases, the mix of diseases that occur in some gay men may also promote unusual health risks. For example, human papilloma virus (HPV) has been found frequently in the rectums of gay men, and over 50 percent of these have had evidence of anal intraepithelial neoplasia (AIN), a precancerous condition. This figure compares with an incidence of AIN in HPV-infected heterosexual men of less than 4 percent, suggesting that HPV alone is not responsible for AIN.[11] Similarly, oral hairy leukoplakia and Burkitt's lymphoma, both associated with Epstein-Barr virus infections, are found fairly frequently in gay men with AIDS, but almost never in other risk groups or heterosexuals infected with Epstein-Barr virus.

Anal intercourse and oral-anal sex (for example, "rimming," the act of sticking one's tongue into the anus of another person) also result in exposure to fecal material. Such sexual acts are highly associated with the development of the set of concurrent infections that is often referred to as gay bowel syndrome (although this syndrome is also found among heterosexuals). The combination of parasitic infections that include amebiasis and giardiasis, along with rectal infections with adenoviruses, *Mycoplasmas*, cytomegalovirus, syphilis, gonorrhea, and so forth, can and often does result in chronic diarrhea. Chronic diarrhea is one cause of malabsorption syndrome, a failure of the intestines to take up the nutrients that are being processed through it. Malabsorption syndrome, in turn, is one of the causes of malnutrition. Indeed, there is direct evidence from clinical studies that patients with chronic giardiasis experience significant immune suppression, even in the absence of other identified risks for AIDS.[12] The same is probably true of the other infections associated with gay bowel syndrome, but apparently no one has performed the relevant clinical studies.

Parasitic infections create yet another potential immunosuppressive risk for gay men, and that is due to the pharmacological agents used to treat these infections. Recall that many antiparasitic drugs have demonstrated adverse effects on immune function. Consider, in addition, that the CDC reports that between 20 and 50 percent of all gay men in major American cities have been treated, in many cases repeatedly, with medications for enteric parasites such as giardia and ameoba.[13] Among the most common of these medications are ketoconazole, metronidazole, and quinacrine, all of which have demonstrated immunosuppressive effects on lymphocyte and natural killer cell activity at therapeutic doses in animals and human beings.[14]

Gay men were aware of their disease susceptibility long before AIDS emerged as a problem. Thus, many volunteered for testing experimental hepatitis vaccines. Others took what appeared to be (and what may have appeared to some of their physicians to be) a more conservative approach. They made chronic use of antibiotics, some prophylactically and some to treat recurrent venereal and other infections.[15] I have been told by a number of gay men that it was not uncommon to take a few antibiotics and sniff an ampule or two of amyl nitrite on the way to the baths or bars for a round of anonymous sex. The only formal survey of this practice of which I am aware is by Linda Pifer and her associates, who interviewed the patrons of a gay bar in Memphis, Tennessee. Over 40 percent of the men surveyed responded that they "routinely" treated themselves with prescription antibiotics. Chronic and high-dose antibiotic abuse can lead to signficant immune suppression.

Some gay men have had additional medically related immuno-suppressive risks. A 1983 survey of gay men in major American cities by the CDC found that between 10 and 17 percent of them had been treated with systemic corticosteroids.[16] Originally a history of such a treatment excluded a person from diagnosis with AIDS, since it represented a known cause of immune suppression sufficient to predispose an individual to opportunistic infection. Corticosteroid use has not been exclusionary since the incorporation of HIV into the definition of AIDS and apparently was never exclusionary in practice, at least in diagnosing homosexual men. Unfortunately, the survey did not explain the reasons for these corticosteroid treatments. One can only presume that they were in response to unusual infections or autoimmune problems, since these are the usual medical uses of such drugs.

Sketchy data indicate that the rate of blood transfusions among gay men is significantly higher than among heterosexual men and women, and lesbians. The Pifer study found that 15 percent of the "healthy" gay men they interviewed reported having had at least one blood transfusion.[17] The reasons for these transfusions were not explored, but at least two possibilities exist. One is related to the use of amyl and butyl nitrites, which was at one time nearly synonymous with gay life in the United States. Nitrite abuse can cause methemoglobinemia, a condition in which the drug causes the hemoglobin to be unable to bind oxygen. The abuser then begins to turn blue from lack of oxygen and may pass out.[18] Massive hemolysis (the large-scale destruction of red blood cells) may also occur.[19] In some cases, these conditions have been treated with blood transfusions.[20] Another possible reason for blood transfusions is surgery following rectal injuries due to practices such as fisting or ulceration due to chronic bowel infections. Again, Pifer found that 19 percent of her study group reported regular episodes of anal bleeding, although none admitted to practicing either active or passive fisting.[21]

The mention of nitrite abuse raises another pair of immunosuppressive specters. Pifer reported that 80 percent of the gay men she interviewed used nitrites at least occasionally and that 30 percent of these men used them more than once a week.[22] In major cities such as Washington, D.C, San Francisco, Los Angeles, and New York, about 95 percent of gay men reported using nitrites, often regularly.[23] Thus, the probability that at least some gay men had some immune suppression due to nitrite abuse is reasonably high. In fact, early in the AIDS epidemic, the CDC seriously considered the possibility that nitrite abuse was the cause of AIDS.

The evil collaborator of the nitrite story appears in the unexpected form of antibiotic abuse. Since both antibiotic abuse and nitrite abuse were extremely common, particularly among very promiscuous men, there is a good probability that some subset of these men took both drugs concurrently. That raises a frightening possibility. Nitrites convert most antibiotics into carcinogens in test tube and animal studies.[24] Possibly this chemical reaction also occurred in many gay men, accounting for the unusually high incidence of both Kaposi's sarcoma and non-Hodgkin's lymphomas that are found almost uniquely in this risk group.[25]

Nitrite abuse is only the beginning of the drug-associated immunosuppressive risks many gay men have added to their virologic, parasitic, and allogeneic exposures. About 10 percent of all homosexual men with AIDS in the United States, Canada, and Britain are intravenous drug abusers. Intravenous drug abuse apparently only scratches the surface of a much broader problem, however. In November 1991, the Lesbian and Gay Substance Abuse Planning Group, working under a grant from the San Francisco Department of Public Health, released results of a survey of more than 700 lesbians, gay men, and bisexuals concerning their drug use. Almost 40 percent of the gay and bisexual men (and 20 percent of the lesbian and bisexual women) reported substance abuse at levels considered by experts to be indicative of chemical dependence or addiction. The drugs most frequently used by gay and bisexual men were, in order from greatest to least use: alcohol, marijuana, amyl nitrite, painkillers, amphetamines ("speed"), cocaine, and tranquilizers. Only 15 percent of the men surveyed reported total abstinence from all alcohol and drug use.[26]

Frank Davis, executive director of 18th Street Services, an alcohol and drug treatment program for gay and bisexual men in San Francisco, called the survey shocking: "I didn't think the rates were up that high. . . . A lot of people surveyed said that they drink and used drugs to avoid or deal with emotional pain. Some gay men use drugs to deal with sexual issues, or to have sex, for that matter." Davis blamed the pressures of living in a homophobic world and possible innate psychological needs for the high rates of drug and alcohol abuse and suggested that it is not a phenomenon unique to San Francisco but a national problem for gay and bisexual men.[27]

Davis is almost certainly right. A CDC survey conducted in 1983 found that a "typical" gay man in New York, Los Angeles, and San Francisco used four street drugs regularly. Those who had developed AIDS by 1983 had a history of increased drug use both in terms of frequency of use and number of different drugs used regularly. Ninety-five percent of the gay men surveyed regularly used inhalant nitrites; over 90 percent smoked marijuana; 60 percent used cocaine; about 8 percent used heroin; over 50 percent used amphetamines; over 20 percent, barbiturates; almost 50 percent, LSD and methaqualone; and about 40 percent had used phencyclidine.[28] Linda Pifer's 1987 survey of gay men in Memphis found slightly lower rates of drug use. Over 80 percent of this group admitted to using nitrites at least occasionally and 30 percent more than once a week. Seventy-four percent admitted to use of other illicit drugs, including marijuana, cocaine, phencyclidine, and LSD, with an average of nearly seven years of "routine use." Eleven percent described themselves as being "heavy drinkers" and another 37 percent as "moderate drinkers."[29] Multiple drug use was the norm among the heavy abusers.

A number of quite unexpected and unusual observations suggest that other immunosuppressive risks may exist for both gay men and other people as well. A. J. Schecter, B. J. Poiesz, and their colleagues at the State University of New York Health Science Center have reported that AIDS patients have significantly increased levels of dioxins and dibenzofurans when compared with asymptomatic HIV-positive individuals and heterosexuals.[30] Both dioxins and dibenzofurans are potent immunosuppressants of cell-mediated immunity in animals studies, but the results in human studies are mixed.[31] Whether these compounds are triggers for AIDS or a consequence of the pathology of AIDS is not yet determined, but at least two possibilities exist. William Helfrich of Michigan State University recently presented evidence that ultraviolet light can transform tryptophan into dioxin agents.[32] Tryptophan is an essential building block of proteins and came into vogue during the last decade as a nutritional supplement. Ultraviolet light is the type of light that causes tanning, whether from natural sunlight or the lights used in tanning salons. Many gay men pride themselves in their physical appearance and health and may have inadvertently created dioxin agents in their skin by tanning after taking tryptophan supplements. Heterosexuals may have the same risk. Ultraviolet light has also been reported to activate HIV in cellular and animal models.[33]

The other possible explanation of the high level of dioxins in AIDS patients is that it is an epiphenomenon of the disease. Because of the high incidences of hepatitis A, hepatitis B, and non-A, non-B hepatitis, chronic infections are common, and these can lead to severely impaired liver functions. Alcoholism is also associated with liver damage. The major function of the liver is to filter the blood of its toxins and then chemically break down these toxins. When the liver is damaged, this detoxification process is impaired. In consequence, compounds that would normally pose little or no health threat may become a severe problem. Thus, dioxin exposure, or dioxins stored in fats that would normally be of no consequence to a healthy individual, may become pathogenic in people, such as AIDS patients, who are already chronically ill. If so, then other chemicals that are normally of little medical interest may also be found to contribute to AIDS pathology.

Another immunological risk that is all too frequent among gay men is malnutrition, particularly of trace elements required for immune function. Pifer noted significant zinc depletion in her non-symptomatic homosexual study group,[34] which has been confirmed by other studies.[35] Similarly, B. M. Dworkin and his colleagues found that selenium deficiency was common in both HIV-positive and HIV-negative men.[36] In all of the studies, there was a correlation between these mineral deficiencies and measurable immune deficiency in HIV-negative men, though whether it was the direct cause of that deficiency could not be determined. It is known that animals stressed by multiple infections often develop zinc and selenium depletion, possibly due to the tremendous loss of lymphocytes that occurs during disease.[37] Thus, the multiple, concurrent infections experienced by many gay men may produce unusual nutritional needs that are not met. In addition, antibiotic abuse is correlated with zinc depletion.[38] Pifer and her colleagues have also hypothesized that nitrite abuse may lead to zinc depletion.[39] A great deal of zinc is lost during ejaculation, suggesting that men who are extremely sexually active may have unusual nutritional needs.[40]

The point to stress is that being homosexual does not, per se, increase a person's risk for AIDS. The risk is conferred by specific activities associated with (but not unique to) male homosexuality. Gay men are not uniformly at risk for AIDS. Heterosexuals who share some or all of these immunosuppressive behaviors also risk AIDS. This conclusion is based almost completely on information available prior to the AIDS epidemic. Responsible physicians and public health officials could have and should have warned promiscuous, drug-abusing, multiply-infected gay men of the damage they were doing to their health. Much pain and suffering would thus have been avoided.

Immunosuppressive Risks of Intravenous Drug Abusers

Joseph Michael Donlon lives (or perhaps, by now, lived) in Newark, New Jersey, one of the world's drug abuse capitals and the AIDS capital of the world when measured by cases per capita. During December 1990, Donlon allowed himself to be interviewed by John Tierney of the *New York Times*. Tierney described Donlon as suffering from a gash on the head delivered by a cocaine dealer swinging a cinder block. He also had a badly infected foot. He had just returned from taking methadone at a local clinic and was preparing to inject himself with a syringe full of cocaine, a ritual he had been practicing with one or another drug for half of his forty-eight years. Donlon had AIDS and, as fate would have it, had lost the prescription for his AIDS medication. He did not seem to mind.

Ceford M., age thirty-nine, another Newark addict interviewed by Tierney, had a similar story. He, too, was taking methadone, had fresh tracks from shooting up cocaine, and had just been diagnosed as being HIV positive. He was not surprised. He had shared needles with friends who had died of AIDS several years earlier. "I brought it on myself," he said. "They told me I got the virus and I've been shooting ever since. There's nothing to live for. I'm going to shoot, shoot, shoot. Maybe I'll get lucky and get a heart attack. I watched my best friend die of AIDS, and it was awful— he got so small at the end. I'm not going to let it come to that. I'll go out first."

Odds are that Mr. Ceford M. and Mr. Donlon will, indeed, go out first. The prognosis for hard-core, intravenous drug abusers has never been good, even before AIDS appeared on the scene. The majority of drug-associated deaths are due to overdose and related causes rather than disease.[1] AIDS has only precipitated the crisis that drug abuse had already initiated. The problem is addiction. For many addicts, the lure of drugs is too strong for even the specter of AIDS to affect. As Ceford M. said to Tierney, "I'd advise anyone thinking about drugs: Stay away from the product, because you're going to like it. Next to sex, ain't nothing like it." Indeed, Mr. M. liked it so well that having kicked his habit once, he threw away a rising career in computers to return to his addiction. Treatment is too often ineffective, and addicts have become so clever that many now sabotage their own rehabilitation to support their habits. Tierney reports that bottles of methadone, supplied to addicts for weekend use, can be sold for thirty-five dollars or two bags of heroin. Similarly, Sustacal, a nutrient drink intended to supplement the all-too-inadequate diets of addicts and which are often supplied free of charge through Medicaid, can be traded for a vial of cocaine. A twelve-can case is worth fifteen dollars on the street, and some pharmacies and grocery stores buy them to resell at below-wholesale prices. The intended cures become the means for continued abuse.[2]

These thumbnail sketches only scratch the surface of what a drug addict's life is like and only hint at the immunosuppressive risks they incur. Most hard-core addicts are like Donlon. They abuse several drugs simultaneously, or one after another, including heroin, cocaine, amphetamines, and barbiturates.[3] Alcoholism is common among addicts.[4] In one group of pregnant women on methadone maintenance, 56 percent were multidrug users during their pregnancies, taking diazepines, opiates, cocaine, barbiturates, amitriptyline, and amphetamines. Fifteen percent of these women imbibed unusually large amounts of alcohol compared with non-addicts.[5] In consequence, in addition to drug addictions, some children of addicts also suffer from fetal alcohol syndrome.[6] Drug-associated immunological abnormalities are the norm for this group.

Drug-induced immune suppression is only the beginning of the story. The fact that Sustecal and other nutritional supplements are given to many addicts as part of their rehabilitation points to another serious source of immunosuppression: malnutrition. The need for nutrient supplements is all too real. Nutritional studies have found that addicts as a group tend to be generally malnourished, to have anemia, decreased appetite, gastrointestinal distress, and emaciation, as well as key nutrient intake below two-thirds the normal levels recommended.[7] Specific problems include suboptimal levels of almost all essential vitamins and minerals. Dr. Jenny Heathcote and Dr. Keith B. Taylor of Stanford University found that vitamin C, vitamin B_6, and serum albumin levels (a measure of available protein) were significantly lower in heroin addicts than in people of the same age and sex who did not abuse drugs. These results are particularly striking since the addicts they studied were not overtly malnourished, as many other addicts are.[8]

Other studies of addicts have characterized a broader spectrum of nutritional problems. Dr. Ahmed el-Nakah and his colleagues at the New Jersey Medical School in Newark found in 1979 that of 149 intravenous drug abusers tested, only 24 percent had normal levels of vitamin B_6, B_{12}, folate, thiamin, riboflavin, nicotinamide, carotene, and vitamin A. Twenty-six percent had a significant deficit in one of these vitamins, 29 percent in two, 15 percent in three, 5 percent in four, and 2 percent in greater than four.[9] Such multiple vitamin deficits usually result only from long-term malnourishment and are well-known correlates of impaired immune function.[10] Deficiencies of vitamins B_{12}, folate, and vitamin A are each sufficient to cause significant immune suppression individually. It is very likely that Donlon and Ceford M. suffered from more or less severe malnutrition in addition to their drug abuse and infections.

But even the drugs and malnutrition themselves are but a beginning. Nearly everyone has had the experience of sitting in a physician's examination room while a nurse swabs down the skin on one arm with alcohol in preparation for a vaccination. The alcohol is used as an antiseptic to kill any bacteria or other infectious agents that might reside on the skin at the site of the inoculation. Many of these organisms are perfectly harmless as long as they remain on the skin, but some become pathogenic when they are introduced under it. In addition, the nurse or physician uses a brand-new disposable sterile needle and, more often than not, a disposable sterile syringe, for the inoculation. The use of antiseptics such as alcohol or iodine to prepare skin for inoculation is rare among drug addicts. Many addicts share previously used, and therefore contaminated, needles and syringes. By the time one combines the immunosuppressive effects of the drugs they inoculate, their failure to observe sterile injection procedures, and the tendency of many of them to share unsterile needles or to reuse their own unsterile needles, the stage is set for medical disaster.

Given these general settings, it is not suprising to find that addicts have an unusual incidence of needle-transmitted infections. Intravenous drug abusers have much higher rates of infection by hepatitis B virus, a number of bacteria including *Staphylococcus* and *Escherichia coli*, candidiasis, and endocarditis (infections of the valves of the heart). They are at higher risk of death from all forms of pathogenic infection, including tuberculosis and many pneumonias, and tend to develop cancers at higher rates than non–drug users.[11] Lymphocytotoxic antibodies (LCTAs) are present in almost all long-term intravenous drug users. Indeed, several studies have found that lymph node and thymus pathology is the rule rather than the exception among drug abusers. Dr. B. Kringsheim and Dr. M. Christoffersen found "enlargement and histological signs of activation in all the lymph nodes investigated" and concluded that these symptoms "are caused by continuous antigenic stimulation due to repeated injections of various antigens. The same may in part be applied to the thymus changes demonstrated."[12] Since lymph nodes and the thymus are integral parts of the system that activates T cells and permits their interactions with other immune system cells, any pathology is a possible symptom of impaired immunity.

The infection risks encountered by some addicts are increased by two more all too common practices: the exchange of sexual favors for drugs and the practice of prostitution to support a drug habit. "Kim Williams's" story is all too common. Kim is (or was) a prostitute in Newark, New Jersey. "First you use a drug," she says, "then it uses you. I never used as a teenager. . . . But a few years ago my boyfriend got locked up, and I got so depressed because nobody was there for me. A friend offered me some coke, and I got hooked." She could not hold a steady job, could not survive on welfare with a child to care for, and could not get enough money to buy the drugs she now depended upon. She turned to prostitution. On the day that Kim was being interviewed by John Tierney, she was on her way to a "freak show" where she was to have sex with another woman while a group of men watched or joined in. One of the men who had organized the freak show was rumored to have AIDS, and nearly everyone there was a drug abuser. Kim had been offered a combination of "crack" cocaine and cash for her services.[13]

As awful as Kim's story is, there are even worse. Tierney was told that some of the cases of prostitution involved girls of eleven or twelve years old, actively soliciting "johns" to support their crack habits.[14] In San Francisco and New York City, some of these addicted children are boys performing homosexual acts for drugs or pay or both.

Promiscuity of the sort required by a prostitute vastly increases the probability that he or she will contract venereal diseases, unless the most stringent safer sex guidelines are obeyed (as does occur in legal prostitution establishments in Nevada and Amsterdam, for example). Safer sex is not a frequent concomitant of the crack house, freak show, or street hooker. Thus, sex with a drug-abusing prostitute increases the risk for the patrons. And many of the patrons are, as we have seen, drug abusers themselves. It is hardly surprising to find that the incidence of all sexually transmitted diseases, including disseminated gonococcal infections and syphilis, is extremely high among drug abusers, compared with all other AIDS risk groups except homosexual men.[15] The resulting open sores increase the probability of infection with HIV and other viruses and may also increase immunologic exposure to semen.

Many intravenous drug abusers are aware of their unusual risk of infection and take prophylactic measures, but these efforts may actually increase their immune suppression. Dr. Scott R. and Dr. Sydria K. Schaffer, who practice medicine in Philadelphia, report that about 60 percent of the drug abusers they treat in the emergency rooms of Temple University and Hahnemann University hospitals admit to using antibiotics without a prescription. A significant proportion of the remaining 40 percent may also use such illicit antibiotics but are unwilling to admit to this illegal practice. Most surprising, every person who admitted to using these antibiotics said they were obtained on the street from their drug dealer. According to the Schaffers, the antibiotic "is reportedly procured via manipulation of routine prescriptions given by physicians to drug users and their friends for unrelated infections."[16] In consequence, drug abuse and antibiotic abuse have become synonymous for many drug addicts, with the inevitable result that many of these abusers are subjected to unusually high doses of antibiotics for extremely prolonged periods of time, amounting to months or years. Some of these addicts use amyl and butyl nitrites and may experience the same carcinogenic risks that were described for gay men. They may also experience antibiotic-associated zinc and copper depletion, accompanied by immune suppression. The irony of the situation is that while these drug abusers may not contract the typical pathogenic bacteria that might otherwise warn them and hospital personnel that these people are in immunological trouble, they are not protected against the opportunistic infections that antibiotics are unable to control. Thus, in protecting themselves against everyday infection, they open themselves up to more exotic, and more deadly, infections.

The result of these multiple immunosuppressive risks is a complex picture of inevitable immunological decline in which no single factor can be identified as being of paramount importance. Dr. P. H. Chandrasekar, Dr. J. A. Molinari, and Dr. J. A Kruse of the Wayne State University School of Medicine in Detroit reported in 1990 that a combination of any three of the following practices was highly associated with the development of AIDS in the drug addicts they studied: (1) illicit antibiotic use, (2) duration of intravenous drug use, (3) visits to "shooting galleries," where needles are often shared, (4) sexual promiscuity, (5) sex with prostitutes, (6) a history of sexually transmitted diseases, and (7) travel to areas where HIV is endemic.[17] Given this wide range of behavioral, nutritional, and disease-related factors, AIDS in addicts is no simple phenomenon and will have no simple solution.

Multiple Immunosuppressive Risks of Transfusion Patients

During 1982, California tax reformer Paul Gann entered the hospital for open heart surgery. During the course of that surgery, he received at least one unit of HIV-contaminated blood. Several years later, he tested positive for HIV infection and subsequently developed the clinical symptoms of AIDS. He died of AIDS in September 1989. More recently, tennis star Arthur Ashe has announced that he, too, contracted AIDS following heart surgeries in 1979 and 1983.

About two years after Gann contracted his HIV infection, a newborn baby named Ariel Glaser also became infected. Her mother, Elizabeth Glaser, had required transfusions as a result of complications during her pregnancy. Both she and her fetus were infected with HIV. Ariel developed symptoms of AIDS in 1988 and died shortly after. Elizabeth Glaser is ill at the time this book is being written, but she is alive and has told her sad story in the excellent book, *In the Absence of Angels*.[1]

Gann, Ashe, and Glaser played prominent roles in bringing the risk of transfusion-associated AIDS to public attention. Gann actively campaigned for recognition of the problem in California, while Glaser, as wife of "Starsky and Hutch" star Paul Glaser, used her Hollywood contacts to reach the Reagan administration heavy-weights who helped to put pressure on the appropriate government agencies in charge of such issues. Ashe reminds us how many years may elapse between surgery and AIDS. From a purely human standpoint, however, Gann, Ashe, and Glaser and her infant are but four representatives of the many thousands of transfusion-associated AIDS cases recorded worldwide. These tragic cases represent about 2 percent of all AIDS cases both in the United States and in other Western nations.

Blood transmission of HIV followed by the development of AIDS is one of the strongest types of evidence mustered by those who believe that HIV is necessary and sufficient to cause AIDS. Some proponents, such as James Curran, director of the AIDS program at the CDC, put the argument quite bluntly: "The evidence just hits you in the face. . . . [Can you] tell me that the transfusion cases are caused by lifestyle? How about the 60-year-old wife of a hemophiliac who gets infected? She's out cruising, too?"[2] But a closer look at the wide range of immunosuppressive risks encountered by transfusion and hemophilia patients suggests that even here the usual story that HIV equals AIDS is too simplistic. Transfusion patients and their spouses may not be out cruising the bars and baths or "hitting the shooting galleries," but their lives are far from free of immunosuppressive risks.

There is little doubt that Gann, Ashe, and Glaser and her daughter were all subjected to much more than simply a unit of HIV-contaminated blood. First, it must be emphasized that not everyone exposed to HIV through blood products becomes HIV seropositive or develops AIDS. It is estimated that about 12,000 transfusion recipients were exposed to HIV through contaminated blood in the United States between 1980 and 1985, when blood began to be screened for the presence of HIV antibody.[3] Recent studies suggest that no more than 60 percent of the patients exposed to HIV through transfusion become HIV seropositive, and one study demonstrated that of 765 patients who received HIV-tainted blood during the early 1980s, only 43 were known to have developed AIDS an average of five years later.[4] These figures suggest that factors other than HIV exposure contribute to who becomes actively infected, who subsequently develops AIDS, and at what rate.

Several factors have been identified. The most prominent is that HIV-seropositive people who developed AIDS had an average of twenty-one units of blood or plasma transfused during their hospitalization, whereas those who did not develop AIDS had an average of only seven units.[5] The average transfusion involves less than five units of blood or plasma.[6] Among those who received HIV-contaminated blood, the disease rate was thirty times higher in patients receiving more than ten units of blood compared with those receiving fewer than ten units of blood. Ann Hardy and her colleagues at the CDC have postulated that "persons receiving transfusions of large volumes of blood may be more susceptible to infection because of severity of underlying disease and other host factors or because they have an increased chance of exposure to other cofactors that modulate the clinical expression of infection with the virus causeing AIDS [HIV]."[7] Other investigators have hypothesized that the type of transfusion may matter, too: patients receiving large amounts of plasma or granulocyte transfusions, in particular, progress to AIDS much more rapidly than patients receiving just a few units of red blood cells.[8] The amount of blood a patient receives correlates with other immunosuppressive risks, such as more severe underlying illness. People who progressed to AIDS after transfusion were hospitalized for significantly longer periods of time than those who did not (nineteen days versus thirteen) and were more likely to have been admitted to intensive care (64 percent versus 46 percent).[9]

Another correlate of the amount of blood transfused is the amount of immune suppression due to the blood itself and the probability of contracting blood-transmissible immunosuppresive viruses, such as cytomegalovirus, Epstein-Barr virus, and hepatitis viruses that the blood may contain. During the late 1970s, it was estimated that nearly 3 in every 100 units of blood contained sufficient cytomegalovirus to result in active infection in the recipient.[10] About the same figure applied to transmission of Epstein-Barr virus. Three in every 1,000 units of blood contained hepatitis B or non-A, non-B hepatitis viruses.[11] In other words, on a purely probabilistic basis, a person receiving fifteen units of blood—as is fairly typical of those developing AIDS following transfusion—had a one in two chance of developing mononucleosis due to cytomegalovirus or Epstein-Barr virus and a one in twenty chance of developing hepatitis. These risks are in addition to the possibility of having received HIV as well.

Actually, the preceding calculation underestimates the risk of infection among those exposed to HIV. As we saw in Chapter 4, people who have active HIV infections are far more likely than the rest of the population to be co-infected actively with other viruses and bacteria. Thus, the probability that an individual will receive an HIV-tainted unit of blood that is co-contaminated with one or more of these other infectious agents is much higher than that of a person who receives HIV-negative blood. Direct evidence supporting this prediction was found by Jacob Nusbacher and his associates at the Canadian Red Cross Blood Transfusion Service in Toronto, Canada. Blood donors self-identifying themselves as being in a high-risk group for AIDS not only had 100 times the probability of testing positive for HIV as other people but were also much more likely to have an active hepatitis infection and antibodies against both hepetitis B virus and cytomegalovirus.[12] Given the roles of cofactors in stimulating HIV infectivity and cell killing, such multiple, concurrent infections could be of great significance for understanding who develops active HIV infections, and who does not, following tainted transfusions, but I can locate not one study of whether HIV-infected transfusion patients also develop simultaneous, concurrent infections. No one has bothered to look.

It is time that someone does look. The incidence of primary and recurrent cytomegalovirus infections following transfusion has been found to vary from about 1 percent to over 50 percent, depending on the patient population studied, the total number of units of blood transfused per patient, and the incidence of cytomegalovirus in the donor population. The greater the incidence is in donors, the higher the incidence of infection among recipients, and the more debilitated the recipient is, the greater is the incidence of infection. Thus, low birthweight infants were at much higher risk of cytomegalovirus infection than previously healthy adults.[13] Non-A, non-B hepatitis continues to be an problem following transfusion, as does hepatitis B virus infection, despite attempts to develop screens for these viruses.[14] Add HIV into the picture, and the disaster of AIDS no longer seems so unlikely.

Blood transfusion patients have a number of risks even beyond the blood they receive and any infectious agents that blood may contain. As Dr. R. Arnold, Dr. S. F. Goldmann, and Dr. H. Pflieger of the Department of Internal Medicine at the University of Ulm noted in 1980, "Sensitization due to HLA antigens is a well-known phenomenon in patients receiving blood transfusion . . . [resulting in] the presence of lymphocytotoxic antibodies after whole blood transfusions, red cell and platelet transfusions."[15] LCTAs can result after almost all types of tranfusions, although the incidence varies depending on whether the transfusion is of whole blood, packed red blood cells, plasma, and so forth. Arnold's group demonstrated that granulocyte transfusions resulted in a particularly high incidence of LCTAs against HLA proteins (MHC class II). By the end of their study, thirteen of twenty-two recipients of multiple transfusions had developed such autoimmunity. Many studies have confirmed that transfusion-associated immune suppression due to autoimmunity is common.[16] Recall that some anti-HLA antibodies mimic HIV antibodies on some ELISA tests.

In addition, transfusion patients almost always encounter a series of further affronts to their immune function. Injuries or surgeries requiring blood transfusions generally require anesthesia, and anesthetics are almost universally immune suppressive. The same sorts of injuries and surgeries are usually followed with the administration of one or more painkillers, often including opiate-based compounds such as morphine or codeine, which are also immune suppressive. Finally, the trauma of gross injury or surgery is itself immune suppressive. While none of these immunosuppressive risks is long term, when added together over a very short period of time such as a few days or weeks, it is hardly surprising that, in some cases complicated by infection and autoimmunity, the immune system never fully recovers.

This complex set of factors is not present in every transfusion patient, so patients have very different risks of developing AIDS following transfusion. Only about half of patients transfused with blood from an HIV-seropositive donor develop antibody to HIV. Even among these, the risk of subsequent AIDS seems to be quite varied. AIDS following transfusion is extremely rare among people aged one to twenty years, perhaps because these tend to be healthy individuals who are hospitalized due to trauma or accident rather than illness. Older adults and infants younger than one year of age are more prone to transfusion-associated AIDS. For example, although infants under one year of age make up only 2 percent of transfused patients, they account for over 20 percent of transfusion-associated AIDS cases. This observation suggests that other, underlying factors are at work in infants, including prematurity, multiple infections, drug addiction, or malnutrition.

The most horrendous aspect of the transfusion story has not been told: Most transfusion patients apparently do not survive the long-term consequences of their illness and surgery whether they develop AIDS or not. Mr. Gann survived for seven years following his open heart surgery and transfusions. Mr. Ashe has survived for thirteen. They are and were both quite fortunate. Dr. John Ward, Dr. Scott Holmberg, and their colleagues associated with the AIDS program of the CDC have compiled the most complete statistics available on mortality following blood transfusions.[17] Their data is truly frightening. They report that "of the 694 recipients [of HIV-contaminated blood] whose medical records were located, 331 (48 percent) had died within one year of transfusion." Mortality continues to mount as the years tick by, so that by seven years, over three-quarters of HIV-positive transfusion patients are dead. Clearly, people who receive HIV-contaminated blood have reason to fear for their lives.

Amazingly, and uniquely in all the studies of transfusion-associated AIDS, Ward, Holmberg, and their associates also provide data for an HIV-negative group of transfusion patients as well: "By comparison, 73 of 146 recipients of components from a random selection of donors not known to be infected with HIV (50 percent) died in the year after transfusion." Fifty percent. The same mortality as HIV-positive patients! If these data are accurate—and they have been validated by several other informal American studies[18]—then it is fair to say that AIDS is the least of the worries that a transfusion patient faces. Most transfusion patients are lucky to live long enough to develop AIDS.

I want to emphasize the importance of these figures. Over the short term—one year—there was no difference in mortality between patients infected with HIV and those who were not. Little can be said beyond that one year. Only a single study exists anywhere in the world looking at the mortality rate of HIV-negative transfusion patients beyond one year. This study was published in 1988 by Gordon Whyte of Canterbury, New Zealand. In his study, 21 percent of transfusion patients were dead within a year of transfusion, and nearly 30 percent by the end of three years. Whyte's study, however is not comparable to the Ward-Holmberg study. The patients in Whyte's study received an average of only 3.7 units of blood, whereas the patients in the Ward-Holmberg study received an average of about 12 units of blood each. The threefold difference surely translates into increased mortality for the American group. For example, Whyte found that among seven patients who received 228 units of platelet concentrates (an average of 33 units each), all but one was dead by the end of two years. The type of transfusion was also found to be important. Nine patients received 34 units of plasma (an average of only 4 units each), and all were dead within two years.[19] The CDC study does not give details of the type of transfusion. All that can be determined from Whyte's study, then, is that mortality continues to increase with time, even among HIV-negative patients. Clearly we need to know much more about this important subject, not only with regard to AIDS but to the long-term risks incurred by all transfusion patients.

In the meantime, there is no basis for maintaining that HIV increases mortality or alters the types of diseases contracted by transfusion patients, since there is no controlled study of a group of HIV-negative transfusion patients with whom to compare. On the contrary, all of the data that do exist suggest that HIV has no influence on mortality and may or may not have any on morbidity. Statements such as those by James Curran and other AIDS researchers to the effect that transfusion patients prove the necessity and sufficiency of HIV as the cause of AIDS are wishful thinking.

The Multiple Immunosuppressive Risks of Hemophiliacs

Hemophiliacs are another group of people at unusual risk for AIDS. Hemophilia is a complex disease with a variety of manifestations. Despite popular myths that hemophiliacs are always men, the disease also occurs in women. One type, in fact, was discovered precisely because it is just as common among women as among men. In 1921 Professor Erik von Willebrand, a physician working in Helsinki, met a young girl named Hjördis S. from the town of Föglö in the Aland archipelago of Finland. Hjördis had had several severe bouts of bleeding from her nose, lips, and following tooth extractions. Four of her sisters died between the ages of two and four of uncontrollable bleeding, and all but two of her other eight siblings also had bleeding problems. Willebrand was fascinated, since bleeding disorders in women are rare, and such a high incidence of such disorders in any family was almost unheard of. He carefully followed Hjördis's progress and slowly began to meet and study her relatives. Altogether he located sixty-six family members, of whom twenty-three had the same symptoms as Hjördis.

Hjördis and her relatives did not have typical hemophilia. Their symptoms differed in some interesting ways. For one thing, bleeding in the joints is very common in typical hemophilia but was almost unknown among the family in Aland. Moreover, typical forms of hemophilia are inherited as a sex-linked recessive trait, whereas the disease he was studying was inherited as an autosomal dominant. This difference explained why so many women were prone to this disease. In more typical forms of hemophilia, a woman can carry the recessive, defective gene on one X chromosome and show no symptoms of disease because her other X chromosome carries a normal gene that compensates. She would have to inherit two defective genes to become a hemophiliac—a rare event. Men, however, have one X chromosome and one Y. The Y has almost no genes on it to compensate for any defects in genes on the X chromosome, so if a man inherits the X chromosome with the defective gene from his mother, he will develop hemophilia. Thus, because of the genetics, typical forms of hemophilia are much more common in men than in women. But in this case, Dr. Willebrand found that just as many women as men were inheriting this particular bleeding disorder. Moreover, if the parent had the disease, their offspring almost always inherited it. This was clearly something new and unexpected.

There was little Dr. von Willebrand could do for his new-found patients. When they bled extensively, he replaced their lost blood by means of blood transfusion, and he also found that transfusions could help to control bleeding when it began, but he had no effective treatments and certainly no cure. Many of his patients died. Hjördis herself died in 1926 at the age of thirteen when her fourth menstrual period failed to end. Willebrand commemorated her death by publishing the first medical description of her disease. It now bears his name.[1]

The various forms of hemophilia, including von Willebrand's disease, subsume a group of genetically inherited conditions in which blood clotting is impaired. Blood clotting is a complicated process beginning with the formation of tangles of a long protein, fibrin and followed by a series of chemical reactions involving about a dozen so-called clotting factors. These factors cause the fibrin net to be properly woven and its loose ends tied together so that a sturdy clot is formed for catching and holding in the platelets and red blood cells, lymphocytes, and the various proteins found in blood serum. These clots control both internal bleeding (hemorrhages and bruising) and external bleeding from cuts and scrapes.

The three most common types of hemophilia are hemophilia A, in which clotting factor VIII is lacking or defective; hemophilia B, in which clotting factor XI (Christmas factor) is absent or defective; and von Willebrand's disease, in which the protein necessary for delivering factor VIII ("von Willebrand's factor") is absent. In all three types of hemophilia, an incomplete fibrin net results following injury, and since the clots are not correctly assembled, bleeding is difficult or impossible to control. If the injury is small, the normal process of tissue healing will eventually correct the problem. Untreated cases of hemophilia result in death when the injury is too great for normal healing processes to compensate for the injury in a reasonable length of time. Even with treatment, hemophiliacs still face significant risks. Half of all deaths associated with hemophilia were attributed to uncontrolled bleeding during the 1970s.[2] The amount of factor necessary to treat a hemophiliac differs with the degree to which blood clotting is genetically impaired. Some hemophiliacs require a great deal of clotting factor and are termed severe hemophiliacs; others rarely require clotting factor treatment and are termed mild cases.

A look at the life and treatment of a modern hemophiliac is revealing from the standpoint of AIDS risks. Perhaps the best such look was made in 1975 by Robert and Suzanne Massie in their best-selling book, *Journey*, detailing their experiences learning to deal with their son, Robert Jr.'s, hemophilia. It is a heart-wrenching book, filled simultaneously with courage, hopelessness, joy, pain, and ongoing victory. Hemophilia is a lifelong struggle both for the individual who inherits this trait from his parents and for the parents themselves. The threat of AIDS adds to the already considerable burden.

Like many other male hemophiliacs, Robert Massie, Jr.'s, problems began with his circumcision. It would not stop bleeding. He is a severe hemophiliac, whose natural clotting is badly impaired. Severe bruises became potentially life threatening. A hard knock on the head had the potential to cause bleeding inside the cranium that could result in brain damage. Lifting heavy objects, walking long distances, or even running for a short distance could break enough blood vessels within his joints to cause them to swell to the size of grapefruits. Untreatable joint damage was always a potential threat following such incidents. To prevent such damage from occurring, Robert wore braces on his legs and began physical therapy at a very young age. The braces have been discarded; the physical therapy has never ceased.

Therapy and reasonable care prevented many of Robert's problems. Nonetheless, sometimes, unpredictably, Robert would develop bleeding within one or more of his joints. "The pain of these episodes was terrible. For all of Bobby's childhood, he lived with it and we lived with it," reported his parents. "Doctors say that the pain of a bleeding joint is one of the worst known to medical science. . . . In the past, seeking to alleviate their agonies, hemophiliacs became drug addicts. Our doctor had seen such hemophilic addicts during his medical studies, and because of this we tried in every way to give drugs to Bobby as sparingly as possible. In any case, there is no drug that can completely alleviate the pain of bleeding joints. Besides, Bobby was very sensitive to any drug. They tended to excite, rather than soothe, him. We tried Darvon. We tried codeine, but codeine only made his eyes grow wild, blurred his vision, and did not take away the pain, only somehow made it pound through a haze of half-conciousness. Under its influence he would sometimes grow hysterical and violent, and scream, over and over, 'No! No more pain! No more pain!' Sometimes when, for no reason we knew, transfusions did not help or only slowed the bleeding, the pain would go on for days and nights."[3]

This short passage encapsulates many of the immunosuppressive risks faced sometimes daily, sometimes much less frequently, but certainly repeatedly, by every hemophiliac. First, there are the repeated transfusions with blood plasma containing the clotting factors the hemophiliac requires. Plasma transfusions are associated with greatly increased mortality. All too often, the blood plasma contains hepatitis viruses, cytomegalovirus, and Epstein-Barr virus. A severe hemophiliac like Robert, Jr., or Ryan White, one of the most visible of American hemophiliacs to die of AIDS, might require dozens or even hundreds of plasma transfusions each year. Many developed and continue to develop circulating immune complexes and lymphocytotoxic antibodies as a result of their repeated transfusions and viral exposures,[4] and nearly all hemophiliacs during the 1960s and 1970s were infected with hepatitis A and B viruses, cytomegalovirus, and Epstein-Barr virus. The average life expectancy of a hemophiliac in 1970 was a mere thirty-three years.[5]

Plasma transfusions were slowly replaced with a concentrated form of clotting factors, called cryoprecipitate, beginning in 1966. Robert Massie, Jr., was one of the first to be treated with these new concentrates. They are much more effective and safer than plasma, as is indicated by the fact that the average life expectancy of a hemophiliac is about fifty-five years today.[6] The fact that hemophiliacs still have about a twenty-five-year shortfall in their life expectancy compared with the general population demonstrates that they still have significant medical risks. Some of these risks, such as the factor concentrates themselves, are associated with immune suppression. Part of this immune suppression is apparently due to repeated exposure to the factors themselves and part is due to the continuing inability to eliminate some immunosuppressive viruses such as non-A forms of hepatitis and their antigens from these preparations. Chronic hepatitis, and the immune suppression that accompanies it, are a problem for about 10 percent of hemophiliacs.[7] Cytomegalovirus infections, whether acquired independently of blood clotting treatments and reactivated or transmitted in the factor serum or during a blood transfusion, have been found to increase the rate at which AIDS develops in HIV-seropositive hemophiliacs about threefold.[8] And despite screening procedures for hepatitis A, occasional transmission of the virus is still recorded.[9]

Blood products, and the viruses they may carry, are only the beginning of the immunosuppressive risks a hemophiliac may encounter. As the Massies recount, the use of painkilling drugs—both opiate based (synthetic morphine or the morphine derivative, codeine) and nonopiates (Darvon)—is a common, and perhaps even necessary, part of the treatment of many hemophiliacs. The more severe the hemophilia, the more likely is the need for such drugs. Thus, a terrible and intractable problem arises. A hemophiliac is most likely to require both multiple factor transfusions and painkillers simultaneously, increasing the likelihood and degree of immune suppression. If, as happened in the case of Ryan White, those factor transfusions also contain HIV (and possibly other viruses as well), then a hemophiliac may be in the worst possible immunological state to fight off the resulting infection. Indeed, a severe hemophiliac who has been given an HIV-contaminated factor transfusion while simultaneously being treated for pain with an opiate-derived painkiller is effectively mimicking the risks of an intravenous drug abuser. There are, however, no data whatsoever on the possible role of painkillers in the development of immune suppression in hemophilia—yet another lacuna in our knowledge.

A hemophiliac's immune suppression may be increased by other medical treatments as well. At one point, it appeared that factor therapy was not working for Robert Massie, Jr.[10] In 5 to 10 percent of hemophiliacs, the factor therapy itself results in an autoimmune reaction against the factor. When this happens, tremendous amounts of factor may be necessary to allow clotting or further factor therapy may be essentially useless. This leaves the hemophiliac at the mercy of his disease unless, using immunosuppressive drugs, the autoimmunity can be brought under control.[11] Such drugs confer additional risks of developing opportunistic infections. Thus, once again, the very complications of hemophilia may require treatments that can activate latent HIV infection and that are themselves conducive to the development of AIDS-associated diseases.

Some hemophiliacs have been subjected to even further immunosuppressive regimes. In 1971, Robert Massie, Jr., had an episode of untreatable joint bleeding that threatened to result in such extensive damage that the joint would become locked in position. The Massies agreed to several experimental procedures out of desperation. The first consisted of injections of osmic acid, the second of injections of radioactive gold salts.[12] These are the same sorts of treatments that have been tried with more or less success for arthritis. Both the osmic acid and radioactive gold are still used by some physicians,[13] and both are known to be immunosuppressive. Happily, in Robert Massie, Jr.'s, case, he survived all of these insults to his immune system, attended Princeton University, and has gone on to other impressive achievements. Not all hemophiliacs, especially in this age of AIDS, are so lucky, as Ryan White reminds us.

Not all hemophiliacs have the same risk of developing AIDS. The incidence of AIDS is six times higher in those with hemophilia A than hemophilia B, in part because type A hemophiliacs are more likely to have severe disease and therefore to need more clotting factor more frequently. Severe hemophiliacs are also more likely to require painkillers, treatment for joint injuries, and associated problems than are mild hemophiliacs. Thus, the incidence of AIDS among people with severe type A hemophilia is three times that of those with moderate hemophilia, and seven times higher than among those with mild hemophilia.[14] These figures once again suggest that there is more to AIDS than just HIV.

An additional factor with an important effect on AIDS risk among HIV-seropositive hemophiliacs is age. A National Cancer Institute study of 1,219 hemophiliacs headed by James J. Goedert reported in 1989 that HIV-seropositive children and adolescents, age one to eleven, developed full-blown AIDS at a rate almost fifteen times lower than the rate at which adults above the age of fifty develop AIDS. As a group, all children and adolescents under the age of eighteen developed AIDS about five times more slowly than did adults over the age of thirty-five.[15] Similar results have been reported from a study of 1,201 hemophiliacs in Britain.[16]

Goedert and his colleagues comment that these data suggest that "older adults either have a smaller reserve of precursor CD4 cells or are perhaps more susceptible to the pathogenicity of HIV-1 because of other conditions that are common in older persons with hemophilia, such as destructive inflammatory synovitis [joint damage] or chronic liver disease."[17] Certainly their speculations match what we have seen here concerning the decrease in immune activity with age, the use of immunosuppressive drugs to treat joint problems and their associated pain, and hepatitis virus–induced immune deficiency. Clearly, the older a hemophiliac becomes, the more likely he or she is to encounter one or more of these acute or chronic problems, any of which may help to trigger AIDS. Older hemophiliacs are also much more likely to have been infected with hepatitis A virus, since many were treated prior to the development of screening procedures for this virus, and some hemophiliacs are chronically infected with it. Just as certainly, the incidence of both acute and chronic infections with various herpes viruses and hepatitis B and non-A, non-B hepatitis viruses also increases with age. What is missing from Goedert's list of suggested factors are additional risks such as the fact that older hemophiliacs lived through a period in which the treatment for hemophilia was very different than it is today. Hemophiliacs age twenty and older may carry over long-term immunological "scars" from the repeated plasma and blood transfusions they received in the preconcentrate era.

In short, hemophiliacs like Robert Massie, Jr., and Ryan White had or have many risks for acquired immune suppression other than HIV. Whether these additional risks set the stage for HIV, promote it once it is present, or are themselves responsible for AIDS is not clear. It is clear, however, that we can learn much from AIDS that may help to extend the life expectancy of all hemophiliacs.

The Immunosuppressive Risks of Infants

Maria Prophet died of AIDS in a San Francisco hospital in 1982. She was only four years old. She is one of the greatest tragedies of the epidemic—a bud that never had the chance to blossom, let alone to reach fruition. She not only represents the potential that AIDS may devastate our entire society—men, women, and children as well—but cases like hers have provided one of the most important bastions for the theory that HIV is both necessary and sufficient to cause AIDS. Anthony Fauci, chief immunologist at the National Institutes of Allergy and Infectious Diseases (NAIAD), has disparaged notions that anything other than HIV may be involved in infant AIDS by asking rhetorically what possible immunologically risky behaviors a fetus could indulge in.[1] That is a good question. But the answer is not nearly as obvious as it may at first seem. Thus, an analysis of why some infants, such as Maria Prophet, develop AIDS is critical, not only for ferreting out the true risk of AIDS to future generations but to understanding AIDS itself.

The usual story concerning AIDS among infants and young children goes something like this. The unfortunate mother of an AIDS infant contracts an HIV infection prior to conception or during pregnancy. The virus sneaks through the protective barriers in the placenta that normally prevent the exchange of anything other than nutrients between mother and child. The HIV invades the fetus's developing lymphocytes and begins to replicate. The infant is born more or less normal, but as he or she grows and the immune system matures, so does the HIV infection. Eventually the child's immune system is wiped out; she or he contracts various opportunistic infections and eventually dies. It is a story portrayed by AIDS researchers, AIDS activists, and AIDS reporters alike as one of the greatest pathos: innocent victims cut down by forces they are too young to even comprehend.[2]

Consider Maria Prophet as an archetypal case. As Fauci asks, what possible risk factors could she have had? She could not inject herself with drugs. She was not sexually active (although 5 percent of AIDS cases in American children under the age of thirteen are due to sexual abuse).[3] Maria may have had various surgical operations or blood transfusions, since these are unusually common among AIDS infants, but there is no evidence on that, and for the moment let us assume that she did not. And she was not a hemophiliac. She was, however, retrospectively found to be infected with HIV. Presumably she contracted her HIV infection from her mother, who also developed AIDS shortly after her daughter. It looks like an open-and-shut case for HIV.

Would that it were true. Would that Maria's story were as simple as that, for then there would be hope in vaccines and retroviral drugs for other children like her. Unfortunately, Maria's story, like the stories of most other AIDS infants, is much more complicated. The reality is equally filled with the pathos of the innocent victim, but Maria was victimized in ways very different than the narrative just presented suggests.

To begin, Maria's mother was a prostitute and an intravenous drug abuser. It is highly probable that although Maria herself did not use drugs, she was repeatedly exposed to them in the womb. Additionally, her mother's prostitution undoubtedly exposed her repeatedly to sexually transmitted diseases. As a result, she was born prematurely and never developed normally. She was so brain damaged that she never learned to talk or to walk. She had chronic problems breathing and eating. She spent much of her life being fed through tubes directed into her stomach and often required oxygen supplied by various forms of breathing apparatus. When she was finally stabilized sufficiently to be taken home, her mother could not be located. Maria had been abandoned. A pediatric nurse at the hospital adopted her at the age of two, but even she found Maria's deteriorating condition too much to cope with and consigned her to a foster care home where she died. As of June 1990, Maria's mother and one of her two siblings had also died of AIDS. The other sibling was still alive but had been diagnosed as being HIV positive.[4]

There is much that a keen mind can read into this medical history. Maria's HIV infection cannot and does not explain her prematurity, her failure to thrive, her brain damage, or her difficulty eating and breathing. Indeed, we now know that simply being born HIV positive is not, for many children, a death sentence. Less than a third of HIV-seropositive infants go on to develop AIDS. Other factors tip the scale.

There is no doubt that there are other actors operating. Nearly 80 percent of infants with AIDS are, like Maria, born to drug-abusing mothers. Another 10 percent acquired their HIV during operations requiring blood transfusion, 6 percent are hemophiliacs, and the remainder have no established risk (although these may be accounted for by child abuse, drug abuse in a noncooperative parent, or other causes). Thus, the vast majority of children who contract AIDS are in well-established risk groups.[5]

The observation that nearly all infant AIDS cases other than those requiring blood products are born to drug abusers is extremely enlightening. The pre-AIDS literature on infants born to drug-abusing mothers contains many descriptions reminiscent of Maria. There is certainly nothing new about intravenous drug use's increasing the mortality of infants born to addicted mothers. During the 1950s, Cobrinik[6] and Goodfriend[7] reviewed the relevant literature back to 1875 and documented greatly increased mortality in such children. Subsequent studies pinpointed specific problems. Nearly 40 percent of such infants were low birthweight (less than 2,500 grams or 5.5 pounds) compared with 7 percent in the general population. Two-thirds of the babies were significantly premature (less than thirty-seven weeks gestation) compared with 7 percent in the general population. And over 30 percent of the babies were small for their gestational age, as compared with 2 percent of babies in the general population. Ten percent of the addict infants were in the lowest tenth percentile for weight, size, and head circumference for their gestational age. Apgar scores, a measure of newborn health parameters ranging from a high of 10 to a low of 1, below 6 were significantly more common (approaching 20 percent in infants of drug abusers), indicating moderate to severe physical distress at birth. In many cases, this distress was due to withdrawal symptoms.[8] Thus, although Maria was not, in the usual sense of the word, an intravenous drug abuser, she was probably born addicted to drugs. She participated, albeit unknowingly, in her mother's addiction.

Other problems plagued these addict infants. Congenital malformations were much more common in children of addicts than in other children. In some cases, these malformations may have been a result of fetal alcohol syndrome.[9] Fetal malnutrition was evident in the hypoglycemia (low blood sugar) present in 7 percent of the infants, and in the hypocalcemia (low level of calcium) evident in 15 percent. Infant malnutrition was associated with significantly higher rates of anemia (iron deficiency) and much lower

weight gain in the mothers than in healthy controls. Five percent
of addicted mothers actually lost weight during their pregnancy,
as compared with no weight losses among unaddicted mothers.
Jaundice, suggesting liver malfunction, was significantly more
common in drug-addicted infants. Some liver problems were due
to viral hepatitis. In over half the cases, the cause of the jaundice
could not be determined.

Not surprisingly, these multiple health problems had their toll.
Mortality rates among infants born to addicts during the 1970s
were three times the national average during the first year after
birth. Nine percent of the infants developed nonfatal pneumonias
during the first year of life (as compared with 1 percent in the
general population), but since the types of pneumonia were unspecified, any possible relationship to AIDS cannot be determined.
It was, however, recognized during the 1970s that *Pneumocystis*
pneumonia (PCP) was most common among low-birthweight and
malnourished infants like those just described. Sixty-eight percent
of infants, who were not congenitally immune deficient, but who
were diagnosed with PCP prior to 1975 were below the tenth percentile for weight at their age, and 75 percent had total serum
proteins and serum albumin levels significantly below the normal
range. PCP was also common in African and Asian children suffering from kwashiorkor.[10] Thus, just those children who must already bear the crosses of being born to malnourished addicts or
desperately impoverished, starving people are those most likely to
be immunosuppressed sufficiently to contract opportunistic infections such as PCP. The fact that this was true even prior to the
recognition of AIDS should be cautionary. Perhaps the only difference between these children and AIDS infants today is that we
have the ability to distinguish those who are HIV seropositive
from those who are not. Other, nonfatal infections that may have
characterized AIDS (such as chronic thrush) were not reported.[11]

Most of the problems just listed are independently associated
with immune suppression in infants. To begin, the mothers themselves are immune suppressed, and the infant inherits a temporary
immune system from the mother during gestation. For the first
three months of life, or longer if the infant is born prematurely,
the infant is dependent on the mother's immune system to protect
him or her from infection. If the mother's immune system is
subpar, the infant's immune system will also be subpar. This state-

ment applies whether the mother's immunosuppression is due to AIDS or any other cause, such as severe malnutrition or a combination of surgery, anesthesia, and blood transfusion. Ariel Glaser, whose mother received an HIV-tainted transfusion, was exposed not only to HIV but also to all of the other immunosuppressive regimens that affected her mother. Thus, even infants who do not become infected with HIV from their HIV-infected mothers nonetheless display subnormal immune responses.[12]

Given what we have found out previously in this chapter about the immunosuppressive risks of drug abusers, it is not surprising to find that studies of drug-addicted mothers demonstrate a wide range of problems that can lead to immune disregulation in their infants. The first, and most common, is undoubtedly the drug use itself. Next comes malnutrition. "The lifestyle of the pregnant addict predisposes her to nutritional problems which may lead to anemia [usually due to iron deficiency], and to higher incidence of infection, particularly venereal diseases."[13] In fact, anemia is twice as common among pregnant women who are addicted as among those who are not addicts. Notably, Dr. Gladys C. Vargas of the Chicago Medical School reports that such anemia may signal a developmental risk for the fetus: "Growth failure, including the brain, has been observed in young animals whose mothers were fed a low protein diet early in gestation. Similarly, human neonates whose mothers were undernourished before the third trimester of pregnancy have shown growth retardation of all organs, including the brain. In neonates of heroin addicts, Naeye, et al. [1973] observed that the subnormal size of the organs was associated with a subnormal number of cells at all gestational ages. An inhibitory effect of heroin on fetal growth has been observed in fetal rabbits who were injected with heroin. Whether maternal protein calorie malnutrition or a direct effect of heroin is responsible for the small size of our infants, including the head, cannot be determined from our data."[14] What has been documented is that small-for-birthweight children and microcephalic children (those with an unusually small head size) have a smaller thymus (the organ necessary to activate T cells) and decreased absolute numbers of T cells, problems mirrored in reduced T cell activity and diminished response to infection. The more premature an infant is or the lower the birthweight, the lower his or her T cell count and the greater the immune suppression.[15]

These exact findings are typical of infant AIDS patients.[16] Prematurity, low birthweight, and microcephaly are common. Anemia in AIDS infants is nearly universal (97 percent).[17] And the neurological deficits and behavioral and developmental difficulties that attend AIDS are much more common in the children of drug addicts than in other children, suggesting that Maria Prophet's retardation may have resulted directly from her mother's addiction and malnutrition rather than from her HIV infection.[18]

Since many drug addicts turn to prostitution, venereal diseases are common in addict infants. "On November 27 [1989], Baby Boy W. was born in Brooklyn four weeks early and with test signs of cocaine and syphilis. He is the youngest of three children whose mother is an unemployed crack addict, who has dropped out of drug treatment. She has not seen Baby Boy W. since his birth."[19] Baby Boy W.'s symptoms are all too common. Physicians observing pregnant addicts noted unusually high rates of syphilis and gonorrhea even prior to the recognition of AIDS.[20] The problem is now worse and may be indicative of higher rates of all sexually transmitted diseases, including cytomegalovirus, hepatitis viruses, *Chlamydia*, and *Mycoplasmas*. These infections pose additional immunosuppressive and health risks for infants. Infants congenitally infected with cytomegalovirus and other viruses often develop clinically evident immune suppression even in the absence of HIV.[21] Sexually transmitted infections may also result in encephalitis, meningitis, or myelitis, all of which can be fatal or leave permanent central nervous system damage such as Maria suffered.[22] Similarly, untreated Chlamydial infections, the most prevalent of sexually transmitted diseases in the Western world, are transmitted 5 to 20 percent of the time and are associated with preterm deliveries and low birthweight. They can cause neonatal chlamydial ophthalmia and trachoma, which can leave the child blind, and neonatal chlamydial pneumonitis, a potentially life-threatening lung infection.[23] *Mycoplasmas*, which are directly implicated as cofactors for HIV infection, selectively colonize low-birthweight and premature infants—those infants at highest risk for immune suppression in any case.[24]

In fact, there is a syndrome that is almost indistinguishable from AIDS in infants, save for the absence of HIV infection, called the TORCHES syndrome. TORCHES stands for congenital infections caused by TOxoplasmosis, Rubella, Cytomegalovirus, HErpes simplex virus, and Syphilis. This syndrome was first described by Dr. Andre Nahmias in 1971 and has been widely described in the children of poverty-stricken and drug-abusing mothers ever since.[25] Among the consequences of TORCHES syndrome are any or all of the following: an infant unusually small for gestational age, cytomegalovirus hepatitis, thrombocytopenia, microcephaly, retinitis, failure to thrive, and impaired mental development.[26] All of these symptoms are found in AIDS infants, who differ only in being infected with HIV as well.

Other largely unsuspected risks may await the infants of those who have or are at risk for AIDS. It is well established that HIV can be transmitted by means of both blood transfusion and breast-feeding. What is equally well established but apparently less well acknowledged is that other immunosuppressive risks can be transmitted by the same means. Transmission of cytomegalovirus via breast milk may, in fact, be the typical route of infection among infants, according to a 1980 study by Dr. Sergio Stagno and his colleagues at the University of Alabama in Birmingham.[27] Since the majority of women with AIDS are also carrying active cytomegalovirus infections, many of their infants contract active cytomegalovirus infections, too. Infants with concomitant cytomegalovirus and HIV infections are much more likely to develop full-blown AIDS and to die by the age of twenty-four months than are infants exposed only to HIV.[28] These data suggest a strong role for cytomegalovirus in AIDS.

In addition to infectious risks, breast-fed children may acquire maternal malnutrition since lack of dietary zinc and selenium in the mother results in breast milk deficient in these immunologically required minerals.[29] As in so much else that concerns AIDS, no studies of the transmission of cytomegalovirus, Epstein-Barr virus, or other immunosuppressive viruses from HIV-positive mothers to children via breast milk have been made, nor has the risk of breast milk or other forms of malnutrition been examined as an AIDS risk. Thus, one reason that many clinicians and researchers continue to insist that HIV is the primary cause of immune suppression is that no one has bothered to look at any other factors.

As sad as Maria Prophet's case was (and as sad as the deaths of all AIDS infants continue to be), it is not clear to what extent her death was due to HIV or to other immunosuppressive and developmental risks she would have encountered even in its absence. What is clear is that Anthony Fauci and other HIV proponents who claim that AIDS infants have no immunosuppressive risks other than HIV, and therefore that these cases prove that HIV is the sole causative agent in AIDS, are completely wrong. To attribute the health problems of Maria Prophet and other children like her solely, or even mainly, to HIV is a mistake. History suggests that such children have existed for as long as drug abuse and poverty have existed. Treating them and their mothers only for their HIV infection will hardly decrease their health risks. At best, eradicating HIV will only allow them to die of virtually identical problems, categorized under other names. That is the true tragedy of AIDS infants.

CHAPTER 7

IMMUNOSUPPRESSION IN THE ABSENCE OF HIV INFECTION

Theoretical Implications of Immune Suppression Among High-Risk-Group Individuals in the Absence of HIV

I have demonstrated that there is much more to AIDS than HIV alone. Cofactors—be they noninfectious agents, infectious agents, or various manifestations of autoimmunity—have a major, if often overlooked, role in the progression of the disease. Moreover, people who do contract AIDS have been shown to have multiple, concurrent risks for severe, debilitating immune suppression, as well as their HIV infection. I will show in this chapter that significant immune suppression is present in large numbers of people in high-risk groups for AIDS in the *absence* of HIV infections. Sometimes the degree of immune suppression is equal to, or even greater than, that experienced by HIV-positive, matched patients.

The existence of significant immune suppression in the absence of HIV necessitates several important conclusions. First, HIV is not necessary for inducing immune suppression in people at risk for AIDS. The causes of the observed immunodeficiencies therefore help to identify some of the specific agents involved. Second, many people in high-risk groups for AIDS have significant immune impairment prior to contracting an HIV infection and are thus more susceptible to both infection and the effects of infection than are immunologically healthy individuals. Third, a significant role for cofactors in AIDS becomes plausible. And finally, it is clear that acquired immune deficiencies do not require the presence of HIV infection. The evidence presented demonstrates absolutely that HIV is not the only significant cause of immune sup-

259

pression in any risk group and cannot be designated the sole cause of AIDS.

Indeed, given the evidence that I have presented in Chapters 2 through 6 regarding the multiple, concurrent immunosuppressive risks encountered by people in high-risk groups for AIDS, there can be little doubt as to why these groups are at risk. But risk does not equate with cause. Prior immune suppression may make people much more susceptible to HIV, and therefore to AIDS. HIV may then be necessary but require prior immune impairment to become pathogenic; it may require cofactors; or it may not be required.

We return to an analogy introduced earlier. When a person dies by drowning, how are we to distinguish the causes from the effects? The factors that all drowning victims have in common— high levels of carbon dioxide in the blood, extremely low levels of oxygen, buildup of certain metabolites, and so forth—are not the causes of death. They are the manifestations of death, and they are virtually identical in all drowning cases. The factors that actually caused the drowning, and therefore triggered all the other reactions, are, on the other hand, as myriad as the ways in which a person can be accidentally or purposefully kept under water until he or she is unconscious. At that point, the outcome of the situation is almost a forgone conclusion. We must therefore beware of the common epiphenomenon of drowning and search for the specific and varied triggers.

Originally I introduced this analogy in order to focus attention on the problem of identifying causal agents in the context of Koch's postulates. Here I emphasize a different aspect of the analogy. One must beware of reasoning that just because all patients of any particular disease have the same manifestations of disease, then they must also have the same primary cause of their symptoms. In the context of AIDS, this warning amounts to recognizing that just because HIV replication is an extremely common concomitant of the progression toward death in AIDS, it need not be a cause of that progression. In fact, cytomegalovirus and Epstein-Barr virus replication are just as frequent concomitants of AIDS as is HIV replication. Viral replication may therefore be a marker rather than a cause of immunological events that are more directly relevant to understanding the origins of acquired forms of immune deficiency. Just as drowning results in common symptoms but has many causes, AIDS, too, may have common outcomes but diverse initiators.

One crucial fact must be reiterated at this point: A significant number of AIDS patients appear to be completely free of both HIV infection and of antibody to HIV. When tests for HIV were first developed, as many as 10 percent of AIDS patients were found to be HIV seronegative, even by Robert Gallo's own laboratory.[1] Even since the adoption of HIV as the preferred diagnostic marker for AIDS, as many as 5 percent of AIDS patients who are tested for HIV infection continue to display no signs of infection in the United States.[2] Thus, there are many people who are developing opportunistic diseases diagnostic for AIDS in the absence of identifiable HIV infections. These data clearly represent the sorts of anomalies that suggest that HIV is not the primary cause of immune suppression in AIDS patients.

In fact, a large body of evidence demonstrates that significant immune suppression occurs in the absence of HIV infection in groups at high risk for AIDS but not among low-risk groups. HIV-seropositive individuals within each identified risk group are no more immune suppressed than those who are HIV seronegative, as long as they do not contract other active infections. Thus, HIV seropositivity is a significant marker for the development of increasing immune suppression only among individuals who have other symptomatic infections. Moreover, other infections, such as cytomegalovirus and Epstein-Barr virus, are just as accurate for predicting the development of AIDS as are HIV-related markers. In short, although nearly all studies of AIDS are predicated on the assumption that HIV is the only necessary marker for AIDS, in fact it is neither the necessary nor sole marker for increasing immune suppression. These facts provide added weight to the plausibility of alternatives to the HIV-only theory.

Immune Suppression in HIV-Negative Homosexuals

Evidence of significant immune suppression among HIV-negative people at risk for AIDS has been reported since the first tests for HIV seropositivity became available. For example, studies by the laboratories of Jerome Groopman[1] and Robert Gallo[2] found immediately that as many as 50 to 80 percent of HIV-seronegative homosexual men and hemophiliacs had significantly reduced T-helper/T-suppressor ratios during 1984. A possible cause of this immune suppression was suggested by Elena Buimovici-Klein, who observed the same phenomenon in 1988. She and her colleagues found that nearly 75 percent of homosexual men in New York City had active cytomegalovirus (CMV), Epstein-Barr virus (EBV), herpes simplex virus (HSV), or hepatitis B virus (HBV) infections, whether they were HIV infected or not. Twenty percent of these men had more than one active infection. These figures stand in stark contrast to heterosexual controls, among whom only 6 percent were actively infected with a single type of virus in semen, saliva, or blood and none was multiply infected.[3] These data suggest that homosexual men differ significantly from heterosexuals in their disease load and that this disease load results in immune suppression unrelated to HIV. All of these results have been confirmed and significantly extended by subsequent investigators.

Another early study demonstrating significant immune suppression in HIV-seronegative homosexual men was performed by W. Lawrence Drew, John Mills, Jay Levy, and their co-workers in San Francisco. They studied 109 homosexual men who were HIV seronegative when tested in 1985 and found that the presence of an active CMV infection was highly predictive of substantial immune suppression in these men. The 42 men who had no evidence of CMV infection had a normal mean T-helper/T-suppressor ratio of 1.92. The 67 men who had active CMV infections had a significantly depressed mean T-helper/T-suppressor ratio of 1.03. More important, 35 of these men (2 from the CMV-free and 33 from the CMV-infected group) had persistently inverted T-helper/T-suppressor ratios of less than 1.0 that lasted an average of nearly a year. Seven of these men had ratios of less than 0.4, a figure that represents extreme immune suppression. Twelve CMV-free men became infected with CMV during the course of the study, and all 12 developed inverted T-helper/T-suppressor ratios.[4] Drew and his colleagues concluded, "Depending on the cutoff value used to specify a normal [T-helper/T-suppressor] ratio, 25 percent to 76 percent of homosexual men may have abnormal ratios. . . . These abnormalities appear to occur before, and independent of, infection by the retrovirus associated with the acquired immunodeficiency syndrome [HIV]. We suspect, therefore that cytomegalovirus infection is an important cofactor in the genesis of the syndrome and that cytomegalovirus infection of lymphocytes may be a necessary precursor for the subsequent full expression of infection by the syndrome-associated retrovirus."[5]

Further evidence supporting both a high incidence of immune suppression among HIV-negative gay men and the importance of CMV infection as a possible cause of this immunosuppression has been presented by Dr. Ann Collier and her collaborators from various hospitals in Seattle, Washington. They studied 180 homosexual men and 26 heterosexual men. Their data are virtually identical to those reported by Drew. Nine homosexual men who had no evidence of either CMV or HIV infections had normal T-helper/T-suppressor ratios of 1.96. Thirty-seven homosexual men who had developed CMV infections but not HIV infections had an average ratio of 1.00. One hundred nine individuals infected with both CMV and HIV had an average ratio of 0.78. Clearly the majority of the immune suppression in this group of homosexual men was due not to HIV but to CMV infection. Notably, three of thirty CMV-infected, HIV-negative men (10 percent) were anergic: they had no immunological response to four separate antigens that are commonly used to assess T-cell function. In other words, their immune suppression was virtually total, although they had no evidence of HIV infection. None of their healthy peers and none of the heterosexual controls studied by Dr. Collier and her colleagues were similarly anergic, and only 34 percent of those coinfected with both CMV and HIV were.[6] Most significant, Collier's data demonstrated that active CMV infection frequently, if not universally, precedes HIV infection. Thus, as Drew and his co-authors commented in 1985, "The timing of these two infections may be critical; the fully developed acquired immunodeficiency syndrome may occur only when retrovirus infection is superimposed on active cytomegalovirus infection of lymphocytes."[7]

Another major study has confirmed that HIV-negative homosexuals are at unusual risk for developing severe immune suppression but without fingering a possible culprit. J. N. Weber, A. J. Pinching, and their colleagues at St. Mary's Hospital and Medical School in London, where Alexander Fleming discovered penicillin more than sixty years ago, began studying a cohort of 170 symptom-free homosexual men in 1982.[8] Of these men, only 48 were HIV seropositive by the end of 1986. The other 122 men re-

mained HIV negative. Surprisingly, Weber found that "seronegative subjects were as likely as seropositives to have abnormal immunological tests."[9] As in previous studies, the average T-helper/
T-suppressor ratio was about 1.0 among HIV-negative men, almost exactly identical among HIV-seropositive but healthy men,
and about 0.8 among those actively progressing toward AIDS. Indeed, about 30 percent of the HIV-seronegative men in Weber's
group were consistently anergic, while about 60 percent of the
seropositive men also lacked immunological response to T cell antigens. In short, a significant reduction in T4 (T-helper) cells was
not specific for either HIV-seropositive or AIDS patients among
homosexual men in this study. Something other than HIV was
destroying some of these men's immune systems. Weber's data
suggested that contracting a sexually transmitted disease or continued drug abuse were both significant possible cofactors in disease progression.

Weber's colleagues, Lesley Roberts, Susan Forster, and A. J.
Pinching, further characterized this group of homosexual men in
1989, confirming that seronegative homosexuals had significantly
reduced T cell and B cell activity as compared with heterosexual
controls and were just as immunosuppressed as symptomless HIV-
seropositive homosexual men. "Thus," they wrote, "in symptomless infection, HIV does not appear to cause more impairment
than that seen in their uninfected peers. The cause of these abnormalities, and of the altered T cell subsets, in uninfected homosexual men is unknown. They could be due to intercurrent sexually
transmitted infections or possibly allogeneic stimulation [by
semen]."[10]

Many other studies have confirmed these results, but only four
more will be summarized here. In 1986, David M. Novick of the
Rockefeller University reported that chronic HBV infections, like
CMV infections, were associated with abnormalities in T cell subset ratios, natural killer (NK) cell activity, and T-suppressor activity in both HIV-seropositive and -seronegative gay men.[11] A year
later, Linda Pifer and her colleagues found borderline immunodeficiency associated with drug use, antibiotic abuse, and an unusually high disease load in 27 "healthy" homosexual men.[12] These
men had T-helper cell counts averaging between 500 and 700
compared with a normal range of about 1,000 to 1,200 in heterosexual controls. Similarly, H. W. Murray and his associates in

the Division of Infectious Diseases at Cornell University Medical College studied over 200 homosexual men and reported that among "asymptomatic homosexuals, abnormalities in T4+ cell number and interferon gamma generation [a measure of T cell activity] were similar irrespective of human immunodeficiency virus seropositivity."[13] In both HIV-positive and HIV-negative groups, T cell numbers and activity were significantly lower than in heterosexuals, although not nearly as low as in HIV-positive men with full-blown AIDS. These results strongly suggest, as do the many other studies summarized here, that some immune deficiency precedes HIV infection in most gay men; that HIV infection alone does not add significantly to this T cell deficiency; but that the subsequent acquisition of additional immunosuppressive factors, many in the form of further active infections and perhaps drug use and immunologic exposure to semen components, trigger a rapid increase in immune suppression.

These conclusions are given added weight by the final study to be summarized here, which consisted of an immunological and infectious disease evaluation of 100 "healthy" homosexual men in Trinidad in 1987 carried out by Robert Gallo, William Blattner, and their colleagues. Nearly all of the men in the study, whether they were HIV seropositive or not, had a significant depletion of T-helper cells. HIV-seronegative homosexual men (43 subjects) had an average T-helper/T-suppressor ratio of 1.0, and ranging as low as 0.3, and some of these men had as few as 220 T-helper cells per cubic milliliter of blood—nearly sufficient to qualify them as AIDS patients under the newest proposed definition. The median number of T-helper cells in the HIV-negative group was 560—well below the 1,000 to 1,200 expected in healthy men. Also surprising was the observation that some HIV-infected men had normal T-helper/T-suppressor ratios, as well as normal, or even elevated, numbers of T-helper cells. Thus, HIV alone did not uniquely signify concomitant immune suppression. The only group that did have significant immune suppression without exception were men concurrently infected with HIV and HTLV-I. Notably these men also had unusually high antibody titers to herpes simplex and CMV, suggesting chronic, reactivated infections with these viruses. The most severely immunosuppressed group were, as in other studies, actively infected with several immunosuppressive viruses. Interestingly, Gallo, Blattner, and their colleagues make no mention in the article of the significance of the finding that even the HIV-negative homosexuals were immune deficient or that some of these men had more severe deficiencies than were found in HIV-seropositive men in the same study.[14] In the light of the question of whether HIV is sufficient and necessary to produce acquired immune deficiency, I believe that these data are highly relevant.

Immune Suppression in HIV-Negative Drug Abusers

Intravenous drug users, like homosexual men, are characterized by immune suppression in the absence of HIV infection. Intravenous drug abuse has been known to be immunosuppressive since at least the 1970s when, at the very most, 2 or 3 percent of addicts were infected with HIV. Despite the rarity of HIV, S. M. Brown and his colleagues demonstrated that several measures of immunological function, including response to various antigens that elicit T cell responses, were all significantly low in the majority of heroin abusers whom they tested in 1973.[1] Additional studies by other laboratories conducted at the same time demonstrated antibody overproduction and the presence of autoimmune factors as well.[2]

Surprisingly, these initial observations of immunologic abnormalities were not followed up until 1979, when Robert J. McDonough, John J. Madden, and their colleagues in the Department of Psychiatry at Emory University in Atlanta, Georgia, began studying local addicts with the help of the CDC and other government-associated laboratories.[3] Blood samples were also collected from addicts in Chicago, Illinois, and Belmont, Massachusetts. Lymphocyte counts were compared for forty-four addicts and twenty-eight nonaddicts. A significant reduction in the ratio of T cells to B cells was observed in the addicts. T cells comprise about 70 percent of the total lymphocyte count in a healthy person but comprised less than 30 percent of the lymphocytes in addicts. This massive alteration of lymphocyte subsets lasted more than a month after total abstinence from drugs in five addicts who began treatment at the start of the study. McDonough also found that, contrary to previous observations, the number of B cells was slightly reduced in opiate addicts and that the number of undifferentiated lymphocytes was tremendously elevated, suggesting that opiates interfered with proper development of stem cells or pre-T and pre-B cells into active lymphocytes. In consequence, the addicts all displayed very poor responses to tests of T cell activity.[4]

Subsequently, other researchers at Emory, most notably R. M. Donohoe, performed more detailed studies of heroin-associated immune suppression. One pair of studies is particularly enlightening. In 1982, Donohoe obtained blood samples from more than a hundred intravenous drug abusers in Atlanta and Chicago. All had significantly depressed ability to respond to foreign red blood cells, a sensitive measure of clinically relevant immune suppression. The longer the period of addiction was, the greater was the T lymphocyte suppression. Moreover, as of 1982, 24 percent of Donohoe's study group had T-helper/T-suppressor ratios typical of AIDS patients (less than 1). Retesting of the same individuals in 1985, however, demonstrated that only 12 percent of the group were HIV seropositive. In other words, at least half of the addicts who were severely immunosuppressed had a cause of the immunosuppression other than HIV. Since the rate of HIV seropositivity among drug abusers skyrocketed between 1982 and 1985, it is probable that many more than half of these severely immunosuppressed individuals were actually HIV free.[5]

The McDonough and Donohoe studies have been amply confirmed and extended. Jenny Heathcote and Keith B. Taylor of Stanford University's Department of Medicine found that heroin addicts failed to produce leukocyte migration inhibition factor (LMIF), a protein necessary to cause lymphocytes and macrophages to accumulate at sites of infection.[6] Italian researchers led by M. A. Brugo demonstrated that T-lymphocyte populations were significantly altered in a group of heroin abusers studied during 1980 and 1981.[7] Notably, this study was carried out before HIV had been documented to have invaded the drug-using community in Italy. Yet another study by M. M. Reddy reported that HIV-negative intravenous drug abusers had significantly lower natural killer cell activity than controls. Reddy's results were extended by David Novick and Mary Jeanne Kreek of the Rockefeller University, who teamed up with other New York City researchers in 1988 to demonstrate that HIV-negative, infection-free heroin abusers have natural killer cell activity half that of methadone-maintained addicts and healthy non–drug using volunteers.[8]

Not only do the intravenous drugs taken by addicts appear to have direct immunosuppresive effects, so do some of the infections they acquire as a result of their addiction. Dr. R. D. de Shazo and his colleagues at the Tulane/Louisiana State University AIDS Clinical Trials Group in New Orleans reported in 1989 that HIV seronegative methadone-treated former drug addicts who had HTLV-I infections had "severely depressed" pokeweed mitogen responses—a common measure of T-helper cell activity—despite fairly normal overall numbers and percentages of lymphocytes. They concluded "that HTLV-I infection is associated with abnormalities in T cell-dependent B cell proliferative responses [that is, failure to synthesize antibody]. Furthermore, both long-term methadone use and HTLV-I infection are associated with abnormalities in the distribution of CD4+ cell subpopulations."[9] These results confirm the Gallo and Blattner study, which indicated that HTLV-I was an important cofactor in producing immune deficiency in homosexual men.

HTLV I infections may be only one of many similarly immuno-suppressing infections. Measurements of the incidence of infection among drug abusers, a surrogate measure for immune competence, demonstrate that there appear to be many causes of immune impairment in HIV-negative people. During 1988, C. R. Horsburgh, Jr., J. Anderson, and E. J. Boyko ascertained the incidence of infections in 270 HIV-seronegative intravenous drug abusers compared with 562 non-drug-abusing people. They found that intravenous drug users "have an increased incidence of infection compared to control subjects." Many of the infections were directly associated with needle use (such as hepatitis) or sexual activity (such as gonococcal infections). Since infectious agents such as hepatitis B can, themselves create immune suppression, each additional infection may contribute to an ever-more vicious circle. Part of this vicious circle is undoubtedly due to specific drug use, however, since Horsbaugh found that "heroin users, but not other [intraveneous drug users], had an increased incidence of infections not thought to be associated with needle use, suggesting impaired immunity."[10]

Intravenous drug abusers have also been found to have alterations in immunological markers for disease. Compared with non-drug-using heterosexuals, drug abusers "had a significantly greater level of antibody titers for antinuclear antibody . . . than did controls." Four percent of New York City drug addicts also have lupus erythematosus antibodies, suggesting serious autoimmune disease. "Abnormal serologic results were not significantly associated with HIV seropositivity."[11] HIV-negative addicts also had unusually high levels and incidence of *Toxoplasmosis* antibody. Notably, antinuclear antibody is a diagnostic sign of autoimmunity, and *Toxoplasmosis* one of the opportunistic infections associated with AIDS. These findings suggest that exposure to multiple infections in addicts has adverse effects on immune function in the absence of HIV.

Direct evidence of opportunistic infections in HIV-negative heroin addicts has also been reported. These include several cases of disseminated or systemic candidiasis[12] and a variety of skeletal infections, including skeletal tuberculosis.[13] These cases prove that the immune suppression experienced by HIV-seronegative addicts can and sometimes does lead to the same disease outcome as is found in HIV-associated AIDS.

In fact, deaths from non-AIDS pneumonias and septicemias among intravenous addicts have increased over the past decade at exactly the same rate as deaths from AIDS-associated infections (from 3.6 per 1,000 to about 14 per thousand), leading the CDC to acknowledge that it "cannot discern . . . to what extent the upward trend in death rates for drug abuse reflects trends in illicit drug use independent of the HIV epidemic."[14] This passage acknowledges that there is no evidence that HIV infection leads to higher rates of disease or death among drug addicts. Their unusually high rates of death are due to other debilitating causes.

Immune Suppression Among Infants in the Absence of HIV

Infants of addicts "inherit" their parents' immunodeficiencies, as we saw in Chapter 6. Thus, not surprisingly, the infants of drug addicts have an unusually high incidence of infection, hospitalizations, and death as compared with infants of non-drug-using parents[1] and display immunological abnormalities, whether their mother is infected with HIV or not.

Between August 1984 and March 1985, Dr. Kenneth W. Culver and his team in the Department of Pediatrics at the University of California, San Francisco, enrolled nineteen pregnant street addicts who admitted using drugs during pregnancy in a study of the immunologic effects of their addiction on their infants. Although only one of the nineteen mothers was HIV-seropositive, all of the drug-exposed infants were severely immunosuppressed at birth. These infants had only slightly more than half the number of white blood cells found in infants who were not exposed to drugs, and the total numbers of T cells and monocytes were correspondingly half those of the control group. Significantly low levels of T-helper cells were still a problem for the drug-exposed infants a year later. In addition, the few T cells that the drug-exposed infants did have responded relatively poorly to antigenic stimulation when compared with the responses of the control group. Cytomegalovirus and HIV infection and malnutrition were ruled out as possible causes of the immune suppression, since CMV was not cultured from any infant and all mothers showed normal weight gain. Only one infant was HIV seropositive. "We conclude," Dr. Culver wrote, "that infants of drug-abusing mothers have significant abnormalities in T cell immunity compared with normal infants and adults, which cannot be explained by either intrauterine infection or malnutrition . . . suggesting a direct toxic effect of the drugs on fetal immunologic development."[2]

Other studies have confirmed the basic thrust of Dr. Culver's findings. In one recent study, the infants of eighty-seven HIV-seropositive mothers, most of whom were intravenous drug abusers, were followed until at least fifteen months of age. Only ten of the eighty-seven infants were themselves HIV seropositive and developed AIDS. Notably, an additional ten infants had a series of nonspecific findings, including failure to thrive, persistent lymphadenopathy, persistent oral candidiasis (thrush), and poor development, all of which are "suggestive but not diagnostic of HIV infection." None of these ten infants was ever HIV-seropositive, however, and analysis of their blood lymphocytes for HIV DNA using the PCR technique(polymerase chain reaction—a test for HIV genes) failed to find any HIV genomic material.[3] Virtually identical results were found by Dr. Marguerite M. Mayers and her associates in the Bronx. They studied infants of addicts who were both HIV seropositive and seronegative. Several of the infants born to HIV-seronegative mothers had T-helper cell counts and helper/suppressor ratios as low as those of HIV-infected infants. The incidence of pneumonias and other respiratory tract infections was elevated for both groups compared with healthy controls. In addition, seven children born to HIV-seropositive mothers developed opportunistic diseases typical of AIDS despite the fact that these infants had no evidence of HIV infection themselves according to any test. Their immune suppression was due to factors other than HIV. Persistent anemia and growth retardation were the most highly correlated factors for opportunistic infection for these infants as well as for HIV-seropositive infants. Congenital infections with CMV, EBV, venereal diseases, and related factors were not studied.[4]

Data such as these make one wonder whether HIV is responsible for the development of AIDS among HIV-seropositive individuals. Perhaps HIV is just an additional burden on an already failing immune system. These data certainly undermine Anthony Fauci's contention that HIV alone is responsible for immune suppression and disease in infants at risk for AIDS: "The public will understand," he says, ". . . the mother who has the virus and gives it to her baby and the baby gets sick. While in the same hospital, the mother without the virus has a baby and there's no way her baby gets AIDS. The public will understand that."[5]

Immune Suppression in HIV-Negative Hemophiliacs

Hemophiliacs, like homosexuals and drug addicts, have chronic sources of immune suppression other than HIV that can lead to significant immune suppression. A notable example was published in September 1986 by Dr. P. M. Mannucci of the University Institute of Internal Medicine in Milan, Italy. An HIV-negative hemophiliac had died from extensive lesions of Kaposi's sarcoma. The patient was a thirty-year-old heterosexual man suffering from severe hemophilia A. He had developed a strong autoantibody response to factor VIII and, terminally, severe anemia. His symptoms required him to transfuse unusual amounts of factor VIII and to undergo frequent whole blood and platelet transfusions. He died of respiratory failure caused by disseminated Kaposi's sarcoma tumors in his lungs. He also had tumors in his brain, intestines, and pancreas. Three separate blood samples taken over a period of five months prior to the man's death were tested for HIV antibody using both ELISA and Western blot techniques. All the tests were negative. Whatever the cause of this man's demise, it was not HIV.[1] This case is similar to the dozens of AIDS-like deaths among hemophiliacs in the United States reported by David Aronson for the two decades prior to 1979.

In fact, evidence of immunosuppression in HIV-negative hemophiliacs is legion, although it is not usually so severe that opportunistic diseases are common.[2] David J. Lang and his colleagues in the Transfusion Safety Study Group of the City of Hope National Medical Center in Duarte, California, performed a long-term study that included 418 people treated with blood factor products for bleeding disorders. Of these people, 131 were HIV seronegative throughout the study. The study group found, however, that 24, or 18 percent, of these HIV-seronegative hemophiliacs had T-helper/T-suppressor ratios of less than 1. The lymphocyte counts and distribution of these 24 people were indistinguishable from their HIV-seropositive peers.[3]

Other studies of hemophiliacs confirm that significant T cell suppression can occur in the absence of HIV infection. Dr. Jin, Dr. Cleveland, and Dr. Kaufman of the Michigan State University Medical School report that some immune abnormalities exist in even untreated hemophiliacs, and their immunodeficiency is exacerbated by factor treatment.[4] They found no difference in the degree of immunodeficiency between HIV-positive and HIV-negative hemophiliacs in their study. Similarly, W. S. Mahir and his colleagues in the Department of Haematology of St. George's Hospital Medical School in London observed that the responses to phytohaemogglutinin, a T cell–stimulating agent, were uniformly low among all of the hemophiliacs whose T cells were tested, regardless of whether they were HIV seropositive. Moreover, some (but not all) of the patients (both HIV seropositive and seronegative) displayed very low levels of activity against CMV and HSV 2 antigens when compared with other patients and with healthy controls.[5] Abnormally poor T cell responses to *Mycobacterium tuberculosis* antigens and poor production of lymphokines (immune-stimulating proteins) have also been observed.[6] These observations clearly demonstrate that T cell deficits in hemophiliacs are translated into measurable defects in lymphocyte activity that are found in both type A and type B hemophiliacs.[7]

The magnitude of the resulting immune impairment has been documented clearly in type B hemophiliacs treated with factor IX.

HIV-negative type B hemophiliacs who used factor IX more than once a week had significantly lower numbers of T-helper cells than did those using factor IX less than once a week (731 T-helper cells per cubic millimeter of blood versus 1,037 T helper cells per cubic millimeter). The same depletion of T helper cells was seen in HIV-positive individuals as well. Those who used factor IX more than once a week had an average of 515 T-helper cells per cubic millimeter of blood, as compared with 726 T-helper cells per cubic millimeter for those using factor less than once a week.[8] What is particularly interesting about these figures is that they demonstrate that frequent factor use by HIV-negative hemophiliacs results in a depletion of T-helper cells that is just as great as the T-helper cell depletion attributed to HIV infection alone (731 cells per cubic millimeter versus 726 cells, respectively). Thus, just as in homosexual men, many hemophiliacs have the same degree of immune suppression in the absence of HIV infection that they do following asymptomatic infection.

Even more interesting are the results of a prospective study of twelve hemophiliacs and six normal males begun in 1983. The baseline values for T cells were established for these men between October 1983 and May 1984. At that time, the six type A hemophiliacs had 452 T-helper cells per cubic millimeter of blood, the six type B hemophiliacs had an average of 505 T-helper cells per cubic millimeter, and the six normal males had an average of 1,157 T-helper cells per cubic millimeter. None of these individuals was HIV positive or had symptomatic AIDS at that time. By August 1988, all but three of the hemophiliacs (who declined HIV testing) had become HIV seropositive, and five of them had developed AIDS.[9] In other words, this study clearly demonstrates that severe T cell depletion preceded the occurrence of HIV infection among these hemophiliacs.

Other indications that factor-related immune suppression may predispose hemophiliacs to unusual infections also exist. Skin diseases are much more common in AIDS patients than in immunologically normal patients. Dr. N. R. Telfer and his colleagues at the Slade Hospital in Oxford, England, undertook a study of hemophiliacs to determine the prevalence of these skin diseases in this risk group. Not surprisingly, they found that HIV-seropositive hemophiliacs had unusually high incidences of seborrheic dermatitis, folliculitis, candidiasis, and fungal infections of the skin and nails. All of these infections have been identified as symptoms of the AIDS-related complex preceding full-blown AIDS among homosexual AIDS patients. What was surprising was that several HIV-seronegative hemophiliacs also had seborrheic dermatitis and folliculitis. No such cases were found in studies of healthy heterosexual control groups.[10]

These studies all agree, that significant immune suppression is present in almost all hemophiliacs, regardless of their HIV status. It is impossible to attribute the immune suppression associated with many cases of hemophilia to HIV. On the contrary, it appears likely that immune suppression was almost always present in hemophiliacs prior to their exposure to HIV and exacerbated such an infection once present. Thus, at an August 1992 meeting of experts convened by the CDC to discuss the origins of HIV-negative AIDS, James Mosely of the University of Southern California argued that hemophilia itself might be the risk, and Anthony Fauci conceded that such cases may have always existed.[11]

Immunosuppression in HIV-Negative Heterosexuals

The same agents that cause immune suppression in AIDS patients can also cause immune suppression in non-AIDS patients. Since surgery, anesthesia, and blood transfusions are all causes of T cell suppression, one would expect to find that patients having all of these risk factors are also prone to opportunistic infections. In fact, according to Gabriel Virella and Hugh Fudenberg, "opportunistic infections with 'nonpathogenic' organisms such as *Staphylococcus epidermidis* and *Candida albicans* are a definite risk after surgery, both in immunosuppressed and in apparently immunocompetent individuals. Cardiac surgery is particularly associated with a significant risk of infective endocarditis and/or septicemia, involving bacteria and fungi. . . . Immunodepression induced by surgery itself leads to a high risk of infection. Other types of invasive procedures, such as intravenous fluid therapy, parenteral alimentation and urethal catheterization, involve a high risk of opportunistic infections, particularly if other predisposing factors, such as malnutrition, are present. Some specific surgical interventions, such as jejunoileal bypass [in which the upper intestines are bypassed to allow, for example, ulceration to heal], can also lead to a depressed immunocompetence that can be reflected by increased incidence of histoplasmosis and tuberculosis. It is likely that malnutrition secondary to malabsorption due to the bypass is the primary cause of immunoincompetence in these patients."[1] Surgery, particularly when accompanied by blood transfusion, carries an increased risk of various immunosuppressive viral infections, including hepatitis and cytomegalovirus. Notably, the presence of cytomegalovirus infections greatly increases the probability of developing opportunistic infections and pneumonia in both kidney transplant patients and heart transplant recipients.[2]

These observations create a conundrum with regard to AIDS. People who develop histoplasmosis, disseminated candidiasis or tuberculosis, or CMV infections after transplant surgery can and do develop all of the symptoms of AIDS but are not classified as AIDS patients unless they also have the misfortune of becoming infected with HIV as well. When such people develop an HIV infection, there is no difference in their symptoms, but they are then considered AIDS patients. Why? In both cases, the primary modes of immune suppression are well known: surgery, chemotherapeutic immune suppression, possibly blood transfusions, possibly graft-versus-host disease, and certainly viral influences. In both cases the outcome is the same: death by opportunistic infection. Clearly HIV seropositivity or antigenemia is not a differentiating factor in the prognosis of these patients. Other viruses, such as CMV, are just as predictive of superinfection.

AIDS may have been mistaken for such multiple-infection induced fatalities and other unusual medical problems in the past. A significant number of syndromes mimic various aspects of AIDS so closely that they can be confused. The presence or absence of HIV is often the only distinguishing feature. For example, Dr. Anthony B. Minnefor and Dr. James M. Oleske reported a 1979 case involving a five-year-old black girl who died of what they diagnosed at the time as being Nezelof's syndrome, a genetically inherited defect of the immune system specifically leading to decreased numbers of T lymphyocytes but not affecting B cell numbers or antibody production. The result is identical to AIDS. Patients, usually very young, develop a series of viral, parasitic, and fungal infections and usually die by their third or fourth year of life. The child Minnefor and Oleske saw fit the description of Nezelof's syndrome exactly, but when AIDS was recognized in 1981, the

physicians tentatively revised their diagnosis in the light of some
previously uninteresting information. The mother of the dead girl
was a sixteen-year-old, promiscuous, intravenous drug abuser suffering from idiopathic thrombocytopenia (autoimmune destruction of her own red blood cells). The girl herself also suffered from
thrombocytopenia. Both mother and daughter had been given
multiple blood transfusions. Because of these multiple risks associated with AIDS, Minnefor and Oleske suspect that both the
mother and their patient were among the earliest AIDS patients,
having contracted the disease in 1974. If that is true, then other
cases of Nezelof's syndrome may also represent either pre-1979
cases or HIV-free cases of AIDS.[3] This speculation is given weight
by the existence of a reasonable number of children, aged two to
sixteen, who have been diagnosed with a syndrome very similar
to Nezelof's, called "benign combined immunodeficiency." "The
syndrome refers to children with typical clinical manifestations of
a severe T cell defect such as oral *Candida*, severe varicella
[chicken pox], and progressive vaccinia, who survive beyond the
age of 2. Most of these children have normal or raised immuno-
globulin levels; a few have low IgA. However, humoral [T cell]
immunity is impaired."[4] The cause of benign combined immunodeficiency has not been established.

Another unexplained phenomenon resembling AIDS is idiopathic Beck-Ibrahim syndrome, diagnosed by the presence of
chronic fungal infections, especially *Candida*, of the fingers, toes,
skin, mouth, and hair. Most cases are due to inherited immune
deficiencies that run in families. However, the late-onset form,
which occurs in adults over the age of twenty, is not genetically
inherited, and its etiology is unknown.

These various diseases tell us at the very least that many as-yet
unidentified causes of immune suppression exist. To what extent
multiple concurrent infections such as those in which combined
Candida and *Mycoplasma* result in severe depletion of lymphocytes are clues to AIDS, and to what extent AIDS may reveal clues
about the etiologies of these mysterious syndromes, remains to be
seen. Whatever the solution to these mysteries turns out to be,
they should remind us of Joseph Sonnabend's admonition that
there may be many causes of acquired immune deficiencies; AIDS
may not have a single cause or be a single disease.

Of one thing only can we be certain: the astonishing extent of our ignorance about what causes immune suppression and who may be at risk for it. For example, malnutrition is one of the best-studied causes of immune suppression. Anorexia nervosa and bulimia, two eating disorders common to teenage girls and more rarely found among women and men of other age groups, is a well-known cause of malnutrition. A comprehensive search of available computerized databases has not, however, revealed a single study of the immune status of anorexic or bulimic patients. This fact does not mean that such studies have not been done, since computerized searches are not infallible, but it does mean, at best, that very little is known about these causes of malnutrition and the possible risks they may confer for the development of AIDS.

Similarly, it is well known in medical circles that immune suppression is common in diabetics. The manifestations include depressed macrophage function and poor T cell–mediated immunity, leading to increased rates of tuberculosis and other infections.[5] One would expect that diabetics might have an unusual incidence of opportunistic infections typical of AIDS. This prediction is, in fact, correct.[6] Diabetics contract various opportunistic parasitic infections (including protozoa and amoeba) at unusually high rates,[7] as well as Kaposi's sarcoma. Although the incidence of diabetes mellitus is estimated to be about 1 percent of the American population, 4 (4 percent) out of 100 patients with Kaposi's sarcoma examined at one hospital in 1973 had concurrent diabetes mellitus.[8] A recent cluster of five elderly, HIV-negative patients with *Pneumocystis* pneumonia included two who were diabetics, another who had had open heart surgery, a fourth with chronic obstructive lung disease, and one man without obvious predisposing factors.[9] Similarly, diabetics share with immunocompromised, severely burned, and intravenous drug-abusing patients an unusually high incidence of fungal[10] and *Pseudomonas aeruginosa* (a bacterium) infections.[11] The medical literature reports dozens of cases of esophageal candidiasis and other forms of disseminated and chronic candidiasis in men whose only reported risk factor was diabetes. These cases have been reported since the 1960s and are often associated with bladder and blood infections of bacteria.[12]

Diabetics are also unusually susceptible to disseminated herpes virus infections. In 1962, Dr. Brian E. Heard and his colleagues in the Department of Pathology of the Postgraduate Medical School in London reported the death of a previously healthy diabetic from combined cytomegalovirus and aspergillus infections of the lungs.[13] A similar case was presented by Dr. Simon Lauzé of the University of Montreal's Hôpital Notre-Dame in 1961.[14] These are apparently the first such cases in the medical literature. Of twenty-eight patients with eruptive herpes zoster infection seen by Dr. I. Neu and Dr. S. Rodiek during the mid-1970s, sixteen were diabetics between the ages of twenty-five and eighty-five years.[15]

These data clearly demonstrate that diabetics have been contracting the same opportunistic diseases typical of AIDS for at least several decades and that they continue to do so today, even in the absence of HIV infections. Thus, diabetics may represent an unrecognized group at high risk for both acquired immune suppression (whether we want to call this "AIDS" or not) and for HIV infection. This risk is magnified by the fact that diabetics use needles for insulin injection and, like drug addicts, sometimes share their injection equipment or fail to sterilize it properly.[16] Amazingly, there are apparently no studies of the incidence of AIDS among diabetics, and no research concerning the possibility that diabetics may be at higher risk for AIDS than other heterosexuals. My guess, and it is no more than a guess, is that if the CDC data are reanalyzed in terms of the additional heterosexual risk factors for immune suppression elaborated here, the majority of the "no known risks" cases will reveal identifiable immunosuppressive risks not yet associated with AIDS.

Non-HIV Markers of Developing Immune Suppression in AIDS

The existence of significant immune suppression in groups at high risk for AIDS raises some important diagnostic issues. Perhaps the most important of these is that HIV may not be the most useful or even a necessary measure of the development of AIDS. HIV antigenemia may be, like the increasing levels of carbon dioxide in the blood of drowning swimmer, a reasonable measure of how close a person is to death, but if AIDS is as complicated as the picture that is being painted here, then there should be other measures of the progress of AIDS as accurate as HIV antigenemia.

T cell levels, increased antigenemia of other infectious agents, degree of autoimmunity, and other factors causing immune suppression should also be indicative of the status of an AIDS patient, just as one can measure the amount of oxygen, time under water, and other factors for a drowning person.

Papers describing HIV-related markers for the progression of AIDS are legion. A very good correlation between the amount of p24 antigen (an HIV protein) and the degree of immune dysfunction found in an AIDS patient has been reported from many laboratories. The more of this protein that is present in AIDS patients, the worse off they are. This fact is often used to bolster the argument that HIV must be the cause of the primary immune suppression in AIDS. That, indeed, is one possible explanation of the data. Another interpretation is suggested by two observations already mentioned. The first is that HIV can be transactivated by many other infectious agents associated with AIDS, including CMV, HTLV-I, and herpes simplex viruses, and it has shown synergistic effects in the presence of Epstein-Barr virus, hepatitis B virus, *Mycoplasmas*, and various drugs. The second observation is that several of these viruses extend the range of cells infected by HIV, so that the proportion of host cells infected increases with the presence of cofactor infections. It follows that the more cells that are infected, the more HIV proteins will be produced. It also follows that the degree of cofactor infection should also rise in direct proportion to increasing immune suppression.

In fact, although very few studies have been performed, cytomegalovirus appears to be as good a marker for increasing immune incompetence as HIV. R. J. Biggar and his colleagues reported in 1983 (prior to the isolation of HIV) that a very good correlation existed between the excretion of CMV in the semen of homosexual men and the degree of their immune suppression.[1] This work has recently been updated by M. Fiala and his colleagues, who report that the amount of CMV present in blood or excreted in urine or semen increases with disease progression in AIDS patients: "The intensity of CMV viremia appears to be inversely proportional to the decline in immune responsiveness of patients."[2] Thus, in CMV-infected AIDS patients (who constitute more than 95 percent of all such patients), the amount of CMV replication is just as good a marker for disease progression as is any measure of HIV replication.

So is Epstein-Barr virus. In 1986, Charles R. Rinaldo, Jr., and his co-workers demonstrated that homosexual men who seroconverted to HIV simultaneously experienced a fourfold increase in antibody titers to EBV VCA antigen (virus capsid antigen). Furthermore, they documented a direct correlation between HIV antibody titer and EBV antibody titer. The higher the one, the higher the other.[3] Similar results have been reported by Sonnabend and his colleagues, who report that nonspecific enhancement of an antibody type called IgA preceded and was predictive of decline in T-helper cell numbers.[4] EBV activation results in just such IgA activation. Whether other viruses associated with AIDS, such as HBV or HTLV-I, are similarly predictive of disease progression remains to be seen, since no one, as far as I can tell, has every bothered to look. This failure to look has left us in the position of assuming that HIV is the only valid measure of disease progression in AIDS, without the scientific benefit of having checked the assumption.

One other notable measure of progressive immune deficiency must be mentioned. Chapter 5 extensively documented data indicating that the presence of lymphocytotoxic antibodies was directly correlated with progressive immune decay, while the absence of such antibodies, even in HIV-seropositive individuals, was indicative of a positive prognosis.[5] The same correlation has been observed between the presence of circulating immune complexes (one sign of autoimmunity) and T-helper cell depletion.[6] These data are once again consistent with a non-HIV-mediated cause of immune suppression and suggest that appropriate measures of the extent of the autoimmune reactions occurring in AIDS patients might prove an excellent indicator of their immune status. We will not know until we look.

CHAPTER 8

WHY AIDS IS EPIDEMIC NOW

AIDS is a Social Disease

Some people are far more susceptible to AIDS than others, and the reasons are far from mysterious: immunological exposure to semen, blood or other alloantigens; multiple, concurrent infections; prolonged medical or illicit drug use; malnutrition; and so forth. None of these risk factors is new, however. Why, then, has AIDS become epidemic only recently?

The recent spread of AIDS can be understood only in terms of one of the most basic principles of epidemiology: diseases that are transmitted by exposure to blood or through sexual means are social diseases. It is impossible to understand such diseases from a purely medical, biological, or laboratory perspective. Indeed, Jonathan Weber, a senior lecturer at the Royal Postgraduate Medical School in London, has recently made the point with regard to syphilis: "*Treponema pallidum*, the bacterium that causes syphilis, affects 14 men to every woman in Britain. The proportion is similar in other parts of Europe. Biology per se does not affect the distribution of this bacterium [that is, it does not have an intrinsic 'preference' for men over women], because *T. pallidum* will readily infect both sexes. Instead, social factors are the cause of the unequal distribution of syphilis between the sexes. The proportion of cases of syphilis between males and females 50 years ago in Britain was about 50:50. The difference today is explained by the near eradication of syphilis in the heterosexual community but the maintenance of a reservoir of syphilis among homosexuals. Sociology, not biology, explains the distribution of sexually transmitted agents."[1] I believe that the same conclusion can be reached about AIDS. To understand AIDS, we must document and understand the sociological changes in homosexuality, drug use, and medical practice that have created the conditions that allowed the syndrome to explode into prominence during the past decade.

Coming Out of the Closet

Social revolutions always have medical consequences. The rise of the first cities and the invention of sailing vessels capable of exploring new continents set the stage for transporting lice-infested rats into crowded urban settings where bubonic plague could devastate medieval Europe. The invention and dissemination of birth control devices in this century has allowed widespread control of procreation but also sexual freedom and a subsequent spread of sexually transmitted diseases. It is well to remember that AIDS was presaged by prior epidemics of herpes simplex, *Chlamydia*, gonorrhea, and syphilis. The Stonewall riots in New York City, the 1969 crucible from which the movement for gay liberation was cast, created another social revolution that is no exception to the medical rule. "Coming out of the closet" has altered not only our social perception of homosexuality but its medical face as well.

The sociological manifestations of homosexuality have changed radically in the recent past. As Jonathan Weber noted, the incidence of syphilis a few decades ago was almost exactly equal between men and women but is now found mainly in homosexual men. Since homosexuality is almost surely as old as humanity and is present in almost every society, the unusually high incidence of syphilis among homosexual men today cannot be ascribed to homosexuality per se but to significant changes in homosexual behavior in the recent past. New expressions of homosexuality concomitant with the gay liberation movement have created an unusual and new disease profile for gay men.

The medical literature is quite explicit about some of these new manifestations of gay male life. Sometime during the mid-1970s "D. F.," a thirty-two-year-old homosexual man, was admitted to the emergency room of the Lenox Hill Hospital in New York City. He complained of severe abdominal pain that had begun some eighteen hours earlier while having sex with his male partner. The pain had begun when his partner had pushed his clenched fist into

D. F.'s rectum to the level of his upper forearm. At that time, D. F. had experienced a sudden sharp pain in his abdomen and referred pain in both shoulders as well. Blood was evident when the partner removed his hand. When the pain had not subsided after several hours, D. F. had sought medical assistance.

D. F. was seen by two surgeons at Lenox Hill, Michael A. Weinstein and Norman Sohn. They determined that D. F. had a slight fever and a tender abdomen, suggesting abdominal infection. A laparotomy, or incision through the abdominal wall, revealed that D. F.'s colon had a ten-centimeter-long tear that had allowed fecal material to contaminate the peritoneal cavity. D. F. was treated with a battery of antibiotics to combat his infections, and a series of operations were performed by Weinstein and Sohn. These included sewing up the colon injury and a colostomy to create an artificial anus in D. F.'s side so that his colonic injury could heal. The colon was later reattached to D. F.'s natural rectum, and he recovered successfully.[1]

Rectal injuries following anal sex of various types, and in both men and women, were not unknown to the medical community prior to the mid-1970s, but they were almost always associated with the improper use of enema syringes, vibrators, dildoes, butt plugs, fruits, vegetables, and other sexual instruments in conjunction with veneral diseases of the rectum.[2] In fact, the first medical report of proctological lesions as a common correlate of homosexual activity came from a physician in New York City in 1964 who noted that among eighteen avowedly homosexual patients, he had observed four instances of anal ulceration associated with trauma (injury caused by a foreign object), five instances of granulomatous anorectal lesions, five instances of nonspecific anorectal lesions, and five cases of perianal condyloma acuminata (an anal wart associated with papilloma virus infection).[3] He described these symptoms as very unusual. A decade later, such symptoms were no longer surprising. Of 260 gay men, comprising 10 percent of a private proctological practice in New York City in 1976, 134 patients had condyloma acuminata, 43 hemorrhoids, 31 nonspecific proctitis, 30 anal fistulas (secondary openings through the anus), 18 rectal abscesses (inflamed infections that have filled with pus), 18 anal fissures, 17 amebiasis, 16 pruritis ani, and 7 "trauma and foreign bodies."[4] It is now accepted that such injuries and infections greatly increase the risk of concurrent infections (HIV or otherwise) and of semen gaining access to the immune system following anal intercourse.

Despite previous experience with these other types of medical problems associated with anal intercourse, rectal or colonic injury following the insertion of a partner's fist was a new observation in the mid-1970s.[5] It was to be only one of many. Dr. Sohn and Dr. Weinstein eventually treated eleven patients for complications similar to those of D. F. between 1973 and 1977 and wrote a benchmark paper on such "social injuries of the rectum." "With the rapid emergence of the new sexual mores and permissiveness in our society," they wrote, "as well as a greater acceptance and understanding of sexual deviation by the general public, the surgeon is now confronted with new problems in diagnosis and treatment of unusual anorectal injuries. One such newly recognized injury is [due to what is] aptly called 'fist fornication.' This refers to the practice of introducing the closed or clenched fist into the rectal ampulla, upper rectum, and sigmoid colon to achieve sexual gratification."[6] All eleven of the patients seen by Dr. Sohn and Dr. Weinstein were homosexual men, and all had previous experience with anal insertion of sexual instruments and the fingers of sexual partners. All reported that their experience with fisting was extensive but of relatively recent inception, which probably explains why the paper by Sohn and Weinstein is the first to describe this practice and its complications to the medical community.

Dr. Sohn and Dr. Weinstein were sufficiently intrigued by this new complication that they enlisted the help of psychiatrist Joel Gonchar to survey the prevalence of fisting among New York homosexuals. A surprising proportion of gay men had apparently taken up this new form of sex. Sixty of 200 homosexual men (30 percent) provided with a questionnaire responded in 1977. All sixty had participated in fisting, most as both the active and the passive partner. Almost all admitted to having used one or more drugs, including amyl nitrite, marijuana, LSD, mescaline, and alcohol, prior to engaging in fist fornication. Subsequent studies by other physicians suggested that the use of amyl or butyl nitrites was particularly associated with anal intercourse and fisting, since these drugs act both as aphrodisiacs and as muscle relaxants that facilitate such sexual acts. Five percent of the survey group had been hospitalized with consequent major complications, and the majority of the responders, despite engaging in this activity, correctly questioned its safety. Sohn, Weinstein, and Gonchar concluded their study with the warning that "the general medical community should be made aware of this potentially dangerous sexual practice so that significant pathologic features are not missed. . . . In addition, practicing homosexuals should be made aware of the inherent dangers of this practice."[7]

Despite the dangers, subsequent surveys performed during the 1980s found that approximately 40 percent of gay men had engaged in fisting their partner and about 15 percent admitted to having had their partner's hand inserted into their rectum. Only about 15 percent did not at all engage in fisting or receptive anal intercourse.[8]

Injuries due to fisting were not the only new medical problem that emerged following gay liberation. Dr. Sohn was also one of the authors of a paper written with Dr. Henry Kazal and several of their colleagues identifying another new and unusual syndrome in 1976. They called this new conglomeration of lower gastroentestinal problems the gay bowel syndrome, since the diagnostic symptoms were found with unusual frequency among homosexual men but with less frequency among people of other sexual preferences. Gay bowel syndrome consists of some combination of condyloma acuminata; pruritis ani; anal syphilis or gonorrhea; actively bleeding hemorrhoids, anal fissures, ulcers, and abscesses (often resulting from fisting or anal intercourse); hepatitis; and intestinal infections of shigellosis, giardiasis, and/or amebiasis. The syndrome was highly associated with oral-genital, oral-rectal, and rectal-genital contact, and the vast majority of the patients were promiscuous individuals between the ages of twenty and forty.[9] Those suffering from gay bowel syndrome were the exact group who would also be identified a few years later as being at highest risk for AIDS. Indeed, studies performed around 1980 revealed significantly higher rates of anal intercourse, fisting, and oral-genital contact among homosexual men in San Francisco, New York, and Los Angeles (cities shortly to develop high rates of AIDS) than in Denver and St. Louis (cities in which AIDS has been relatively rare).[10]

These observations of new syndromes associated with a very active male homosexual life-style suggest that both the type of sexual activity and the extent of promiscuity associated with it changed markedly during the 1970s. I must emphasize that this is a gross generalization that does not apply to every homosexual man. Some homosexual men are monogamous or have had only a handful of sexual partners during their lives. Some have never engaged in anal intercourse of any type. Many who have engaged in anal-genital intercourse have never engaged in fist fornication. Nonetheless, there was a documentable increase in risky behaviors among gay men immediately preceding the explosion of AIDS.

Much of this new-found sexual liberation and promiscuity was fed by new institutions. According to J. Weeks, who has studied the emerging social acceptance of homosexuality, "The 1970s did witness an explosion of what has been described as 'public sex' amongst gay men, with the appearance in most of the major American, Australian and European cities (with the partial exception of Britain where the laws remained restrictive) of such facilities as bath houses, backroom bars and public cruising areas where casual, recreational sex with multiple partners became the norm. . . . It clearly represented some form of decoupling of sex and intimacy, and a normalisation in a new way of sex as recreation and pleasure."[11] It also appears that for many gay men, establishing the gay liberation movement also meant developing uniquely homosexual types of sex (such as fisting) and specific places, such as bath houses and sex clubs, in which this liberation could be explored. Sex became, in a sense, a political statement.[12] Indeed, Michael Callen recalls hearing a lecture by Edmund White, a co-author of The Joy of Gay Sex, in which White proposed that "gay men should wear their sexually transmitted diseases like red badges of courage in a war against a sex-negative society." Callen recalls thinking, "Gee! Every time I get the clap I'm striking a blow for the sexual revolution!"[13]

The extent of this sexual revolution is revealed by the figures. A large study of homosexual men in San Francisco performed during the mid-1980s found that no one reported having had a single lifetime sexual partner; 8 percent reported between two and ten lifetime sexual partners; 17 percent reported between eleven and fifty lifetime sexual partners; and 75 percent reported more than fifty lifetime sexual partners.[14] Homosexual AIDS patients have often been—though not always—among the most promiscuous of their brethren, sometimes reporting thousands of lifetime sexual partners.[15] The figures for heterosexual lifetime partners contrast markedly. Eight percent of the heterosexual men interviewed in Atlanta, Georgia, during the early 1980s reported having only one lifetime sexual partner; 48 percent reported having between two and ten partners; 32 percent between eleven and fifty partners; and 12 percent more than fifty. Heterosexual women were even less likely to experiment with sexual partners. Twenty-two percent reported having had only one partner; 55 percent between two and ten; 20 percent between eleven and fifty; and only 3 percent more than fifty.[16] Although very public figures such as Magic Johnson and Wilt Chamberlin rival the most promiscuous of homosexual men, the figures clearly show that such heterosexual promiscuity is far rarer than is homosexual promiscuity. (See Figure 16.)

Although gay men have not always been promiscuous (Kinsey and his colleagues reported in 1948 that homosexual men had few sexual partners),[17] by the 1980s many gay men were having sexual relations with several, sometimes anonymous, partners each week, especially in major cities such as New York, San Francisco, and Los Angeles.[18] This transformation of gay life is mirrored in medical records of venereal diseases.

The incidence of venereal diseases has long been recognized to be a sensitive indicator of levels of promiscuity.[19] Rates of venereal diseases began a noticeable climb during the mid-1950s, as ad-

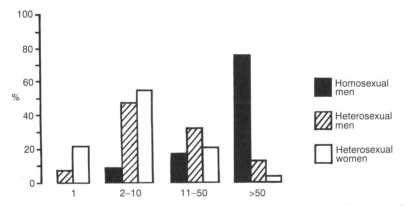

FIGURE 16. Number of lifetime sexual partners of heterosexual men and women in Atlanta, Georgia, and homosexual men in San Francisco.
SOURCE: Nahmias AJ, Lee FK, Beckman-Nahmias S. 1990. Sero-epidemiological and -sociological patterns of herpes simplex virus infection in the world. *Scand J Infec Dis* Suppl 69:19–36, fig 5.

vances in birth control became widely available, and they skyrocketed during the 1970s. Whereas the increase was found
among both men and women during the 1950s and 1960s, the
vast increase in new cases of venereal diseases during the 1970s
was found almost entirely in homosexual and bisexual men and
has been directly attributed by the medical community to the consequences of gay liberation.[20] The title of an article in the *Journal
of the American Medical Association* in 1977 by Dr. S. Vaisrub
said it all: "Homosexuality—a risk factor in infectious disease."[21]

Analysis of the increases in specific venereal diseases provides a
detailed look at the growth of homosexual promiscuity. Studies of
the prevalence of syphilis around 1960 demonstrated that in Los
Angeles and Vancouver, at least 70 percent of cases among men
were associated with male homo- or bisexual activity.[22] When the
rate of syphilis among white males in the United States increased
by 351 percent between 1967 and 1979, it was found to be due
in very large part to increased homosexual activity.[23] Cases of gonorrhea increased from 259,000 in 1960 to 600,000 in 1970, to
over 1 million in 1980.[24] Again, health officials attributed this
growth in large part to increased homosexual activity. The number of hepatitis B cases also increased dramatically, from 1,500 in
1966 to 8,300 in 1970 to 19,000 in 1980 and 25,000 in 1987.[25]
Although some of this increase is undoubtedly due to intravenous
drug use, many AIDS patients have both homosexuality and drug
abuse as risks, and evidence from Europe and America suggests
that the increase is mainly due to disease spread among homosexual and bisexual men.[26] A similar increase in hepatitis A during
the same period was also noted among homosexual men but in no
other part of the population.[27]

Other sexually transmitted diseases showed similar huge increases among homosexual men but only small increases among
heterosexuals. For example, Dr. André Nahmias of the Emory
University School of Medicine in Atlanta, Georgia, has reviewed
the incidence of herpes simplex viruses between 1960 and 1980.[28]
Among his striking observations are that the incidence of diagnosed genital herpes (herpes simplex 2) in the general population
of both the United States and England rose sevenfold from about
1 in a 1,000 patients in 1965 to nearly 7 in 1,000 patients in 1980.
During the same period of time, the incidence of genital herpes
only doubled (from about 0.5 in 1,000 patients to 1 in 1,000 pa-

tients) in women. That means that the huge growth in genital herpes was mainly due to increased promiscuity among men not engaged in intercourse with women. In fact, by 1978 the incidence
of HSV 2 infections was ten times higher in homosexual men in
the United States than in comparable women, and the incidence
of infection among gay men nearly doubled during the next seven
years.[29] By 1985 homosexual men in their early twenties had an
incidence of genital herpes infections eight to twenty times that
of heterosexual males of the same age and socioeconomic status
throughout Western Europe and North America.[30]

Data on increases in other viruses are more difficult to detail
definitively, but a comparison of reports from the beginning of the
AIDS epidemic with ones from a decade later suggest that infections such as cytomegalovirus (CMV) also spread rapidly. A 1980
study of the prevalence of CMV among homosexual men in San
Francisco revealed that only 14 of 190 (7 percent) homosexual
men were excreting CMV in their urine and 27 of 190 (14 percent)
had CMV cultured from their blood, semen, or urine.[31] A 1988
study in New York found that the rates of active infection had
almost tripled. Thirteen of 100 homosexual men were excreting
CMV in their urine, and 41 had it cultured from one or more
secretions.[32]

Some gay men became unwitting guinea pigs for the elucidation
of how various diseases were transmitted. Diseases such as amebiasis, shigellosis, and giardiasis were not known to be transmitted
sexually prior to 1970. Their sexual transmission was first documented in gay men, and they are now known to be associated with
anal intercourse and anal-oral contact. Once again, these diseases
therefore provide measures of the increases in these types of gay
sex.

All of these diseases were rare in the United States and England
prior to the 1970s, with outbreaks always associated with
fecal contamination and poor public hygiene.[33] This picture
changed dramatically in the aftermath of gay liberation. Cases of
amebiasis reported to the CDC rose from a relatively constant
level of fewer than 2,500 prior to 1972 to more than 7,300 in
1982 (an increase of 320 percent).[34] The bulk of this increase was
among men, most, if not all, of whom were gay; what had been
just a handful of gay-related cases blossomed to many thousands
in only a decade. Schmerin and his colleagues at the New York

Hospital, for example, reported that amebiasis was identified in the stools of ninety-eight patients between 1971 and 1976. All of the forty-two women for whom information existed acquired the disease from travel outside the United States or were congenitally or therapeutically immunodeficient. Fifty-six patients were male. Thirty of these male patients had contracted the disease while visiting endemic areas outside of the United States. No information was available for six male patients. The remaining twenty male patients were all homosexuals who had not traveled outside of the New York metropolitan area. There were no cases of nontraveler amebiasis in 1971 in their sample; the number grew yearly thereafter. Almost half of these homosexual patients also had concomitant shigellosis, giardiasis, syphilis, gonorrhea, and/or hepatitis or a history of these diseases.[35] By 1977 physicians recognized that the presence of any enteric pathogen such as amebiasis or giardiasis in a man who had not traveled out of the country was likely to have resulted from homosexual activity.[36] What had been an extreme rarity had become a common enough symptom to be used diagnostically.

Some physicians saw what was happening even as it happened. Dr. H. Most and Dr. B. H. Kean, for example, noted that the Manhattan homosexual community had begun to display the unusual disease profile typical of "a tropical isle" or Third World country beginning about 1968.[37] Nearly 60 percent of all cases of shigellosis in the New York metropolitan area were among gay men by 1976[38] and an epidemic of the disease was noted in the gay community in San Francisco in 1974.[39] Schmerin and his collaborators found that nearly 90 percent of giardiasis cases among nontravelers living in New York were avowedly homosexual in 1977.[40] In the same year, it was reported that "physicians with homosexual male patients in New York City, San Francisco, and Boston have also noted a marked increase in enteric protozoal infections occurring in those who practice analingus."[41] And much the same picture began to emerge in Great Britain at the same time.[42] Not surprisingly, the recognition of an epidemic of lymphadenopathy in gay men has also been traced back to the decade prior to the recogniton of AIDS, but it went unnoticed at the time.[43]

These data demonstrate definitively that the gay liberation movement resulted in a great increase in promiscuity among gay men, along with significant changes in sexual practices that made rectal trauma, immunological contact with semen, use of recreational drugs, and the transmission of many viral, amoebal, fungal, and bacterial infections far more common than in the decades prior to 1970.[44] The same data strongly suggest that recent changes in sexual and drug activity played a major role in vastly enlarging the the homo- and bisexual male population at risk for developing immunosuppression. Since promiscuity, engaging in receptive anal intercourse, and fisting are the three highest-risk factors associated with AIDS among gay men[45] and since each of these risk factors is correlated with known causes of immunosuppression,[46] they represent significant factors in our understanding of why AIDS emerged as a major medical problem only after 1970.

One other very important change in homosexual activity also took place during the 1970s, at least in the United States: Male prostitution grew tremendously. During the second half of the 1960s, about 7,000 men were arrested each year on charges of prostitution—about one-quarter of the number of women arrested on similar charges. By 1976 the number of men had grown to 15,000 per year,[47] and by 1979 there were 25,000 arrests. More than 35,000 men were arrested on prostitution charges in 1987.[48] Thus, the number of men arrested yearly on prostitution charges increased by a factor of five over a period of two decades. Male prostitutes are now almost as common as female prostitutes.

Greatly increased male prostitution is important to understanding AIDS. Studies of male prostitutes suggest that they are among the most promiscuous of male homosexuals and that they are more likely to engage in high-risk types of sex without safer sex measures. Male prostitutes also have extremely high rates of drug use, even compared with other homosexual men, with as many as 75 percent in cities such as San Francisco being addicted to intravenous drugs.Male prostitutes may represent one of the most important reservoirs of all sexually transmitted diseases in the homosexual community and one of the greatest AIDS threats.[49] (See Figure 17.)

If we now go back and ask why AIDS emerged as a problem for gay men only in the past decade or so, despite the acknowledged antiquity of homosexuality itself, the answer becomes clear: AIDS became a problem for homosexual men only when rampant pro-

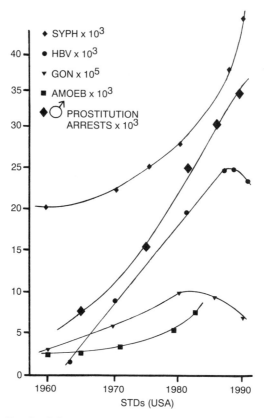

FIGURE 17. Graph of the increasing incidence of sexually transmitted diseases (STDs) and male prostitution arrests in the United States, 1960–1990. Cases of syphilis (SYPH ♦) in thousands; hepatitis B (HBV ●) in thousands; gonorrhea (GON ▼) in hundreds pf thousands; amoebiasis (AMOEB ■) in thousands; and male prostitution arrests (♦) in thousands.

miscuity, frequent anal forms of intercourse, new and sometimes physically traumatic forms of sex, and the frequent concomitants of drug use and multiple concurrent infections paved the way. As Mirko Grmek has concluded, "American homosexuals created the conditions which, by exceeding a critical threshold, made the epidemic possible."[50] This conclusion stands regardless of whether one wishes to interpret the social revolution of gay liberation as the means by which HIV has been spread, the vehicle for transmitting HIV with all of its necessary cofactors, or the direct cause of the immunosuppressive habits that have medically debilitated so many gay men.

The Explosion of Drug Abuse

Another factor that has created the social context of AIDS is drug abuse. Intravenous drug abuse, like promiscuous homosexual behavior in men, is a recognized AIDS predictor.[1] Significantly enough, both AIDS and intravenous drug abuse are found predominantly (70–80 percent) among men aged twenty to forty-four.[2] Virtually all measures of opiate- and cocaine-related drug abuse indicate an escalating problem during the past three decades.

The growth in intravenous drug abuse has depended on social changes just as significant as those that have occurred in the homosexual community. Very few people used intravenous drugs just three or four decades ago. Drug addicts then, like most addicts today, sniffed, snorted, smoked, or ingested their drugs. Syringes and needles were too difficult to come by. In fact, the widespread modern use of syringes and needles to deliver pharmaceutical drugs dates only since World War II. For more than a century before this, needles and syringes were expensive instruments made of metal and glass that had to be sterilized before each use. They were not readily available to any but medical practitioners. The invention of the cheap, dispensable plastic syringe in 1970, with its replaceable metal needle, changed all that. Syringes became available in huge quantities, to be thrown out after each use. Hospitals and laboratories no longer kept close watch on how many needles and syringes were used by whom. It became much easier to obtain, illicitly or otherwise, the paraphernalia necessary to "shoot up."[3]

The results of the new technology made an immediate impact. In England, the number of registered heroin addicts rose from 10 in 1960 to 509 in 1966[4] and grew exponentially thereafter. Similarly, the U.S. Bureau of Narcotics reported that the total number of registered heroin addicts increased from 45,000 in 1960 to 68,000 in 1970 to 99,000 in 1973 (an increase of 120 percent).[5] There are now believed to be over 200,000 intravenous drug abusers in New York city alone. Estimates of the actual number of addicts in the United States in 1973 were between 300,000 and 500,000.[6] About half of these addicts were to be found in New York City; other urban centers associated with AIDS, such as Los Angeles and San Francisco, were also among the cities reporting the most addicts.[7]

Other measures confirm that drug abuse expanded explosively. The number of arrests for breaking narcotic drug laws grew exponentially in the United States: 29,000 arrests in 1962, 100,000 in 1967, 190,000 in 1969, 400,000 in 1971, 530,000 in 1977, and 690,000 in 1986.[8] A total of 1.4 million arrests were made on drug charges of all kinds in 1989.[9] During these arrests, drug enforcement officials confiscated 500 kilograms of cocaine in 1980 and 100,000 kilograms in 1990.[10] Of particular significance for understanding the risk for AIDS among young people is the fact that about 70,000 teenagers under the age of eighteen are now arrested each year in the United States on narcotics-associated charges.[11]

This epidemic of intravenous drug abuse is also evident in information collected by the federal government's Drug Abuse Warning Network (DAWN). DAWN data indicate that in virtually every major city in the United States, heroin overdoses seen at hospital emergency rooms quadrupled between 1976 and 1985.[12] The use of cocaine, which may also be taken intravenously, increased nearly 1,000 percent in all major cities during the same period.[13] In 1983, cocaine was mentioned 7,194 times in DAWN reports. This figure increased to 13,501 times in 1985 and 46,331 times in 1987.[14] (See Figure 18.)

Not surprisingly, individual hospitals have encountered increasing numbers of drug-addicted patients. Dr. Peter Dans recently published a report on the total number of addicts seen among patients at the Johns Hopkins Hospital in Baltimore. Between 1983 and 1988 the number of patients seen at the hospital remained almost constant, yet the number of patients with opiate-related symptoms increased from an average of about 55 per year to nearly 300 (a sixfold increase). Cocaine-related symptoms exploded from a couple per year to 250 (a 100-fold increase). And babies affected by one or more addictive drugs at birth skyrocketed from 1 or 2 to 200 per year (a 100-fold increase).[15] Such figures are typical of what has been observed throughout the United States and Canada at hospitals in most major cities.

As was the case with gay men, the medical literature reflects the growing problem of drug abuse with the recognition of unusual immunologic and infectious complications in abusers. The medical community in the United States first commented on the "epidemic" of drug addiction during the late 1960s, when physicians

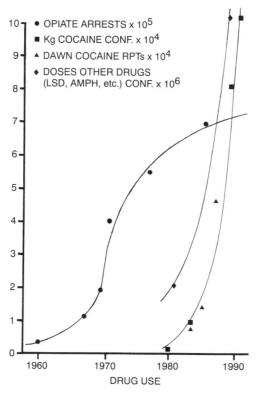

FIGURE 18. Use of addictive drugs in the United States, 1960–1990. Opiate arrests (●) in hundreds of thousands; kilograms of cocaine confiscated (■) in tens of thousands; Drug Awareness Warning Network (DAWN) cocaine reports (▲) in tens of thousands; doses of other drugs (LSD, amphetamines, barbituates, etc.) confiscated (♦) in millions.

in all major cities began regularly encountering addicts among their patients. The high mortality and vastly decreased life span of opiate addicts was studied.[16] Severe systemic infections with gram-negative bacteria, *Candida*, and other opportunistic organisms began to be reported.[17] Disseminated tuberculosis (another diagnostic indicator of AIDS) was identified as a major complication of drug addiction in 1973.[18] At the same time, lymphadenopathy was recognized as a common and often misdiagnosed or undiagnosed problem in drug abusers,[19] with concomitant immunological dysfunction being identified in such patients.[20] Malnutrition was recognized as an immunosuppressive risk associated with intravenous drug abuse by 1976.[21] The inversion of T cell subsets accompanying opiate use had been identified by 1980.[22]

Studies undertaken since 1980 have verified that the huge explosion of drug abuse has been accompanied by a corresponding explosion of all of the infectious diseases associated with drug behavior, including hepatitis A, hepatitis B, and non-A, non-B hepatitis. The incidence of each increased over fivefold between 1980 and 1990 among addicts.[23] Septicemia, a general term for infections of the bloodstream, also increased fourfold over the same period of time.[24] Of particular concern is the fact that intravenous drug abuse appears to be highly associated with prostitution, the trading of sex for drugs, and high-risk sexual behaviors, including promiscuity. Thus, drug abuse is only one aspect of a complex problem that includes both the direct immune suppression of ever-larger numbers of people and the vastly increased incidence of sexually transmitted and other diseases in the same group.[25]

As drug abuse has become more prevalent, so have the numbers of babies born to addicts. From 1950 through 1953, only three babies were born to heroin addicts at the Vancouver General Hospital in Canada. During the next four-year period (1954–1957), ten babies were born to addicts; from 1958 through 1961, eighteen such infants; thirty-one between 1962 and 1965; forty between 1966 and 1969; and forty-seven between 1970 and 1973. In brief, the number of infants born to drug-addicted mothers skyrocketed by a factor of eighteen (1,800 percent) in a mere twenty years.[26] Half of these mothers admitted to prostitution as a means to support their drug habits. The same scenario has been played out in many other cities as well. In New York City, for example, the number of babies born to addicts was about 800 per year in 1978. It is now over 5,000—an increase of over 600 percent in a single decade.[27] The stage was clearly set for infant AIDS long before it was recognized.

The message is clear: AIDS—particularly what the CDC and the press are inaccurately calling "heterosexual" AIDS—is a problem that cannot be dissociated from drug abuse. AIDS has clung to the coattails of drug addiction like a demon from hell intent on claiming its victims. And like all such demons, it has spread its crop of disease and destruction all along the way.

The Implications of Homosexuality and Drugs
for the Blood Supply

It is important to emphasize several points about the dramatic changes in homosexual behaviors and drug abuse over the past two decades. One is that this information is crucial for understanding the growth of AIDS, whether HIV is the sole necessary and sufficient cause of AIDS or not. Just as it is impossible to understand the current uneven distribution of syphilis in the general population or the vastly increased risk of homosexual men and highly promiscuous heterosexuals for this disease in the absence of social factors, it is impossible to understand the distribution of AIDS, HIV, or any other AIDS-associated infection without taking into account the same sociology.

A second crucial point is that the spread of AIDS from high-risk individuals to hemophiliacs, blood transfusion patients, and their sexual partners must also be put in the context of this overall pattern. Occasional cases of AIDS induced via blood transfusions and factor treatments would have been expected—and indeed, some of these have been documented in previous chapters—prior to the AIDS epidemic, since non-HIV immunosuppressive agents, including cytomegalovirus, Epstein-Barr virus, and hepatitis viruses, were all present in the blood supply. When we consider the sudden and tremendous increase in sexually transmitted diseases, including all of the viruses just listed, in the 1970s and 1980s, it is hardly surprising to find that the incidence of AIDS among people receiving blood products should also go up until the risks were observed and countered. More and more blood donors were multiply infected with many viruses—HIV undoubtedly included— and screening was done for none of these infections except hepatitis A until 1984 or after. Thus, not only did the incidence of HIV in the blood supply grow tremendously, so did all the other infections associated with AIDS.

Finally, just as sociological factors are at work in who is at risk for AIDS among homosexuals and drug addicts, the same is true for transfusion patients. Countries such as the United States and France that allow blood donors to be paid for their donations have had much worse problems with transfusion- and hemophilia-associated AIDS cases than have countries such as Great Britain, which has an entirely voluntary blood donation program. The reason for the difference is simple. In countries permitting blood donors to be paid, addicts and others on the "street" can transform their blood into the cold, hard cash they need for food or drugs. Just those people who are most likely to be carrying multiple, concurrent infections are the people who have the greatest incentive to donate their blood. The victims of this practice are the innocent, and unsuspecting, medical patients who receive that blood. The "gift" of blood in this case places two people in intimate contact who otherwise would probably avoid one another at all costs. The social forces governing this gift giving have also played their role in creating the context for AIDS.

CHAPTER 9

SOURCES OF ACQUIRED IMMUNOSUPPRESSION IN HETEROSEXUALS

A COMPARISON OF NORTH AMERICANS AND EQUATORIAL AFRICANS

AIDS Among Heterosexuals in Africa and Western Nations

Not everyone is at equal risk of contracting AIDS. That much is evident. Can we nonetheless expect AIDS to become rampant among heterosexuals? Many medical experts think so. In April 1987 Stephen Jay Gould wrote a one-page article for the *New York Times Magazine* proclaiming heterosexual AIDS a "natural" and therefore inevitable phenomenon. He recounted how, in 1984, he and other scientists had received a manuscript from bio-physicist John Platt arguing "that the limited data on the origin of AIDS and its spread in America suggested a . . . frightening prospect: we are all susceptible to AIDS, and the disease has been spreading in a simple exponential manner. . . . Most of us were incredulous," recalled Gould, "accusing Platt of the mathematical gamesmanship that scientists call 'curve fitting.' After all, aren't exponential models unrealistic? Surely we are not all susceptible to AIDS. Is it not spread only by odd practices to odd people? Will it not, therefore, quickly run its short course within a confined group?" The answer to these questions, Gould assured us, is a resounding "no." Reconsidering in the light of 1987 figures, Gould proclaimed that "the AIDS pandemic . . . may rank with nuclear weaponry as the greatest danger of our era." AIDS, he was con-

vinced, had remained in high-risk groups initially "only by accident of proximity, not by intrinsic susceptibility. Eventually, given
the power and lability of human sexuality, it spreads outside the
initial group into the general population. And now AIDS has begun its march through our own heterosexual community." Some-
day in the near future, Gould predicted, AIDS in Western nations
would resemble AIDS in "African areas where it probably originated, and where the sex ratio of afflicted people is 1-to-1, male-
female." Those afflicted will be "our neighbors, our lovers, our
children and ourselves. AIDS is both a natural phenomenon and,
potentially, the greatest natural tragedy in human history."[1]

Certainly the specter of AIDS that the mirror of Africa holds
out is horrendous. Over 60 percent of all the estimated HIV infections in the world—perhaps 10 million infected people—are concentrated in sub-Saharan Africa. Oddly, neither northern nor
southern Africa is experiencing epidemics of AIDS, suggesting that
there is something unique about the tropical region lying between,
a point to which I shall return shortly.[2] Within the epidemic region, the ratio of men to women with AIDS is nearly one to one.
That is, for every man who is dying of AIDS, a woman dies, too.
This picture contrasts sharply with figures from the United States
and Europe, where almost twenty men die for every one woman.
In some cities, such as the West African city of Abidjan, on the
Ivory Coast, over 35 percent of all cadavers were found to be HIV
seropositive, and AIDS accounted directly for the deaths of 14
percent of the population in that city.[3] Figures such as these truly
raise the apparition of the Black Death. Fortunately, Abidjan is
unique even among African cities. In most endemic areas, including Kinshasa, Zaire, and various cities in Uganda, the rate of
death from AIDS is about 1 in 1,000 people, and about 5 percent
of the population are HIV seropositive.[4] These rates of death are
still hundreds of times higher than in Western nations, however.
If Africa is a barometer predicting the future of AIDS elsewhere in
the world, whether AIDS eventually accounts for only 1 percent
or 15 percent of all deaths is irrelevant; the consequences will devastate both our health care system and society. The question is
whether the extrapolation from African to Western nations is justified.

The facts are not in question. Our answer depends instead on
the assumptions that underlie the reasoning. Implicit in any com-

parison between the risk of AIDS in heterosexuals in Europe and the Americas and the risk of AIDS in equatorial Africans is the assumption that these two populations do not differ in any essential way. If, however, it can be demonstrated that heterosexuals in Africa are not comparable healthwise with heterosexuals in developed nations, then the type of argument made by Gould and his medical colleagues is no better than the argument that people who use logic are infallible because logic is infallible. We know better. Similarly, anyone who has looked carefully at the health status of the typical heterosexual in Africa and in Europe or North America knows better, too. They are not comparable.

The Unusual Immunosuppressive Risks of African Heterosexuals

AIDS in Africa cannot be used as a model for AIDS in Western nations because typical sub-Saharan Africans are not comparable to Western heterosexuals in their disease load, their nutritional status, or their immunological functions.

Many factors determine the disease load of a population. One is the degree of promiscuity typical of the culture. Sub-Saharan Africans tend to be much more promiscuous than heterosexuals in Western culture, both pre- and postmaritally.[1] This fact is mirrored in an extremely high rate of female sterility, rising as high as 30 to 50 percent in some sub-Saharan countries, and directly linked to a very high incidence of sexually transmitted diseases (STDs).[2] Notably, the "infertility belt" in Africa corresponds almost exactly to the area in which AIDS is most prevalent, suggesting that AIDS in Africa, as in Western nations, is most prevalent among people who are highly promiscuous and who contract multiple STDs.[3] African countries such as Nigeria and South Africa, in which sexually tranmitted diseases are relatively rare, also have a very low incidence of both HIV and AIDS.[4]

No direct measures of promiscuity exist for Africans, but as we have seen previously, the incidence of sexually transmitted diseases is considered to be a sensitive marker. Huge differences in rates of sexually transmitted diseases certainly exist between Equatorial African and Western heterosexuals (Figure 19). Gonococcal infections, which affected about 4 in 1,000 Americans in 1981, were estimated to be present in one in every 10 people in

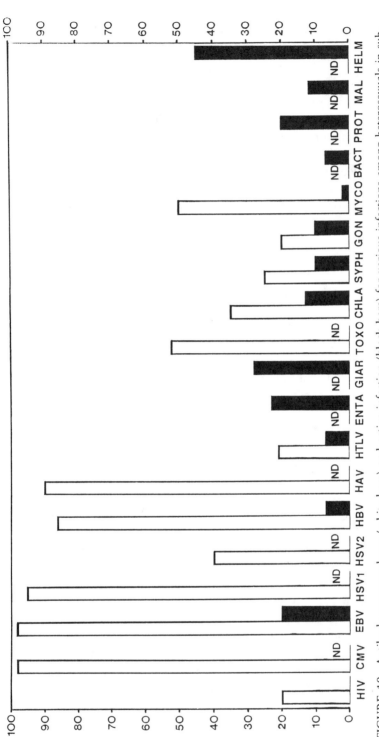

FIGURE 19. Antibody seroprevalence (white bars) and active infection (black bars) for various infections among heterosexuals in sub-Saharan Africa. Compare with Fig. 3. The infections are, listed from left to right, HIV, cytomegalovirus, Epstein-Barr virus, herpes simplex virus 1, herpes simple virus 2, hepatitis A virus, hepatitis B virus, HTLV I and II, *Entamoeba*, *Giardia lamblia*, *Toxoplasmosis*, *Chlamydia* (all species), syphilis, gonorrhea, *Mycobacteria* (both tuberculosis and atypical forms), pyogenic and septicemic bacteria (*Pseudomonas*, *Escherichia coli* in the bloodstream, *Staphylococcus*, etc.), protozoa (including *Pneumocystitis carinii*), malaria, and helminths.[5]

Uganda, and rates of syphilis were similarly elevated.[6] *Chlamydial* infections are also very common, with 25 percent of postpartum women in Ghana being antibody positive and 5 to 8 percent actively infected. Three percent of these women were also infected with gonorrhea.[7] Again, these figures are much higher than in Western nations. It is estimated that fewer than 5 in every 1,000 people in Britain and the United States are infected with *Chlamydia* each year.[8]

Other sexually transmitted diseases point to much greater promiscuity among heterosexuals in Central Africa than in Western nations. Like gonorrhea and syphilis, HSV 2, the sexually transmitted form of herpes simplex virus, is much more prevalent in Central Africa than it is in North America or Europe. Whereas the seroprevalence of antibody to HSV 2 is about 20 percent among adult white Americans and Europeans, the incidence of antibody among adult heterosexuals in Zaire is 40 percent, and in the Congo, 70 percent. These seroprevalence rates compare with homosexual men and female prostitutes in North America and Europe. Given the immunosuppressive effects of HSV infections, it is also noteworthy that about 85 percent of all HIV-seropositive heterosexuals in Central Africa are HSV 2 seropositive as well.[9] Indeed, A. J. Nahmias and his colleagues have suggested that HSV 2 acts as a cofactor for HIV in AIDS among African heterosexuals.[10]

The incidence of HTLV infections is much higher among equatorial Africans than it is among comparable groups in Western nations. Studies suggest that HTLV I is endemic in this region and may affect as many as 5 of every 100 sexually active people.[11] This is dozens of times higher than among heterosexuals in Western nations. Indeed, the increased risk of infection persists even among Africans who have emigrated to Western countries such as Great Britain and the United States.[12] The incidence of hepatitis viruses is similarly elevated. Whereas the number of hepatitis B virus carriers among blood donors in northern Europe, North America, and Australia is significantly less than one-tenth of 1 percent (1 in a 1,000 or less)—including homosexual men and drug abusers—in some parts of Africa and Asia, 20 percent or more (two of every ten people) are actively infected with the virus.[13] Not incidentally, it is estimated that 20 to 30 percent of sexually active adults in the age bracket twenty to forty years are

HIV seropositive in sub-Saharan Africa,[14] an indication of just how immune suppressed this population may be.

Immunosuppressive viruses are only the beginning of the infectious problems encountered by inhabitants of central African and other tropical countries. Tuberculosis, almost unknown in Western nations except among alcoholics, drug abusers, and the poverty stricken, is rampant in Africa. The World Health Organization and the International Union against Tuberculosis and Lung Disease estimate that as many as 10 million new cases of tuberculosis occur each year, accompanied by 2 to 3 million deaths, mainly in equatorial nations. The highest rates of tuberculosis in the world occur in Africa, where about half of the sexually active adults have been infected.[15]

A very high association of HIV with active tuberculosis has been reported by many investigators in countries in Africa and in Haiti.[16] Approximately one-third of HIV-seropositive people are co-infected with *Mycobacterium tuberculosis*. Conversely, a third to half of the people who suffer from active tuberculosis are seropositive for HIV in countries such as Zaire, Uganda, and the Central African Republic.[17] Extrapulmonary tuberculosis, a diagnostic indicator of AIDS in Western nations, represented 8 to 10 percent of all tuberculosis cases in Kenya and Tanzania in 1970 and has remained at that level or slightly higher ever since.[18] Thus, tuberculosis represents a very real infectious risk for peoples in equatorial Africa that is not shared by most Western heterosexuals.

Another disease, malaria, represents a risk unique to developing nations. In tropical Africa, AIDS and HIV-seropositivity are virtually synonymous with regions in which malaria is endemic.[19] Malaria may therefore represent an important factor predisposing people to AIDS. It is certainly anything but a benign disease. Of 20 million children born each year in Africa, nearly 700,000 will die before their fifth birthday of malaria or its complications.[20] Those who survive often develop malaria-associated anemia. They are, as a consequence, iron deficient and immune suppressed. The most common treatment for their anemia is blood transfusion. Transfusion is also a frequent treatment for sickle cell anemia, a genetically inherited defect in hemoglobin that is common among the black population of equatorial Africa. Sickle cell anemia has a detrimental effect on oxygen transport but is protective against malaria. The prevalence of malaria-associated anemia and sickle-cell anemia can be gleaned from the fact that between August 1985 and July 1986 almost 70 percent of nearly 13,000 transfusions performed at the Mama Yemo Hospital in Kinshasa, Zaire, were given to children with malaria. Lack of money prevents most central African hospitals from screening donated blood for HIV or other pathogens. Thus, the more blood a child receives, the greater is his or her chance of contracting all blood-transmissible viruses and the greater is the probability of AIDS.[21] These blood transfusions can induce lymphocytotoxic antibodies, some of which may falsely mimic HIV antibodies.[22]

The malaria confers AIDS risks beyond blood transfusion. As noted in Chapter 4, malaria and other parasitic infections such as schistosomiasis and filariasis can cause immune suppression themselves and may induce the formation of lymphocytotoxic antibodies against B cells.[23] Studies in Africa and Venezuela have found that malaria often results in false-positive HIV tests and may be a risk factor for HIV infection independent of blood transfusion.[24] Moreover, as noted in Chapter 3, many antimalarial and antiparasitic drugs are also immunosuppressive.[25] Indeed, inappropriate use of injectable agents to treat malaria "was identified as an expensive and potentially harmful practice" in Togo during the early 1980s.[26] Even when oral chloroquine, the preferred treatment, was instituted, less than 30 percent of the prescribed doses were sufficient to be of therapeutic value.[27] Thus, between the presence of rampant parasitic diseases and the side effects or inefficacy of their treatments, immunosuppression due to chronic malaria is far from rare.

Combinations of disease agents take their toll on immunocompetence and lives. Both aggressive Kaposi's sarcoma and Burkitt's lymphoma have been endemic at high rates (nearly 10 percent of all cancers) in central African countries such as Uganda for as long as records are available (mid-1950s).[28] This evidence suggests that AIDS has been present in Africa for as long as the diagnostic techniques have been available to detect it. Even more striking is evidence of high rates of Kaposi's sarcoma in African children during the 1950s. Malaria and other infections have been fingered as possible cooperative culprits. Dr. A. C. Templeton reviewed all cases of Kaposi's sarcoma in Uganda between 1961 and 1970 and records of similar cases back to 1947. Among the atypically aggressive and quickly fatal cases in young men and women, fourteen of thirty-four cases suffered from malaria. An additional eight individuals suffered from sickle cell anemia and may have required frequent blood transfusions.[29] Similarly, John Sixbey and his colleagues have noted that AIDS patients resemble the "inhabitants of the Burkitt's lymphoma-endemic regions of Africa, where malaria is hyperendemic. . . . Such patients have depressed T cell function and intense polyclonal B-cell stimulation, factors that could preferentially affect dissemination and replication of type B Epstein-Barr virus in these populations." Indeed, several investigators have suggested that a combination of Epstein-Barr virus with malaria may be the cause of Burkitt's lymphoma, although this suggestion remains controversial.[30] Notably, Burkitt's lymphoma has also been endemic in Africa for decades.

Helminth parasitism, trypanosomiasis, schistosomiasis, amoebiasis, and giardiasis are all prevalent in tropical countries where AIDS is endemic. Not only can these parasites cause significant immune suppression, they have also been linked to increased risk of developing anemia in pregnant women. The resulting anemia puts the babies of such women at increased risk of low birthweight and malnutrition. Moreover, these parasites can and often are transmitted congenitally, placing the unborn in double jeopardy.[31]

The incidence of infectious diseases is exacerbated in many African (and other Third World) nations by failure to observe the standards of medical practice maintained in Western nations. "Inappropriate drug use seems to be the rule rather than the exception in developing countries. . . . Drugs that are not essential to recovery, of proven efficacy, acceptably safe, or reasonably priced are widely used and are just as likely to be bought from a grocery store or a travelling drug pedlar as from a pharmacy. Potent medicines that should be available on prescription only are often bought over the counter, and diarrhoea is commonly treated with drugs instead of oral rehydration therapy."[32] These are the summary conclusions of a recent report from Health Action International based on a review of published and unpublished studies. The implications of such self-treating of disease, in terms of failing to diagnose or cure infections and in terms of potential chronic abuse of pharmaceutical agents, translate into increased risk of immune suppression for some Third World people. In addition, many Africans employ for their medical care "injection doctors," untrained traveling practitioners who inject over-the-counter antibiotics and other drugs. Their equipment is often either unsterilized or improperly sterilized and therefore represents a possible mode of transmission of numerous infectious agents.[33] Although injection doctors have not been implicated directly in the transmission of HIV, they certainly pose a transmission risk for other diseases such as hepatitis and syphilis.

Dangerous medical procedures are not limited to untrained practitioners. One survey of hospital procedures in Togo found that during immunizations, a sterile needle was used only 75 percent of the time and a sterile syringe only 20 percent of the time.[34] It is not known how prevalent such unsterile practices are in sub-Saharan Africa or in other developing nations, but if they are at all common, then they clearly represent yet another infectious risk that is almost unknown in Western nations outside of intravenous drug users.

Among the other factors chipping away at the immunological competence of central Africans is acute undernutrition or malnutrition, extremely prevalent in the developing countries where AIDS is epidemic. Between 1 and 4 out of every 100 people, including children, have been found to have clinically evident malnutrition in sub-Saharan countries.[35] Such malnutrition has been clinically demonstrated vastly to increase susceptibility to parasitic infections and to exacerbate their pathological effects.[36] This is hardly surprising, given the pronounced immune suppression that can accompany such deficiencies and the remarkable increase in mortality that even a slight deficiency in nutrients such as vitamin A has been shown to confer during infectious disease.[37] Consequently, one could hardly expect anything but the observation that both AIDS and AIDS-associated infections are extremely prevalent in all portions of African populations in which malnutrition is present.

In a controlled study of 1,328 patients in Uganda, amenorrhea (abnormal absence of menstrual flow or failure to have a period) among women was the most specific predictor of HIV seropositivity.[38] Since amenorrhea is frequently a concomitant of severe malnutrition, abnormally low body fat, and anemia, this observation suggests a significant link between susceptibility to HIV and immune suppression due to starvation or malnutrition. Further evidence of the immunosuppressive effect of recurrent malnutrition comes from a study of birthweights of babies in rural Zaire over a ten-year period. Birthweights were found to fluctuate seasonally and significantly according to the energy expenditure of the women, variation in food availability, and incidence of malaria. No variation in birthweight was evident over a five-year period in rural or urban hospitals in Canada that were studied as controls.[39] Like the infants of drug addicts in Western nations, infants of HIV-seropositive mothers in Zaire were more frequently premature, had lower birth weights, and experienced a higher death rate during the first month than did children of HIV-seronegative mothers.[40] As in infants of intravenous drug abusers, it would be a mistake to attribute the observed problems completely or even mainly to HIV.

Added to the problems of disease, malnutrition, and inappropriate or dangerous medical practices, many African heterosexuals are also prone to drug abuse. Despite widespread denials by African governments and even by some medical personnel who have visited Africa, there can be little doubt that some central African nations experience as much drug abuse as Central American countries and the United States. There is no way around the fact that more than 45 percent of all heroin seizures made at U.S. ports of entry during 1991 involved Nigerian couriers.[41] Nigerian drug smuggling rings handle a large percentage of the worlds's heroin supply. It is inconceivable that none of the drug finds its ways into the hands of African abusers or that these heroin rings are not just as active within Africa as they are in America and Europe.

Social and political revolutions are also taking their tolls on African health. As Daniel B. Hrdy has pointed out in his essay, "Cultural Practices Contributing to the Transmission of Human Immunodeficiency Virus in Africa," the human face of Africa is rapidly changing. "Population movements in Africa contribute to the 'sexual mixing' of various African groups and may be related to the spread of AIDS. The entire Central African area (and indeed the whole of sub-Saharan Africa) is experiencing large shifts of population. Some patterns have existed for long periods, such as the movement of Arabic and Nilotic peoples into the northern part of Central Africa. The long-term movement of rural populations into urban areas is also continuing. Other more recent trends include the movement of migrant workers from Zaire and Rwanda to neighboring countries (e.g., to the copper belt in Zambia), the movement of armies on the Uganda-Tanzania border, and the presence of large numbers of refugees, especially from Uganda. Ugandan prostitutes are a major focus of HIV seropositivity in Kenya. It is probably significant that AIDS cases seem to have been present in large numbers in Africa only since the 1970s—a time frame that correlates with the intensification of urbanization and population shifts."[42] Interestingly, historian William McNeill has argued convincingly that such population shifts and the mass movement of armies and refugees have been associated with epidemics of new diseases since at least the time of the first Crusades in Western Europe through the time of World War I.[43] Perhaps there is nothing new about AIDS in this sense but its particular manifestation.

Just as social revolutions affect the transmission of diseases, so do political ones. War creates havoc, all too often manifested in the disintegration of health care facilities, food distribution services, and agriculture. Tens of millions of refugees have been forced from their Haitian, central African, and Asian homes during the past decade. These refugees are often consigned to camps where severe malnutrition, poverty, and disease are rampant. Civil war has repeatedly disrupted Haiti, Uganda, and Sudan, reducing tax revenues and crippling health care budgets. To quote C. P. Dodge, who works for UNICEF in Khaka, Bangladesh, and who has written extensively on the problems of Asian and African nations, "Normal health service delivery systems were broken down forcing doctors, nurses, and other health professionals into towns, cities, or neighbouring countries in search of peace and employment. Scores of hospitals, health centres and dispensaries were abandoned, destroyed, or looted, rendering even the limited physical facilities useless. Preventive public health services such as immunization and provision of potable drinking water were discontinued leaving huge populations susceptible to controllable infectious diseases and epidemics."[44] These mass dislocations and deaths also produced an estimated 1 million orphans in Uganda alone, each of whom was considered to be vulnerable to increased morbidity and mortality.[45] These conditions are not duplicated in any Western country.

Europe and America Are Not Africa

This comparison of the immunosuppressive risks encountered by the typical sub-Saharan African heterosexual and those of the typical American or European heterosexual undermines any attempt to predict the future of heterosexual AIDS in Western nations based on the history of AIDS in Africa. African heterosexuals in areas where AIDS is endemic are not comparable in general health, disease load, nutrition, or sexual activity to heterosexuals in the West. No reasonable basis exists for arguing that any Western nation will develop an endemic heterosexual AIDS problem of the magnitude currently emerging in Africa. On the contrary, the natural history of AIDS in both Western and African nations strongly implicates non-HIV causes of immune suppression as necessary elements predisposing people to AIDS. African heterosexuals are therefore properly compared with promiscuous homosexual men and intravenous drug abusers in the United States and Europe.

As surprising as this conclusion may be, it has been reached by other investigators who have taken the time to compare the relevant statistics. A decade prior to the recognition of AIDS, physicians in New York City, Boston, and San Francisco had already begun to liken the prevalence and types of diseases appearing among homosexual men and drug abusers to those usually found only in Third World nations and on tropical islands such as Haiti.[1] This similarity was validated in 1987 by Dr. Thomas C. Quinn, Dr. Anthony Fauci, and their colleagues at the National Institutes of Health, who conducted a major study comparing the incidence of various infections in Americans with those of Africans in Zaire. Their conclusion is worth quoting at length:

> Our serologic studies, as well as others, demonstrate that [heterosexual] Africans are frequently exposed, due to hygienic conditions and other factors, to a wide variety of viruses, including CMV, EBV, hepatitis B virus, and HSV, all of which are known to modulate the immune system. Exposure to these infections was also observed in the present study and in other studies among American homosexual men, probably secondary to sexual exposure. In contrast, the prevalence rates of these infections was much less frequent among the American [heterosexual] controls within this

study. Furthermore, the Africans in the present study are at an additional risk for immunologic alterations since they are frequently afflicted with a wide variety of diseases, such as malaria, trypanosomiasis, and filariasis [a type of worm causing, among other things, elephantiasis and heart disease], that are also known to have a major effect on the immune system. The frequent exposure to these multiple microbial agents could act collectively or individually to result in immunologic modulations rendering a host more susceptible to HIV infection or by influencing disease progression by increased viral replication and cytolysis of T-positive cells. Our immunologic data on the Africans in this study support these latter concepts.[2]

African heterosexuals, in other words, have the same range of immunosuppressive factors as other identified AIDS risk groups in other countries. The only difference is that nearly everyone in these African countries is exposed to these factors, whereas exposure in Western nations is severely limited.

The inevitable conclusion that must be drawn from these data is that sub-Saharan African heterosexuals are not a reasonable model for heterosexuals in developed nations. Far from presenting us with a look at the future of AIDS in North America and Europe, African heterosexuals simply confirm the fact that AIDS is a problem only for individuals who have multiple causes of immune suppression prior to, concomitant with, or independent of HIV exposure. AIDS will never become the major health threat to Americans and Western Europeans that it has become for Africans. AIDS will be a continuing problem only for individuals whose life-styles, medical histories, or socioeconomic conditions predispose them to immune suppression in general.

Indeed, careful consideration of African AIDS confirms that the usual story about AIDS originating Africa cannot be correct. Recall that HIV is supposed to have originated in Africa. Recall that AIDS in Africa predates AIDS in Western nations. Indeed, studies of the incidence of Burkitt's lymphoma and Kaposi's sarcoma suggest that AIDS-like cases existed as a significant proportion of all cancers in all people by the 1950s in sub-Saharan Africa. Note that the people born in a country with a high proportion of heterosexual AIDS retain an increased risk of AIDS when they emigrate (indicating chronic perinatal or childhood infections) and that sex with such a person with AIDS is listed as a high-risk activity by the CDC.[4] Now consider these facts in the light of the notion that AIDS is caused by nothing more than HIV and that HIV is a sexually transmissible agent. One expects on these bases that AIDS would have been diagnosed first among black Africans who had emigrated to European and North American countries or among Europeans and North Americans who had lived for a time in Africa. Furthermore, the transmission of AIDS into Europe and North America should have begun no later than the 1950s, since diseases diagnostic for AIDS were already endemic in Africa by that time. And since homosexual activity in Africa is reportedly rare and AIDS was already equally distributed among heterosexuals several decades ago, it follows that AIDS would have been introduced first into the heterosexual community in Europe and North America. Yet the facts are quite otherwise. Something is seriously wrong with the standard analysis of what causes AIDS, how it originated, and how it has spread. Those problems begin with our view of AIDS in Africa.

Heterosexual AIDS in North America and Europe

Heterosexual AIDS in North America and Europe is, and will remain, rare. This fact was recognized as long ago as 1987 by key individuals at the CDC and NIH: "AIDS is not spreading at the anticipated rate among non-drug using heterosexual Americans, and medical officials here at the Centers for Disease Control and elsewhere are generally agreed that they see no evidence the disease will reach epidemic proportions, except among homosexuals and intravenous drug users. As a consequence, there is a growing consensus among leading medical scientists that the threat of AIDS to the wider population, while serious, has been exaggerated. . . . In areas such as California and New York where AIDS is epidemic among homosexuals and drug users, 'very little crossover to the mainstream heterosexual population has occurred,' said Dr. Harold W. Jaffe, the centers' chief of AIDS epidemiology."[1] If anything, the subsequent five years have served only to reinforce this clairvoyant prediction. Discounting acquisition of AIDS from an intravenous drug addict or bisexual male, less than 3 percent of AIDS cases in the United States, Canada, and England are listed as heterosexual transmission. In fact, since 1981, there have been only eleven supposed cases of transmission of AIDS from a female to a male in New York City, out of nearly 31,000 documented AIDS cases. Germany reports four such cases, and the numbers are equally small in England and Canada.[2]

In fact, the chances that a healthy, drug-free heterosexual will contract AIDS from another heterosexual are so small they are hardly worth worrying about. One statistician has compared them to the probability of winning a state lottery game or being struck by lightning.[3] Similarly, a report in the journal *Science* states that "the chance of becoming infected with the human immunodeficiency virus (HIV) after one sexual encounter with someone who has both tested negative for HIV and who has no history of high-risk behavior is 1 in 100 million. If the same couple uses a condom, the risk plummets to 1 in 5 billion, say the epidemiologists. Even having sex with someone whose HIV status is unknown, but who does not belong to any high-risk group, yields a calculated risk of 1 in 5 million, or 1 in 50 million per sexual episode, depending on whether or not a condom is used."[4]

But wait a minute. There are heterosexual cases of AIDS in North America and Europe, aren't there? Certainly the media and the CDC itself list such individuals. Indeed, shortly after Earven "Magic" Johnson announced that he was HIV seropositive, the CDC in the United States launched a television advertising campaign predicated on the statement, "If Magic Johnson can get AIDS, anyone can get AIDS." We are all, the message would have us believe, still at risk.

Unfortunately, public policy is being made on the basis of incomplete or even inaccurate information. In the first place, Magic Johnson does not have AIDS—he is HIV seropositive, but he has no symptoms of AIDS, and he may never develop the syndrome. Second, no one knows what risk factors Johnson did or did not have for contracting HIV other than extraordinary promiscuity. We have only his word that he contracted HIV from a woman. He has never directly stated that he never engaged in homosexual activity or used intravenous drugs. Even if we accept his implied denial, we still need to know how many sexually transmitted diseases he had prior to contracting HIV. What drugs was he treated with for his infections? Did he ever use nonintravenous recreational drugs? Was he treated with corticosteroids or painkillers for athletic injuries? Was his nutritional intake adequate to ensure maximal immune response? Did he have any subset of multiple, concurrent risks such as chronic use of antibiotics, cortisone, steroid creams, or other prescription pharmaceutical agents such as tranquilizers or sleeping pills? Did he ever use anabolic steroids or other ergogenic aids to increase his physical performance? What was his nutritional status at the time he became infected and thereafter? Unless we have detailed information about these issues and others, we cannot conclude that Johnson was risk free. And if Johnson was not risk free, then is he a reasonable model for even the most sexually promiscuous heterosexual?

A variety of other cases touted by the government and media as heterosexually acquired AIDS cases are similarly suspect. Steven Murrell, for example, who came from a well-off Nob Hill home in San Francisco, contracted AIDS sometime during the late 1980s and was diagnosed at the age of sixteen. He made several videos with his mother, a film producer, before he died, and he appeared in television commercials in the San Francisco area. He visited schools to promote AIDS awareness. His message was simple and clearcut: "I contracted AIDS through sex with a girl. I contracted it by not using a condom." As important as his message was, it is not clear how Steven Murrell contracted AIDS. It turns out that he lived on the streets of San Francisco on and off for several years. He rarely attended school. He admitted to using drugs from a very early age. At one point, he ran away to New York City, where it is thought that he continued his drug use. No one knows how he earned his keep there—dealing drugs, working as a prostitute, or more legitimate means. Given his age and refusal to attend school, though, it is hard to imagine Steven having had a regular job. So when he said, "I was never going to get AIDS. I bet my life on it. Use a condom. I didn't," one can only wonder how much more there is to his sad story.[5]

Even the widely publicized case of Kimberly Bergalis, the young woman who died of AIDS following her infection with HIV, presumably at the hands of dentist David Acer, is less clear-cut than it is usually presented. She specifically denied ever having had sexual intercourse or using intravenous drugs. The CDC investigators in charge of her case did not attempt to verify these assertions. More interestingly, Bergalis did not explicitly deny using drugs other than intravenous types. Did she, then, ever use nonintravenous drugs? If so, what types and how often? What was her nutritional status at the time she was infected with HIV and following it? Was she bulemic, anorexic, or anemic? What other infections did she have at that time (an infected tooth, hepatitis, mononucleosis)? What infections besides HIV did she acquire from Acer? If she acquired nothing but HIV from Acer (a statistical improbability), then how did she acquire the other infections that resulted in her diagnosis of AIDS? Was she using antibiotics, steroid creams, tranquilizers or other prescription drugs at the time she became infected and during the period following her exposure? Did she ever have dental surgery, such as the removal of impacted wisdom teeth? Did the operation require general anesthesia and opiate painkillers? All of these questions might have revealed additional immunosuppressive risks that Bergalis might have had. Yet none of the relevant questions was asked nor tests performed. Significantly enough, independent investigators who are reanalyzing the CDC data linking Bergalis and four other AIDS cases to Acer have concluded that Acer may not have been the source of their infections.[6] There is room for doubt.

Insufficient data is the problem with all so-called heterosexually acquired cases of AIDS. Who are these people really? How did they actually contract AIDS? And are they truly free of immunosuppressive risks? More specifically, does the existence of supposedly nonrisk heterosexual cases such as Magic Johnson, Steven Murrell, and Kimberly Bergalis in North America and Europe disprove my conclusion that AIDS is more than HIV? Let us see.

AIDS Risks of Heterosexuals in North America and Europe

"He was claiming me, clearing everything else away. I felt him slide in me, like a dream I was having of how it could always be like this. We pushed against each other. I reached to touch him past the bundled clothes, our limbs. 'Ahhh,' he said, and held away from me to let me. And then I placed him again, lower this time, by my ass, and pushed a little. His face above me was startled, foolish with unbelieving eagerness, like a high school boy's. He was still. He let me push. It hurt, nearly too much, and I thought I wanted to stop, no matter what. Through the drug and my fatigue, I could feel some deeper consciousness, some other pain, returning. But I eased against it, against him, opened to him, and then slowly, it didn't hurt anymore. It was easier and easy, and Leo pushed too, but gently. He slid the nightgown over my head and looked at me, my legs swimming in the air around him. Then he tilted back into the purplish light and I couldn't see his face any more. We moved slowly, and when he began to cry out, I felt myself shake from somewhere in my spine, I felt only him, only him, it was a different thing; and then the same; and then he bent gently towards me, he lay on me and kissed my neck."[1]

This is Sue Miller's description of her fictitious heroine, Anna, having her first experience with anal intercourse in the novel *The Good Mother*. Not everyone has as pleasant an initiation. One woman wrote to sex adviser and professional hooker, Xaviera Hollander, that her husband repeatedly requested anal sex. "I know Tom engaged in it before we were married. Evidently, he liked it very much, or else he wouldn't want it so badly from me. The only problem is, I can't bring myself—again—to let him 'bugger' me, as they say. Yet I'd like him to. . . . He fingers the crack of my ass, and I must admit I rather enjoy it. Once we were both feeling really good because of all the kissing, touching, and loving, and he asked me to let him try entering my rear just once. I agreed. The pain was so bad that I made him stop, and I haven't let him try since, even though I'd like him to. But to make Tom think there's still a chance I let him tongue my anus. . . . Tom's licking and fondling has at least proved to me that there is a pleasurable feeling in the rear."[2] Nonetheless, this woman could not overcome her fear of the pain she had first experienced, and various other letters written to Hollander express similar reactions. "K.B." wrote to ask "if there are any serious medical side-effects from having anal intercourse? My husband and I had a little session the other night, and since then my body has been doing abnormal things: painful cramps, passing gas every time I move, and not being able to pass any bowels. Is this normal?" Hollander wrote back: "Very similar things happened to me when I first had anal intercourse. In my case, however, I had to move my bowels immediately after having had sex. For about two days my bowel system was rather shaken, and I did not know whether I really liked the entire experience."[3]

Despite an apparent reluctance to talk about such matters outside the bedroom, at least until recently, both analingus (licking or sucking on a person's anus) and anal intercourse, the insertion of a man's erect penis into the anus of his partner, are far from rare among heterosexuals. Several studies show that between 25 and 40 percent of all American and Danish heterosexual women have experimented with anal intercourse at least once during their lives. As many as 10 percent regularly engage in anal intercourse for pleasure.[4] A Kinsey Institute report found that 25 percent of sexually active women at one midwestern university had engaged in anal intercourse at least once during their college years.

These figures are similar to those reported for university students in Canada.[5] During 1988, 5,514 first-year college students were surveyed. Of the 69 percent of the women who were sexually active, 19 percent had participated in anal intercourse at least once, and 5.5 percent reported a previous sexually transmitted disease. Among the 8.6 percent of women who had had ten or more partners, however, 35 percent had engaged in anal intercourse and 24 percent had had one or more sexually transmitted diseases. Women engaging in anal intercourse also have an unusual risk of developing other infections, such as cytomegalovirus, *Chlamydia*, shigella, and amoebiasis.[6]

These data are relevant to understanding risk for immune suppression, since several studies have demonstrated that heterosexual anal intercourse and STD history but not promiscuity per se confer the same increased risk of AIDS as has been found among homosexuals. The European Study Group reports that a survey of 153 HIV-seropositive men and their wives resulted in the identification of three factors that predisposed the women to acquisition of HIV infection: a history of one or more sexually transmitted diseases in the woman during the past five years, full-blown AIDS in the male partner, and anal intercourse.[7] Seropositivity among women with none of these risk factors was 7 percent, whereas it was 67 percent among women having two or more risk factors. Earlier study of HIV transmission from hemophiliacs to household members found similarly that only one of thirty-five regular sexual contacts of fourteen HIV seropositive hemophiliacs was HIV seropositive. The young lady in question was seventeen years old, belonged to no identified risk group, but "had had vaginal, oral, and anal sex with her hemophiliac friend approximately three times per month during the previous year. Throughout their relationship she had used oral contraceptives but no barrier methods."[8]

Anal intercourse was also found to be the probable basis of a recent heterosexual AIDS "scare" in Birmingham, England. The facts of the case were reported in June 1992 by local health authorities who documented the transmission of HIV from a male hemophiliac to four women, one of whom died of pneumonia. The transmission was attributed to "straightforward heterosexual intercourse." Investigative reporter Neville Hodgkinson of *The Sunday Times,* however, was told by the hemophiliac's girlfriends that he repeatedly had anal intercourse with them, and in the case of one girlfriend, "more often than not."[9]

These findings are amply confirmed by other studies. The European Study Group on Heterosexual Transmission of HIV found that HIV transmission was more than five times as common among heterosexuals who participated in anal intercourse than those who did not, regardless of other risk factors, and Nancy Padian and her colleagues in San Francisco have reported that the two highest risk factors for heterosexual transmission of AIDS are repeated anal intercourse and the occurrence of bleeding during intercourse.[10] Another study of 1,115 women who attended a genitourinary medicine clinic in west London during 1987 also found a high correlation between anal intercourse and the acquisition of HIV among nondrug abusers. Three of the women were HIV seropositive. Two of these women completed a questionnaire concerning their sexual history. Both women had had repeated bouts of sexually transmitted diseases; both were in the upper fifth percentile for lifetime sexual partners; and both were unusual in having engaged repeatedly in anal intercourse.[11] Immune competence has not been measured in any of these studies, nor has any attempt been made to determine if antisperm antibodies, lymphocytotoxic antibodies, or circulating immune complexes are also associated with anal intercourse as might be expected from studies of homosexual men who have developed AIDS.

There are, however, independent studies of the prevalence of antisperm antibodies in women with AIDS. They indicate that, as in homosexual men, immunologic exposure to sperm is associated with development of immune suppression. In a 1982 study of twenty-five prostitutes seen in New York City brothels, Dr. Joyce Wallace reported that she found only a single prostitute with a T4/T8 ratio below 1.0. This woman's helper/suppressor ratio went from 0.26 to 0.08 over the course of two months. She was also the only prostitute in the study to display evidence of abnormal levels of antisperm antibody in two separate tests. Among female intravenous drug abusers, on the other hand, Wallace found that over 50 percent (of whom 31 percent were former prostitutes) were not only HIV seropositive, but also had a history of venereal diseases, hepatitis B infection, and antisperm antibodies.[12] In another study, 30 percent of women with AIDS were found to have abnormally high levels of antisperm antibodies against a range of specific sperm proteins.[13]

Victor Lorian, a physician at the Bronx Lebanon Hospital Center in New York, believes that the take-home message of these studies is that the risk for AIDS "has more to do with sexual practice than sexual orientation. Homosexual men are singled out as being at high risk. Indeed they are, but not because of their homosexuality per se, but because they practise anal sex. Heterosexuals who practise anal sex are at the same high risk. . . . Warnings that anal intercourse is dangerous are valid, can be understood by most lay people, and could save many lives."[14] Dr. David Bolling and Dr. Bruce Voeller concur. "Physicians (especially those in obstetrics and gynecology) and sex counselors should set aside their discomfort in discussing anal sex, learn to draw out their patients, and counsel them about risks and prevention."[15]

Indeed, the same could be said for all of the risk factors associated with AIDS. Sexual preference and history of intravenous drug abuse are not the only things that may predispose people to AIDS. There are, and always have been, a small number of people who develop AIDS who do not fit into the neat categories of gay men, intravenous drug abusers, hemophiliacs, and blood transfusion patients that account for the vast majority of cases. However, the number of heterosexuals without identified risks is extremely small, and it is often found that when sufficient evidence is forthcoming, people "without risk" for AIDS in fact fit into the usual categories.

The CDC's own data make clear how unusual so-called heterosexual AIDS is. The number of AIDS cases attributed to "heterosexual contact" is only 6 percent (10,011 cases as of June 1991) in the United States and slightly less in Canada and Great Britain.[16] Notably, every case of transmission of AIDS by "heterosexual contact" has involved sex with a person in a high-risk group: an intravenous drug abuser, bisexual male, hemophiliac, transfusion patient, or person from a country in which AIDS is widespread among heterosexuals. Over 10 percent of heterosexual transmission cases involve bisexual men, perhaps reflecting that a bisexual male is more likely to ask for anal intercourse with a female than is a heterosexual male. Anal intercourse with *any* individual increases the risk of developing immune suppression. If the partner is a male intravenous drug abuser infected with HIV (and other diseases), the probability of transmission increases significantly. Sex with an intravenous drug abuser accounts for over two-thirds of heterosexual transmission cases. In this instance, the partner may have additional risks of his or her own. One would suspect that partners of intravenous drug abusers are themselves unusually likely to use nonintravenous "street" drugs, to be similarly malnourished, and therefore to have concomitant immune dysregulation. Unfortunately, there are no data whatsoever on the matter—another glaring oversight in our epidemiologic studies.

Another "heterosexual" risk group identified in AIDS studies are the sexual partners of people who have contracted AIDS through blood transfusions or clotting factor therapy for hemophilia. These account for about 1 in 1,000 AIDS cases. Three identified risk factors account for the vast majority of these cases. First is having unprotected sexual contact with a person with full-blown AIDS. Such AIDS patients are multiply infected and almost always actively secreting not only HIV but cytomegalovirus, Epstein-Barr virus, hepatitis viruses and their opportunistic infections as well. Thus, the risk to their sexual partner is not simply HIV. Second, anal intercourse is also a risk, as it is among every other group. Finally, age is an unusual risk for heterosexually acquired AIDS. Thomas Peterman and his associates at the AIDS Program of the CDC found that HIV-positive spouses of people with transfusion-associated AIDS were significantly older than spouses who did not acquire HIV seropositivity. The median age of the husbands who contracted HIV infections was seventy-two years, and of the wives, sixty-two. These ages compare with fifty-three years among the HIV-seronegative spouses.[17] Given what what was said about the correlation between increasing age and decreasing immune competence in Chapter 3, this result is not unexpected. Peterman and his colleagues speculate that HIV transmission (and, by analogy, transmission of other viruses as well) may be due to "increased likelihood of mucosal disruption during vaginal intercourse in older women."[18] Older women do not produce as much vaginal mucus as younger women. Moreover, none of the women in Peterman's study used contraceptives, which can act as lubricants during intercourse. Thus, the probability of microscopic damage to the vagina or even overt bleeding is increased among older women.[19]

In addition to cases listed as heterosexual transmission, the CDC has a category called "undetermined risk" or "no identified risk." This category has caused a great deal of confusion for the public, since such cases are often reported in the lay press in such a way that it sounds as if these individuals developed AIDS "out of the blue." This is simply not true. People with "undetermined risk" for AIDS currently represent about 4 percent of all AIDS cases, or, as of June 1991, a cumulative total of 6,655 people in the United States.[20] This number is only about half of all the people who have been designated "without known risk" at one time or another.

Altogether 13,993 AIDS cases have been initially reported with no apparent risk factor between 1981 and December 1991. The reason that the vast majority of these people were so classified is that little or no information was available concerning their lifestyles, sexual habits, or previous medical histories. Indeed, 6,836 of these 13,993 people were subsequently interviewed in detail or other follow-up information was obtained that resulted in 6,318 cases (92 percent of those reexamined) being reassigned to existing AIDS risk groups. Four of the remaining individuals represent the only known cases of AIDS associated with contact with the blood of an AIDS patient in a health care setting; their risk of exposure is, in fact, known. Of the remaining 514 cases interviewed (8 percent), the mode of exposure to HIV could not be determined. These are the figures that constitute the only reasonable evidence that people with no known risk for AIDS may contract the disease.

There remain about 7,200 people also listed by the CDC as having no known risk for AIDS, but for these people, the records are incomplete. Of these, 1,770 have died, refused to be interviewed, or were lost to follow-up. We will never know what their risk factors may have been. The rest of these cases are under investigation. Presumably, more than 90 percent will eventually be reassigned to known risk groups, as occurred in previous reexaminations, leaving about 8 percent without identified risk. In short, when all is said and done, of the nearly 14,000 cases of AIDS initially listed as having no identified risk, fewer than 1,000 will probably remain so identified after the relevant information has been obtained. These 1,000 cases represent one half of 1 percent (5 of 1,000) of AIDS cases.[21]

The actual number of nonrisk group heterosexual cases of AIDS is probably significantly less than even the 1,000 or so that the CDC has on record. Physicians have learned to suspect that such patients are not always candid for a variety of reasons.[22] For example, a significant proportion of people listed by the CDC in the nonrisk group are active military personnel. In one study of 114 HIV-seropositive members of the U.S. Air Force, "no patient admitted to drug abuse, and very few admitted to being homosexual."[23] This sort of evidence—or, rather, lack of it—is, at best, suspect, since both drug abuse and homosexuality can be grounds for dishonorable discharge from the military. Not only would these individuals lose their jobs, their retirement pensions, and their reputations, they would also lose their medical insurance and access to medical services as well. Even the most honest person would think twice about admitting his or her risk factor in such circumstances. Thus, it is hardly surprising to find that in one report on 61 AIDS patients, of the 5 listed as having "no known risk," all were members of the U.S. military.[24] Military personnel are human, however, and do get court-martialed for drug use and homosexuality on occasion. Indeed, more than 13,000 people have been discharged from U.S. armed services over the past decade for being homosexual.[25] Thus, the CDC announced as early as 1987 that "more careful study of widely publicized instances of heterosexual transmission—such as those that reportedly had occurred among new Army recruits—have been found to be unreliable."[26]

In some cases, the problem of identifying risk may be compounded by the fact that the person at risk is not aware of the risk or does not wish to divulge it. For example, sex researchers such as David Bolling report that female patients are likely to withold information about their experience with anal forms of sex: "We find that such information is obtained from most women only after repeated personal interviews and development of strong trust in the interviewer."[27] Others are reticent to discuss the sharing of sexual toys such as dildos and butt plugs that may also represent modes of transmitting sexual diseases. Moreover, it is estimated that about 10 to 15 percent of homosexual men engage in bisexual activity and the majority of these neither self-identify as gay nor inform their female partners of their homosexual activity.[28] Women married to or who have a brief affair with such a man may not know of his other sexual activities, and such a man may be unwilling to admit his AIDS risk to a family physician. Similarly, high school and college athletes and bodybuilders who abuse anabolic steroids may not be aware that they can contract many of the microbes associated with AIDS by sharing needles, just as addicts do.[29] We must seriously consider the possibility that some fraction of the "no known risk" group actually have identified risks but do not or cannot reveal them.

In other cases, such as physicians and health care workers, identifiable immunosuppressive risks may not have been factored into the AIDS equation. Physicians and dentists, for example, supposedly have the highest rates of drug abuse of any other professional group. Most of this abuse does not involve intravenous delivery of the drugs but often involves the same sets of drugs associated with AIDS: cocaine, morphine, amphetamines, barbiturates, and so forth. Certainly, anesthesiologists and many surgeons and nurses are exposed to repeated and prolonged doses of anesthetics. Imagine that such a drug-abusing or anesthesia-exposed health care worker cuts himself during an operation on an AIDS patient or stabs herself on a needle used to inoculate such a patient. The combination of drug abuse or anesthesia, exposure to HIV, and probable transmission of other infectious agents associated with AIDS, such as hepatitis or cytomegalovirus, would be no different from that encountered by an intravenous drug abuser using a dirty needle. In the case of the health care worker, however, the drugs and infectious agents have been delivered separately rather than together. As far as the CDC or NIH is concerned, the physician would not appear in an identified risk group, having never used intravenous drugs. As far as the immune system is concerned, however, the end result is identical.

The same sort of concomitant risk of HIV, other infectious agents, and drugs may be present in people other than health care professionals. For example, Dr. Rand Stoneburner of the New York City Health Department reported to the *New York Times* in 1987 that two identified patients who were HIV seropositive denied belonging to any risk group: "Researchers expressed doubts that the two patients were honest about their sex and drug activities."[30] Further details of these two men's lives, however, suggest that they may have been telling the truth. Rather, the AIDS experts may not have recognized the risks inherent in their activities. Both men had, in addition to HIV, hepatitis B infections. One man admitted having traded sexual favors for a variety of street drugs (suggesting he may also have been impoverished and malnourished) but denied homosexual relations or intravenous drug use. The other man had frequent group sex with a large number of anonymous women and regularly used heroin in noninjectable forms. Notably, a recent study in London found that heroin abusers who had never injected, and heroin abusers who injected but had never shared their needles, both had HIV seroprevalence levels hundreds or thousands of times higher than the general population, and only slightly lower than intravenous abusers.[31] Thus, heroin abuse is a risk factor for heterosexuals independent of needle use. It is probable, based on studies by the CDC, that both men had had repeated bouts of sexually transmitted diseases— another acknowledged risk factor for HIV acquisition—since promiscuous drug abusers are at high risk for multiple STDs.[32] In short, both men had multiple causes of immune suppression independent of HIV, which could have acted synergistically with it. The fact that they do not fit into neat risk groups can as easily indicate that our risk groups need to be adjusted as it does that the individuals might be lying. We will not know until we take their stories seriously.

Another hidden cause of immune suppression may also play a role in promoting "heterosexual" AIDS: malnutrition. This may be a significant risk in U.S. ghettos, barrios, and inner cities. As R. K. Chandra has written, "Undernutrition, when sought, is found in the poor of every community. A study of 300 randomly selected preschool children of impoverished black families in Memphis [in 1970] showed that approximately one sixth of the children were below the third percentile of standards of height and weight and that one fourth were anemic. The prevalence of retarded skeletal development [suggestive of inadequate calcium and vitamin D intake] and low levels of vitamin A were 27 percent and 44 percent, respectively. A surprisingly high prevalence of mild to moderate nutritional deficiencies has been revealed in random population surveys in the United States and Canada. Among hospitalized children and adults, protein-energy malnutrition is even more prevalent."[33]

Consider these findings in the light of some others. First, deficiencies of the sort described by Chandra are often accompanied by significant immune suppression. Second, such deficiencies are a function of poverty. Poverty affects minorities, in particular blacks and Latinos, at rates approximately three to five times the rates they affect whites and Orientals in the United States. Third, drug abuse is most common among impoverished people. Fourth, impoverished people have the poorest health care, the greatest rates of promiscuous sex, and the highest rates of venereal diseases of any other group of people save homosexual men. And finally, consider the fact that blacks and Latinos in the United States have nearly four times the likelihood of developing AIDS than do whites.[34] Is it really imaginable that these five observations are not intimately linked? And yet, shockingly, not one study has ever been performed to determine whether AIDS risk is related to malnutrition, increased disease load, or poverty among minorities.

We need to know. We need to know because of ghetto children such as "Alex," a two-month-old child wracked with diarrhea, covered with a violent rash, and seriously dehydrated from frequent vomiting when he was brought to the office of pediatrician Karl Hammonds in 1988. The mother, a teenage black woman, had fed Alex little besides Koolaid, chewed-up table scraps, and cow's milk since he was born—no diet for an infant and yet one that my pediatrician friends tell me is all too common. "A lot of the health problems for this age group come from whacko feeding patterns," Hammond commented afterward. "Mostly it's young mothers who just don't know better. They love their children; they just don't know how to take care of them. Just like they don't know about what help is available."[35] Hammond went on to make a truly frightening comparison. He contends "that practicing medicine in the inner city is a lot like practicing medicine in the Third World. Doctors encounter similar diseases in both places, he says, and in both places the health-care system often intimidates an uneducated or uninformed patient population. And, in both places, human and material resources are limited."[36] Africa may not be the model for all heterosexuals, but it may be the correct model for our ghettos.

This, then, is the true risk of heterosexual AIDS: that it will become indigenous among the impoverished, ill-educated, malnourished, drug-tempted people of inner cities who are struggling to get out of their ghettos and their barrios and who are now faced with yet another obstacle to their goal. We are not all at risk, but too many of us are.

CHAPTER 10

ALTERNATIVE HYPOTHESES
FOR EXPLAINING AIDS

Elaborating Possibilities

If anything is clear from the preceding nine chapters, it should be that AIDS is not a simple syndrome. We have now seen that the HIV = AIDS tragedy is riddled with inconsistencies in its plot, has lacunae in its dialogue, and masks outright contradictions in its premises. HIV has not proved sufficient to explain AIDS according to any of the standard criteria of microbiology. In some key populations, such as drug addicts and transfusion patients, there are no data demonstrating that HIV per se increases morbidity or mortality compared with HIV-negative controls. Some key investigators, including Luc Montagnier and Robert Gallo, have invoked cofactors either as necessary for HIV activity or at the very least as adjuvants, or stimulators, of HIV. And a very large number of immunosuppressive agents other than HIV are associated with AIDS. They await offstage for their roles to be redefined as the script is being rewritten. The problem for the writers of this tragedy is that the plot itself is not yet clear. They do not yet know what to do with such a vast and varied cast.

There is only one thing to do. We must elaborate possibilities. In science, as in theater or fiction, the tension of the plot is produced by the alternative resolutions we can imagine. A plot that unfolds without surprise is boring. Similarly, in science, research that can reach only one conclusion is hardly worth performing; it has no potential to yield discoveries. We want a plot that proffers alternatives. HIV has been set up as the villain of this piece, but is it not still possible that we have been led a merry chase away from the real culprits? Might HIV be in league with other ne'er-do-wells? Who (or in this case what) else might we suspect? What kinds of information need we attend to in order to test the integrity of our imagined alternatives?

As fanciful as this elaboration of possibilities may sound, it is the basis of all good science. In fact, I have previously defined scientific discovering as a process of elaborating all imaginable explanations for a phenomenon, constrained by an ever-increasing body of observation and experiment. The resulting recursive interplay of imagination and reality assures us that we have explored every alternative before we conclude that we have reached the correct answer. This is an evolutionary view of how scientific theories are built. Diverse theories, like ideas, compete against one another for survival in a world just as cut-throat as a wild jungle. Just as an animal or plant must make best use of available resources, so must a theory explain all of the available data or lose out to a more efficient competitor. Just as living things must reproduce and thrive, so must theories be transmitted from mind to mind in forms that grow into active research projects in order to be fecund. And just as only some individuals and some species survive in the wild, only some theories can survive in the realm of the intellect.[1] That is as it should be. The purpose of generating diversity is to make sure that our selection includes the best answer. We do not, after all, want to make the mistake that Aristotle made when he asserted that heavy objects fall faster than light ones. The first reasonable answer is not necessarily the best—or the final—answer

AIDS research has not yet explored all of the possible explanations of the syndrome. We adopted the first reasonable answer before comparing it with all its possible competitors. Now we must do what should have been done a decade ago: elaborate alternatives and systematically compare and test them.

HIV Is Necessary and Sufficient to Cause AIDS

The question of whether HIV is necessary and sufficient to cause AIDS must be answered logically and through experiment and observation. There is no doubt that HIV is highly correlated with

AIDS. Correlation is not, however, proof of causation. An analogy will help to explain why it is not.

Millennia of observation have verified the theory that sexual intercourse between a man and a woman often results in a baby. For much of that time, it was assumed that the semen deposited in the woman's womb by the man was all that was necessary and sufficient to produce the baby. No one even suspected the presence of an egg prior to the eighteenth century, nor could one have been seen without a microscope in any case. Therefore it was logical that sperm was all that was necessary. The facts certainly appeared to support the case. No man; no baby. No intercourse; no baby. No semen; no baby.

We now know that the woman makes an equally important contribution to the process. An egg must also be present in the woman's womb, and it must be fertilized by a sperm cell in order to produce a viable fetus. In other words, the semen is necessary, but it is not sufficient to produce a baby. An egg is also necessary. Having made that discovery, we also found that intercourse was no longer necessary. Test tube fertilizations are possible. Even more striking, these test tube studies have demonstrated that simply bringing the semen and egg together may not lead to fertilization. Sometimes the sperm motility is too low or the egg is genetically damaged, and no fetus is formed. And even when fertilization has taken place, the embryo may not implant in the uterus, or it may not develop properly thereafter. In other words, even when the necessary components are brought together, they may not be sufficient to produce a baby. Even more surprising, scientists have discovered a process called parthenogenesis, in which eggs may be induced to form embryos in the absence of fertilization under very precise chemical conditions. (The resulting embryos rarely develop properly.) In short, although sperm is almost 100 percent correlated with formation of embryos, it is neither absolutely necessary, nor is it ever sufficient to result in a baby.

If we compare HIV and AIDS to sperm and babies, it is clear that the very high correlation between HIV and AIDS does not mean that HIV is the sole necessary and sufficient cause of AIDS. HIV may be to AIDS what sperm is to babies—only a part of a more complex equation in which other factors are necessary and in which the appropriate conditions must be satisfied for those factors to be effective. Or perhaps the sperm analogy is not correct. Perhaps HIV is actually an epiphenomenon, to resurrect an earlier analogy, like carbon dioxide in drowning victim—present but not a necessary or sufficient cause of death at all. Logic tells us, in short, that the presence of HIV in AIDS can mean many things other than that it is the cause of the syndrome.

The entire case for HIV as the cause of AIDS rests upon epidemiologic correlations not even as good as those linking semen to conception. Thus, despite repeated statements by government officials that the cause of AIDS is known and that it is HIV, I can no longer find any major investigator in the field of AIDS who will defend the proposition that HIV is the only immunosuppressive agent involved in AIDS. Even Robert Gallo, one of the staunchest defenders of the HIV-only hypothesis, has written that "although infection by HIV-1 has been implicated as the primary cause of AIDS and related disorders, cofactorial mechanisms may be involved in the pathogenesis of the disease. For example, several viruses commonly detected in AIDS patients and capable of transactivating the long terminal repeat of HIV-1, such as herpesviruses, papovaviruses, adenoviruses and HTLV-I have been suggested as potential cofactors. Another candidate is human herpesvirus-6 (HHV-6, originally designated human B-lymphotropic virus), which has not only been identified in most patients with AIDS by virus isolation, DNA amplification techniques and serological analysis, but is also predominantly tropic and cytopathic *in vitro* for CD4+ T lymphocytes. . . . These observations indicate that HHV-6 might contribute directly or indirectly to the depletion of CD4+ cells in AIDS."[1] Statements such as this one suggest that even mainstream HIV researchers are beginning to consider the possibility that HIV may not be sufficient to cause AIDS. They do not doubt that it is necessary.

The question for Gallo and many of his colleagues seems to be whether cofactors are necessary to developing AIDS. In other words, is HIV capable of producing AIDS by itself in the complete absence of cofactors in at least some cases? Is the role of cofactors one of simply enhancing or speeding up HIV activities that would take place in any event? A definitive answer to these questions has not emerged. Conservative researchers, such as immunologist David Klatzmann, one of the people who helped demonstrate HIV activity shortly after the virus was isolated, maintain that HIV is still all that is needed: "I'm not sure we really need major cofactors to explain HIV pathology."[2] NIH's chief immunologist and head of the National Institute of AIDS (NIAID), Anthony Fauci, apparently agrees. When asked his view of cofactor theories, he responded that "critiquing a dubious theory would take time away from more productive efforts."[3] James Curran, head of AIDS research at the CDC stated as late as August 1992 at the international AIDS conference in Amsterdam, "There is not AIDS without HIV."[4] Thus, the official view of AIDS remains, as it has been for the past eight years, that HIV is both necessary and sufficient to cause the syndrome. The role of cofactors, if any, is to make more virulent an infection that would lead to AIDS in any case. These cofactors are not worthy of investigation in and of themselves.

The proof of this position has not yet been forthcoming, either through studies of HIV-infected chimpanzees or of HIV-infected human beings. Despite my oft-stated challenge, no one has been able to document a single AIDS patient, tested for the range of risks outlined in this book, whose only immunosuppressive risk was HIV. I will not, therefore, spend any more time on what may be called the HIV-only theory here. It remains unsubstantiated. Besides, numerous general books and articles exist on the subject, including many *Scientific American*, *Science News* and *Discover* articles, and innumerable government publications. My purpose is not to survey these publications but to draw out some of the key implications of the HIV = AIDS theory so that it can be compared with its alternatives.

Two of the most important implications of the HIV-only theory of AIDS are that all risk groups should develop AIDS at approximately the same rate following HIV infection and that the symptoms they manifest should be, on the whole, the same. The virus alone should control disease progression. One logical implication is that the immunological status of an infected person should be irrelevant to susceptibility to contagion or to the progression from infection to disease. Acquisition of the retrovirus infection should be the sole factor determining whether an individual develops AIDS. Everyone should be at equal risk for AIDS, just as everyone is at equal risk for hepatitis B virus, syphilis, or measles. Clearly, according to this theory, prophylaxis against HIV or drugs designed to control or eliminate existing infection is the only reasonable approach to treating AIDS. Yet all of these predictions have been called into question. Risk of AIDS differs by identifiable sexual, drug, and medical practices. The onset of AIDS following HIV infection varies tremendously. And it is impossible to attribute all of the immune suppression in AIDS to HIV.

Several other implications also follow from the HIV-only theory. First, the HIV-only theory plainly predicts that immune suppression in people at risk for AIDS should not be present until they contract an HIV infection. This HIV-induced immunodeficiency will then lead to increased susceptibility to other infectious agents. Thus, the clear sequence of events according to the HIV-only theory of AIDS is that HIV comes first, followed by the other infections that characterize AIDS patients, such as cytomegalovirus, Epstein-Barr virus, toxoplasmosis, and so forth. Furthermore, since the primary cause of immune deficiency is HIV, AIDS should be as old as HIV, and no older. It should be present only where HIV is present, and HIV-negative cases of AIDS should not exist. All of these predictions have been contradicted by numerous studies. Significant immune suppression often, if not always, precedes HIV infection. HIV itself does not appear to cause immune suppression in the absence of other immunosuppressive agents. And AIDS-like cases existed prior to the identification of AIDS and continue to appear in apparently uninfected individuals.

In short, significant anomalies plague the HIV-only theory of AIDS and bring its validity into question.

HIV Is Necessary But Requires Cofactors

The official view of AIDS has been the focus of increasingly frequent revisions. More and more investigators are considering the possibility that although HIV is very highly associated with AIDS and may be necessary to cause it, it is not sufficient. Thus, HIV alone should not lead to AIDS. The strongest evidence for cofactor theories of AIDS comes from the studies of prostitutes that were summarized in Chapter 2. Although intravenous drug-abusing prostitutes worldwide have a very high incidence of HIV infection, neither their drug-free colleagues nor their drug-free clientele have a significant rate of infection. Such data led Shiokawa to suggest that among both female prostitutes and homosexual men, prior immune suppression is necessary for contracting HIV.[1] In other words, HIV may cause AIDS once it has taken root in a person, but only people with preexisting or concomitant immune suppression may be prone to infection.

A stronger version of Shiokawa's hypothesis has been proposed by Dr. Linda Walliford Pifer and her colleagues at the Health Science Center of the University of Tennessee in Memphis. During 1986 and 1987, Pifer studied a group of twenty-seven healthy gay men and found that they had "borderline" immunodeficiencies. After studying a wide range of possible factors that might contribute to immune deficiency, including recreational and addictive drug use, pharmaceutical agents, types of sexual activity, and nutritional status, Pifer and her colleagues concluded that no single cause of immune deficiency could be found. "Contributory factors might include frequency of use and/or abuse of various 'street' and prescription drugs and frequent diarrhea and accompanying fluid loss." All of these factors were significantly more common in the gay men than in heterosexual men and women or homosexual women in their study. They also found that general insults to the health of these men included various sexual practices, most specifically receptive anal intercourse that frequently led to bleeding, and an unusual incidence of sexually transmitted diseases. They concluded that "these findings may be of significance in attempting to understand why approximately three fourths of AIDS cases to date have occurred in male homosexuals. When these factors are assessed independently, they do not appear to present a substantial insult to the overall immune defenses. When viewed in combination, however, they may present a sufficient impairment to place male homosexuals at a small but significant disadvantage in attempting to maintain immune defenses against a number of infectious agents, including the etiologic agent of AIDS. [Thus] borderline immune deficiency in conjunction with the high frequency of anal intercourse and mucous membrane exposure to antigens in semen may have created optimal conditions for the rapid spread of HIV."[2] This view of cofactors is that they create the conditions for HIV to cause AIDS.

A very different view of cofactors is suggested by the work of Montagnier, who argues that HIV is not sufficient to cause AIDS by itself, no matter what the predisposition of the immune system. The virus is, to use Peter Duesberg's term (which Montagnier oc-

casionally adopts himself), a "pussycat." But for Montagnier, unlike Duesberg, HIV is a pussycat that can be transformed into a
tiger. The cofactor that transforms HIV into a tiger according to
Montagnier is that smallest of self-replicating germs, the *Mycoplasma*. Just as the joining of viable egg and sperm can create a
fetus under the appropriate conditions, so the joining of HIV and
a *Mycoplasma* may produce AIDS under appropriate conditions.
Indeed, research in Montagnier's laboratory, that of American microbiologist Shyh-ching Lo, and a variety of other laboratories,
confirms both the frequent presence and cofactor activity of HIV.[3]

But as we have seen, many cofactors besides *Mycoplasmas* have
also been proposed (Table 2). Montagnier himself has recently admitted in an interview that, "Maybe I am wrong in saying that it
[*Mycoplasma*] is the only one, but *Mycoplasma* is a better candidate, because it is present everywhere."[4] Dr. Shyh-ching Lo and
Colonel Douglas J. Wear of the Armed Forces Institute of Pathol-

TABLE 2. IDENTIFIED COFACTORS FOR HIV

	In Vitro Evidence	Epidemiological Evidence
Viruses		
Adenoviruses	+	
Cytomegalovirus	+	+
Epstein-Barr virus	+	+
Hepatitis viruses	+	+
Herpes simplex 2	+	+
Herpes simplex 6	+	
HTLV I	+	+
Bacteria		
Mycoplasmas	+	
Mycobacteria		+
Drugs		
Opiates	+	+
Cocaine	+	+
Inhalant nitrites		+
Malnutrition		+

Source: Root-Bernstein RS. 1992. HIV and immunosuppressive cofactors in AIDS. *EOS.
Revista immunologia ed immunofarmacologia* 12 (4). Table reprinted with permission of
the publisher.

ogy also caution that their "attitude in dealing with AIDS rsearch is to explore all possible avenues. To make any conclusion lightly or prematurely, such as ruling out any possible role of microbes in AIDS, or to commit oneself exclusively to a particular agent and completely rule out any other possible role of a different microbe, may all result in a greater loss of AIDS victims."[5]

Thus, many possible cofactors are being explored. To review, the primary viruses suggested to stimulate HIV activity are human T cell lymphotropic virus, type I,[6] herpes simplex type 2,[7] herpes simplex type 6,[8] hepatitis B virus,[9] Epstein-Barr virus,[10] cytomegalovirus,[11] and genital ulcer disease.[12] These infectious agents may promote HIV activity by increasing the probability of transmission and active infection, expanding the range of cells that HIV can infect, transactivating HIV when present in the same cells, and helping to induce autoimmune processes. Much more research is clearly needed to pin down these possibilities or to exclude them.

Only one thing is certain: Many scientists have concluded that HIV is not a soloist in AIDS. For example, Dr. Elena Buimovici-Klein and her co-workers, having noted the extremely high viral load associated with homosexual men (whether HIV seropositive or not), suggest that "such infections together with rectal immunization by alloantigens [sperm and lymphocytes in semen] may be important in inducing T4 lymphocyte activation, which is an absolute requirement of HIV replication. . . . The results [also] suggest that the high rate of virus shedding among HIV-negative homosexual subjects might be a factor in the development of AIDS in this high-risk population." They are of the opinion that "HIV is an essential, but not necessarily by itself a sufficient factor in the development of [AIDS]."[13] Dr. E. Fernandez-Cruz and his colleagues in Madrid, Spain, have reached a similar conclusion from their studies of addicts: "Heroin and/or pre-existing infections may be important co-factors in the establishment and evolution of HIV infection in heroin addicts."[14]

As Dr. Fernandez-Cruz makes clear, the possible cofactors for AIDS are not limited to infectious agents. Some scientists view noninfectious agents such as recreational and addictive drugs as playing significant roles in the disease pathology as well. Harry Haverkos, for example, has written that "a multifactorial model can be postulated to explain the various manifestations of AIDS. We suggest that the natural history of AIDS begins with immune

dysfunction resulting from HIV infection of T-helper lymphocytes. One or several cofactors, present in some but not necessarily
all patients, then determine which, if any malignancies or opportunistic infections the patient manifests. Our analysis suggests that
a promoter or cofactor for Kaposi's sarcoma is the use of large
quantities of nitrite inhalants."[15] Similarly, R. M. Donohoe and
Arthur Falek of Emory University have suggested that "the fact
that various abused drugs are themselves immunomodulatory and
immunocompromising has led [us] to the suggestion that such
drugs, in particular opiates, may directly alter susceptibility to
HIV-1 and exacerbate the complications of AIDS."[16] Many other
physicians studying AIDS in addicts have reached similar conclusions.[17]

A different set of factors may be at work in hemophiliacs and
blood transfusion patients. Dr. J. L. Sullivan and his co-workers
found that HIV was not sufficient to explain AIDS in their patients: "Exposure to factor VIII concentrate was significantly correlated with decreased percentages of T-helper/inducer cells, decreased   T   helper/suppressor   cell   ratios,   and   decreased
proliferative responses to plant mitogens [a measure of cellular
immune competence]. . . . Epstein-Barr virus and cytomegalovirus infections acted in a synergistic manner with [HIV] to produce immunoregulatory defects. The authors conclude that hemophiliacs receiving commercial factor VIII concentrate experience
several stepwise incremental insults to the immune system: alloantigens in factor VIII concentrate, [HIV] infections, and herpesvi-
rus infections."[18]

The problem faced by those investigating AIDS is that the number of possible cofactors is so large that it may not be possible to
identify any single one. Dr. A. R. Lifson, Dr. G. W. Rutherford,
and Dr. H. W. Jaffe in their recent review article, "The Natural
History of Human Immunodeficiency Virus Infection," make a
similar point: "Why some HIV-infected persons develop disease
and others do not is not completely understood. The role of cofactors for disease progression needs additional investigation. There
may be no one universal cofactor for progression but, rather, various agents that cause immune stimulation and reactivation of latent HIV. Therefore, exposure to a variety of infectious or environmental agents (such as through sexually transmitted diseases
or injection of intravenous drugs) may accelerate progression to

disease in HIV-infected persons."[19] The fact that at this late date we do not know exactly what cofactors HIV may require, or what their effects in AIDS may be, shows how badly needed are the elaboration and testing of possibilities.

The HIV-plus-cofactors theory can be tested in terms of three important differences between it and the HIV-only theory. First, the HIV-plus-cofactors theory predicts that the probability of infection will be altered by the immune status of the person exposed to HIV. The more immune suppressed the individual is, the more easily infection will occur. Infection should still be possible in the absence of immune suppression, however. Second, the rate of disease development following infection should be much faster in (and may, in fact, be dependent upon) the presence of other immunosuppressive cofactors. Again, such cofactors will not be required for disease progression. They only determine the time course of the disease. Third, the specific manifestations of AIDS in different risk groups should be determined by the specific cofactors unique to that risk group. Thus, one should not expect a hemophiliac, a drug addict, and a homosexual man to develop the same set of opportunistic diseases because the specific factors involved in immune suppression will not be identical. As we have seen, all three of these predictions have been amply verified. Cofactors are necessary for AIDS, and they definitely alter its course.

Cofactors may also determine who develops HIV infections and AIDS. Yale University's Alfred Evans believes that what he calls "clinical promotion factors" determine which individuals, among all those infected with a virus, actually develop clinical symptoms.[20] It follows that an HIV-seropositive person who is able to avoid major immunosuppressive risks might never develop AIDS. Thus, one might be able to control the development of AIDS by controlling exposure to cofactors.

One final set of implications also follows from the HIV-plus-cofactors theory. As in the HIV-only theory, AIDS should be no older than HIV, it should occur only in HIV-seropositive people, and HIV-seronegative cases should not exist. A major difference between the two theories is, however, that whereas immune deficiency is triggered by HIV alone in the HIV-only theory, HIV-positive individuals may experience no immunodeficiency in the absence of cofactors according to the cofactor theory. Thus, the relationship between HIV infection and development of immune suppression is quite different. As we have seen, significant immune impairment is, in fact, present in many people at risk for AIDS prior to or in the absence of HIV infection. Thus, existing data indicate that HIV itself may be an opportunistic agent.

AIDS Is a Multifactorial, Synergistic Disease

A somewhat more radical revision of the HIV-only theory is the possibility that AIDS is a multifactorial, synergistic disease resulting from a combination of many immunosuppressive factors (Figure 20). The major difference between the multifactorial, synergistic theory and the HIV-plus-cofactors theory is that in the former, although HIV is often present in AIDS, it is not necessary to cause

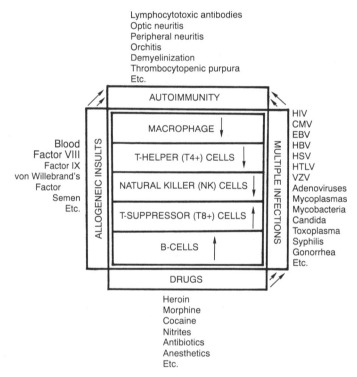

FIGURE 20. A general multifactorial, synergistic model of AIDS. (See also Sonnabend, JA. 1989. AIDS: An explanation for its occurrence among homosexual men. In: Ma P, Armstrong D, eds. 1989. *AIDS and Infections of Homosexual Men.* 2d ed. Boston: Butterworths, 449–470.)

AIDS. In the multifactorial, synergistic theory, HIV is no more than a frequent, opportunistic infection associated with, but not necessary, to cause AIDS. Its high correlation with AIDS may result either from the great difficulty with which it is transmitted (making it an excellent marker for other immunosuppressive agents that are more frequently encountered) or because it is a widespread, nonpathogenic virus that becomes activated under conditions of extreme immunologic stress (that is, HIV may be an opportunist).

The principal exponents of multifactorial, synergistic theories of AIDS are Dr. Joseph Sonnabend of New York City, R. E. Harris of the American Health Foundation in New York, Dr. Eleni Papadopoulos-Eleopulos of Australia, Dr. Luis Benitez Bribiesca of Mexico City,[1] Professor Zh. Zhelev of the University School of Medicine in Pleven, Bulgaria,[2] Dr. G. Mathé of the Hôpital Paul Brousse in Villejuif, France,[3] Peter Duesberg and Harry Rubin of the Department of Molecular Biology at the University of California, Berkeley,[4] Charles J. Hoff and his collaborators at the University of South Alabama College of Medicine, and myself.[5] To quote Dr. Benitez Bribiesca, "We and other investigators have noted a number of unexplained facts that cast many doubts over the simplistic hypothesis of a unique retroviral etiology of AIDS. It is likely that many other factors act synergistically to induce immunosuppression in these patients, i.e.: a) repeated and chronic infections; b) use of drugs: c) alloantigenic stimulation by blood and/or semen; d) anaesthetics; e) antibiotics and f) malnourishment."[6]

Just as those who advocate a role for cofactors as necessary triggers for HIV do not agree on the specifics of which cofactors are necessary, so, too, do those of us who are propounding multifactorial, synergistic theories differ in the details of our proposals. Joseph Sonnabend is, perhaps, on the most conservative end of the spectrum. He was also the first major AIDS investigator to propose that HIV is not a sufficient cause of AIDS and to bear the brunt of that criticism by losing funding and various positions on editorial and research boards. In 1983 Sonnbend proposed that AIDS in homosexual men is caused by a vicious cycle of ever-escalating immune deficiencies, caused, in turn, by a continuous bombardment of immunosuppressive agents. In the earliest stages, sexual promiscuity leads to repeated infections with a variety of disease agents known to cause transient immune suppression, such as cytomegalovirus, Epstein-Barr virus, toxoplasmosis, syphilis, and hepatitis B virus. Because exposure to these agents is almost continuous in high-risk homosexual men and reinfection or activation of latent infection very common, some of these diseases become permanently activated. The immunosuppressive effects of these other infections on the immune system become chronic. HIV also plays a role in this vicious cycle of infections, but Sonnabend does not know whether its effects are any more important than the other infections that are present. What begins to happen is that each infection lowers resistance to infection with other disease agents, so that a vicious cycle of immunological impairment is initiated.

Correlated with this increased disease load, Sonnabend observes, the immune system produces continuous, high levels of interferon, a molecule secreted by the immune system itself that is known to down-regulate T-helper cell activity. Interferon also has a number of pathogenic effects when present at high levels for long periods of time, as occurs in AIDS and various autoimmune diseases. In addition, these multiple infections result in nonspecific stimulation of B cells and autoimmune processes directed at T cells. Immune responses to semen (antisperm antibodies that also recognize T cells as targets) are also elicited in most homosexual men, amplifying the autoimmune aspect of the ever-increasing immune deficiency. This autoimmunity is correlated with the presence of circulating immune complexes, which additionally tie up healthy T cells into inactive clumps. At some point these diverse processes cross a threshold beyond which the immune system can no longer sustain its activity, and AIDS enters a self-sustaining decline into complete destruction of immune function.[7] As Sonnabend notes, the basic reason that this theory remains less substantiated than HIV-directed theories is that no major studies of AIDS patients have been set up specifically to examine the cooperative roles of each of these factors in the process that is AIDS.[8]

The major drawback to Sonnabend's theory is its limitation in explaining AIDS in homosexual men only. He has not attempted to explain AIDS in other risk groups, such as drug abusers or hemophiliacs. Thus, his general approach to AIDS has been rediscovered by other researchers attempting to understand the natural history of the syndrome in nonhomosexual AIDS patients. One such researcher is Charles Hoff. He and his collaborator Raymond

Peterson have argued that although the theory that exposure to semen and blood antigens caused AIDS dropped from favor after the discovery of HIV, evidence continues to mount in favor of causative roles for these factors. They argue that their alloantogenic challenge model provides reasonable explanations for a large number of unanswered questions about immune profiles in HIV antibody negative gay males and in HIV infected individuals who are also exposed immunologically to blood or semen. They cite, in particular, evidence that HIV seropositive gay men and blood transfusion patients often have normal T cell counts and remain healthy for over a decade in the absence of other immunosuppressive challenges.[9] On the other hand, antibodies against semen and lymphocyte antigens are found at unusually high levels in many gay men, some women, and transfusion patients, both HIV seropositive and HIV seronegative, and the presence of these antibodies is highly correlated with immune suppression in all groups. Cross-reactivity between these antibodies and T helper cells explains how these cells are selectively eliminated during AIDS, while the concomitant increase in T suppressor cells that also characterizes AIDS is a symptom of the immune system attempting to suppress this and associated autoimmune reactions. Thus, they argue,

> . . . immunosuppression associated with challenge by lymphocyte alloantigens enhances HIV infection after exposure, promotes survival of the virus, and augments the action of HIV and other reactivated viruses which have immunosuppressive effects. As a result, it is possible alloantigenic challenge can increase both the risk of significant infection and progression to manifest AIDS.[10]

R. E. Harris of the American Health Foundation in New York City and his collaborators have also suggested a wide set of synergistic interactions in drug abusers analogous to those proposed by Sonnabend for homosexual men. Harris and his associates observed a group of forty new patients admitted to a methadone clinic in 1990. They noted that the amount of drugs and number of injections per week had increased significantly since 1984; that very high rates of promiscuity were prevalent (80 percent of the men were sexually active with more than one woman concurrently); condom use was low; malnutrition was very common; and the average body mass of all the individuals was subnormal. In short, these patients were at increased risk for immune suppression due to their drug habit, high-risk sexual practices, multiple sexually transmitted diseases, and malnutrition. "Synergism of these factors should be considered in the manifestation of HIV infection among IV drug abusers," they concluded.[11]

Eleni Papadopoulos-Eleopulos agrees with Harris and Sonnabend about the range of immunsuppressive factors involved in the induction of AIDS but differs in the specific mechanism by which she imagines these to be working.[12] While Sonnabend sees direct immunosuppressive effects of cytomegalovirus, Epstein-Barr virus, hepatitis B virus, and other viruses on autoimmune processes, and Harris the additional direct effects of drugs on immune system cells, Papadopoulos-Eleopulos believes that these effects are indirect. She cites evidence that each of these agents can induce immunological responses that are characterized by oxidation reactions. Such oxidation reactions (which are chemically similar to those that cause bread to go stale and hydrogen peroxide to have astringent qualities) can be highly destructive to immune system cells and other tissues as well. Interestingly, antioxidant compounds such as vitamin C and vitamin E are known to be immunostimulatory[13] and other antioxidants, 2-mercaptoethanol and N-acetylcysteine, have been found to improve T cell proliferation in test tube studies of HIV-infected cells from AIDS and AIDS-related complex patients.[14] Whether these observations are epiphenomena of AIDS or central to its pathogenesis remains to be determined.

Without a doubt, the most radical of the synergists is Peter Duesberg. He has maintained since 1987 that HIV cannot cause significant immune impairment. He believes that the cause of AIDS in all risk groups is drugs: addictive drugs such as heroin and cocaine; recreational drugs such as inhalant nitrites, barbiturates, marijuana, quaaludes, and many others that are prevalent among drug addicts and homosexual men; and pharmaceutical drugs used to treat viral infections, such as AZT and DDI. These promote the infections that lead to death. He and Bryan Ellison have argued strongly for the position that HIV is merely an epiphenomenon of AIDS—an opportunistic "pussycat" of a virus that has no role in producing the initial stages of immune suppression.[15] Whatever the merits or demerits of his case (and there are many of both), he has played the most important role in the AIDS debates by keeping the door to the AIDS mansion open long enough for a few other intrepid souls to sneak in and begin investigating the rooms overlooked by those searching for (and finding) only HIV.

I believe that Duesberg is wrong in ignoring the role of HIV in AIDS. It is certainly highly correlated with the syndrome (even given the methodological sleight of hand involved in defining the syndrome by the presence of the putative causative agent prior to definitive demonstration of causation). It cannot be any less important than, say, cytomegalovirus or Epstein-Barr virus, which are similarly universal among AIDS patients. Even if HIV is normally the "pussycat" he says it is, so are all the other opportunistic infections that take advantage of immune suppression in AIDS patients. I posit that at the very least, HIV, even if it is nothing more than an opportunistic infection that follows the initiation of AIDS, can have just as serious and potentially as deadly effects as cytomegalovirus, toxoplasmosis, or *Pneumocystis carinii* pneumonia—all "pussycats" in healthy human beings. What makes AIDS patients different is the immune suppression they have prior to developing all of these other active infections, including HIV. It is just as big a mistake to ignore the potential role of HIV in AIDS as it is to ignore the roles of all the other immunosuppressive agents that afflict AIDS patients.

In any event, the specific implications of the synergistic theory of AIDS differ from those of the cofactor theory on several major points. Once again, one would expect different risk groups to develop AIDS at different rates and to manifest different symptoms as a result of differing immunosuppressive risks. This prediction is validated by the observation that hemophiliacs develop different manifestations of AIDS than do homosexual men.[16] One would also expect previously immunosuppressed individuals to be most likely to contract the various infections typifying AIDS, including HIV. Thus, the time course of AIDS development should vary depending on exposure to cofactors (as we have seen that it does in Chapter 3) and should involve signficant immune suppression preceding or concomitant with HIV infection in all cases (Chapter 6). The synergistic theory differs from the cofactor theory in that HIV can *never* produce AIDS by itself. Moreover, AIDS may occur in the absence of HIV infection if a sufficient concoction of other immunosuppressive agents is encountered simultaneously or serially. According to this theory, then, HIV is just one of several dozen players, one that may even be replaceable, on the AIDS stage. A significant body of evidence supports these predictions.

The multifactorial, synergistic theory of AIDS has several implications that differ radically from the HIV-only and HIV-plus-cofactors theories. One is that AIDS should predate HIV—as we have seen that it does. It follows that HIV testing may not be the best means of identifying those at risk for AIDS. Treating all AIDS patients in an identical manner, particularly treating them exclusively for their HIV infection (when present), is a tremendous mistake. Many other modes of treatment and prophylaxis should be effective, specifically protection against or elimination of as many immunosuppressive agents as possible.

AIDS Is Caused by Autoimmunity

As if the question of AIDS were not already complicated enough, some investigators, including me, have proposed that another mechanism—autoimmunity—adds to and may actually cause the immune suppression underlying AIDS.[1] Once again, as with the previous hypotheses, there is significant disagreement about the role of HIV in autoimmunity in AIDS. Dr. Rubert Holms, Dr. John Habeshaw, and Professor Angus Dalgleish of the London Hospital Medical School have formed a company called HIVER (HIV Epidemic Research). They contend that AIDS is an autoimmune disease in which HIV infection of T cells and subsequent T cell killing is essentially irrelevant. Habeshaw and Dalgleish cite many of the same problems summarized in previous chapters of this book: "Why do chimpanzees not become ill after HIV infection? why do other viruses that kill T cells (such as human herpes virus-6 and measles) not cause AIDS? and why do some individuals progress to AIDS relatively quickly after infection whereas others never develop the disease?" Their HIVER hypothesis "is that AIDS is not caused by HIV killing T cells, but by a massive activation of the immune system (similar to that seen in graft-versus-host disease) which is triggered by HIV."[2]

Graft-versus-host disease (GVH) is a problem that first surfaced in human beings following transplants, particularly of bone marrow. Such patients are chemically immune suppressed to accept their transplant, and therefore their T cell activity is very low. An inevitable component of many transplants (particularly of large organs such as the liver) are donor lymphocytes. These lymphocytes continue to act as if they are in the donor. Finding themselves placed inside the body of the transplant recipient is, consequently, like finding themselves inside the biggest immunological challenge ever devised. Everything is foreign, everything is a target for elimination. These donor- (or graft-) derived lymphocytes mount an attack on the recipient's body at the same time the recipient's immune system is attempting to eliminate these foreign intruders. The result is GVH. Both graft and host lymphocytes proliferate at very high rates, stimulating increased antibody production and causing significant T cell killing. GVH is characterized by fever, swollen lymph nodes and greatly enlarged spleen (sites of lymphocyte aggregation in the GVH civil war), chronic diarrhea (probably due to destruction of Peyer's patches in the intestines, which normally protect the bowels from infection), and a measles-like rash. GVH can usually be brought under control within a few weeks of onset. Otherwise it is fatal.

Habeshaw and Dalgleish have proposed that HIV triggers a GVH-like response to cause AIDS. Since the HIV envelope protein gp160 resembles MHC class II proteins, they propose that HIV is able indiscriminately to turn on MHC-regulated immune functions, in particular those controlled by MHC class II–bearing cells. As a result, these cells do not respond properly to other foreign antigens. Since properly activated T-helper cells are necessary to both the activation of macrophages and the production of specific antibodies during disease, improperly activated T-helper cells would fail to protect the body against infection.

The HIVER scientists point out that their theory is akin to that proposed by John Ziegler and Daniel Stites in San Francisco, and Geoffrey Hoffmann and his colleagues in Canada, whose theories were discussed in Chapter 5. Hoffmann's theory is essentially that HIV looks very much like MHC class II. When the immune system responds to an HIV infection, it produces antibody against these MHC class II–like proteins. The resulting antibodies are thus, in Hoffmann's language, anti-MHC image (or AMI). These AMI antibodies in turn elicit anti-idiotype antibodies, or MHC-image (MI) antibodies. The combination of MI and AMI antibodies mimics the MHC class II interaction with CD4 protein of T-helper cells, and the antibodies directly attack these cells. The result is immune suppression by T lymphocyte killing.

Various problems remain with both the HIVER hypothesis and the Ziegler-Stites-Hoffmann theory.[3] None of these theories can explain the latency period between HIV infection and AIDS. None addresses why autoimmunity is nearly universal among AIDS patients but so rare among the general populace. And all assume that HIV is sufficient to trigger graft-versus-host disease or autoimmune response. The role of HIV in the induction of autoimmunity has, however, been called into question, ironically enough by Hoffmann's own research. He and his collaborator, Tracy Kion, found that mice with systemic lupus erythematosus appeared to have antibody to HIV, although they had never been exposed to the retrovirus. Hoffmann and Kion concluded that autoantigens or alloantigens may be responsible for these antibodies. E. J. Stott and his collaborators then demonstrated that, indeed, macaques exposed to human lymphocytes develop anti-HIV antibodies, even in the absence of HIV.[4] As John Maddox, editor of the journal *Nature*, commented, "None of this would imply that HIV is irrelevant to AIDS, but that an immune response to foreign cells, most probably lymphocytes, is also necessary."[5] Indeed. Joseph Sonnabend and his colleagues were trying to point this out some seven years ago, although their work was focused on the lymphocytes and sperm cells found in semen, not lymphocytes infected with HIV. Only time will tell whether the difference is important.

I have suggested yet another theory by which autoimmunity directed not only at T cells, but at other tissues as well, might be induced in AIDS patients. This theory, developed with Sheila Hobbs, does not deny HIV a role in AIDS but rather insists that HIV is not sufficient to provoke an autoimmune response on its own.[6] We believe that autoimmunity in AIDS is no different from autoimmunity in any other well-studied animal model and hence requires a combination of appropriate antigens. HIV may be one. It is not a necessary element, however. The important point from our perspective is that the most logical way to explain the universal incidence of antibodies directed at T-helper cells and macrophages in AIDS patients (and in many people at risk for AIDS) is to acknowledge the very high degree of similarity that exists between the proteins of lymphocytes, sperm, blood factor concentrates, cytomegalovirus, Epstein-Barr virus, hepatitis B virus, adenoviruses, and other extremely common infectious agents, and the proteins of T-helper cells and macrophages. We believe that these infectious agents, blood proteins, and sperm antigens, not HIV, provoke the autoimmune response against CD4-bearing cells. The role of HIV, and perhaps of other cofactors in AIDS, is to act as the "adjuvants" necessary to provoke this autoimmunity.[7]

To summarize, autoimmunity may be the primary cause of AIDS, and it may be induced directly by exposure to HIV, or autoimmunity may be a secondary phenomenon in AIDS induced by immunological synergism of some subset of diverse infectious agents. In either case, autoimmunity can and does cause significant immune suppression independent of HIV and by means of a very different mechanism than has previously been supposed. The difference in mechanism translates into the need for different treatments as well. If AIDS is an infectious disease, then vaccination or antiretroviral agents may be effective; if AIDS is autoimmune, direct manipulation of the immune response or lymphocyte subsets may be necessary once the syndrome has been initiated. Trying to treat HIV infection following the induction of autoimmunity may be as useless as treating a drowning person by trying to cleanse the carbon dioxide out of his bloodstream. First, you need to get the person out of the water.

The specific version of the autoimmune theory of AIDS that one chooses also has implications for how old or new AIDS may be. If HIV is a new disease agent and is required for induction of lymphocyte-directed autoimmunity, then AIDS must be a new disease. If HIV is not necessary, then AIDS should be a much older disease. Similarly, HIV-negative cases are impossible in the former case, but necessary and expected in the latter. Once again, I believe that the existing data support a multiple-antigen-mediated cause of autoimmunity in AIDS.

Implications and Testing of These Alternative Hypotheses

The most important point that must be made concerning these alternative explanations of AIDS is that we are still ignorant of the causes of this disease. As shocking as the admission of this ignorance may be, it is by far better than pretending to know more than we do. Informed skepticism is still a necessary component of good research. Dr. Beverly E. Griffen, director of the Department of Virology at the Royal Postgraduate Medical School in London, has urged that we not accept HIV as the cause of AIDS just yet. "It will surely lead to a scientifically healthier society," she says, "if the burden of proof for HIV as a deadly pathogen is returned to where it belongs—to those who maintain that HIV causes AIDS—and others are allowed to pursue alternative approaches in the battle for eradication of the disease."[1] Similarly, John Maddox, editor of the eminent journal *Nature*, has written that he should have given critics of the HIV theory, such as Peter Duesberg, room to express their concerns.[2] "I'm not for a minute saying Duesberg is right in all points," Maddox says, "but I feel sorry that *Nature* has not done more to give his view prominence. It would have hastened the process by which the scientific community is coming around to the view that the pathogenesis of AIDS is more complicated than the baby-talk stories we were all given a few years ago."[3]

And yet the vast majority of investigators and clinicians have uncritically accepted such "baby-talk stories." The HIV-only theory of AIDS has held total dominance in the research and therapeutic arenas since 1984, to the near exclusion of alternative considerations. This is a fact that can be verified by searching through the literature on topics such as AIDS and semen autoantibodies, AIDS and nitrite abuse, AIDS and cytomegalovirus, and so forth. Each of these topics was actively studied, funded, and written about prior to 1984, and each of them nearly died out thereafter. The "cause" of AIDS was elucidated; why pay attention to these irrelevant asides? The answer, which has come eight years later, is that all of these "asides" have turned out to be fundamental processes in AIDS. AIDS, no matter how one looks at it, is more complicated than the scientific community has been admitting to itself or to the public.

The mistake that Maddox made as an editor of a preeminent science journal—the mistake he has now admitted—was to close the door on alternatives too soon. It is a mistake that was shared as well by the editors of other scientific journals, the committees that oversee AIDS policy and funding, and the vast majority of AIDS researchers themselves. We have been operating like burglars who enter a mansion and plunder the first room that has any treasures worth our notice. Without first finding out how many rooms there are in the house and doing a quick inventory of what is in each room, we risk being satisfied with taking a few silver spoons and gilt candlesticks when next door hang a whole gallery of Rembrandts, Van Goghs, and Picassos, unnoticed. When the owners of the mansion return, they may be upset to find their silverware missing, but they will breathe a huge sigh of relief that their true treasures remain.

AIDS may be a complex disease with as many facets as a mansion has rooms. We may have missed some of its key treasures. This is not to deny that HIV is one of these treasures. What is at issue is whether HIV is so important that all the other rooms in the AIDS mansion should be ignored. The inventory of the last few chapters suggests that, at the very least, other treasures worthy of scientific plundering do exist in the house. It is even possible that no single room contains the complete treasure that is the understanding of AIDS. AIDS may be the mansion itself, as complex and varied as its contents. Yet, so far, the only room in the AIDS mansion that has been fully examined is the HIV room. The others have been ignored, possibly at the cost of human lives.

It is time that we do something about this gross and inexcusable oversight. Specific research protocols need to be set up to test the alternative theories and to compare their relative merits and limitations. Whether this research finally provides proofs of the HIV-only theory or casts it aside once and for all is irrelevant. The scientific point is that neither proof nor disproof exists at present; any assumption on our part can be fatal. I am willing to assume nothing when human lives are at stake.

CHAPTER 11

PREVENTING AND TREATING AIDS

THE FUTURE OF AIDS RESEARCH

How Could So Many Scientists Be So Wrong?

In conclusion, several crucial questions remain to be addressed. To begin with, how could so many scientists be so wrong? How could the vast amount of information concerning non-HIV immunosuppressive agents have been overlooked for so many years? Assuming these agents do play a role in AIDS, what are the medical, research, and health policy implications? What can be done to prevent the spread of AIDS that is not being done now? What can be done for people with AIDS that has not yet been tried?

There is no mystery about how so many scientists could be so wrong. Science, despite its elusive grail of objective truth, is just as human and just as fallible as any other human endeavor. Martin Kamen, a biochemist whose invention of radioactively labelled compounds allowed several Nobel Prizes to be won for elucidating the nature of cellular metabolism and viral replication, tells a revealing tale in his autobiography. During the early 1950s, Arthur Kornberg, a future Nobel laureate, arrived at Washington University in St. Louis where Kamen was then working. "Arthur inaugurated a 'brown bag' lunch seminar in which we met at regular intervals to discuss papers of interest. These sessions were great learning experiences for me. Whoever was assigned to lead the discussion would prepare a short typed summary of relevant data provided in the paper. We would examine these without prior consideration of the author's summary and contentions and decide what they showed. Then we compared our conclusions with the author's. It was surprising how often authors of great prestige in the biochemical community failed this test."[1] Surprising, and yet predictable, since almost all data have more than one interpretation and fit more than one pattern.[2] That was the lesson that Galileo taught Aristotle. It was the lesson that Pasteur taught the medical community, that Darwin taught biologists, and that Einstein taught Newton. It is what makes scientific revolutions possible.[3] Perhaps it is a lesson that we must learn again with AIDS.

The question of why non-HIV immunosuppressive agents in particular have been overlooked in AIDS has several answers, each of which sheds light on a different aspect of how science is performed. One answer is that data take on different meanings in light of different theories. Another answer stems from the nature of scientific logic. Scientists are taught to obey Occam's razor: the simplest theory that fits all of the facts is best. Clearly the simplest theory of AIDS is that it has a single infectious causative agent. All other theories are more complex. Moreover, infectious diseases are easily treated by vaccination or antibiotics, whereas multifactorial diseases, such as heart disease or cancers, are difficult to prevent, treat, and cure. I cannot fault my scientific colleagues for trying to place AIDS in the monocausal category. But we must now recognize that HIV is not the entire answer to AIDS. Simplicity has become oversimplification. As Einstein is supposed to have cautioned, "Make it as simple as possible, but no simpler."

The reasons that so many scientists have participated in the oversimplification of AIDS hinge on other human limitations. One is overspecialization. If AIDS is as complex and multifaceted as I have portrayed it here, then few people have the skills to address anything but a small part. The virologist, the immunologist, the biochemist, and nutritionist, the sexually transmitted diseases expert, and the pharmacologist are like the blind men and the elephant, each thinking that his or her part is the whole. Just as naturally, each tends to ignore or overlook data that are discordant. In fact, there is a saying in science ("Maier's law") that if some of the facts do not fit your theory, throw out the facts—they may be wrong. Indeed, they may be. But "Agassi's law" also warns us that not all data supporting a theory are to be believed. We must be skeptical both of what fits and what does not fit our preconceptions, and broadly trained enough to pay heed to both.[4] Thus, my approach has been to try to see the forest for the trees and to focus on those elements of AIDS that are all too easily overlooked because they should not be there: the anomalies, paradoxes, and contradictions that spell out problems for the HIV-only theory. Unfortunately, anomalies are almost always overlooked, until after a new theory has been proposed that singles them out as significant.[5] Perhaps in the light of the new possibilities raised here, the anomalies of AIDS will take on new significance, and the shape of the elephant will begin to emerge.

A related human foible—gullibility—has also played a role in the oversimplification of AIDS. Carl Djerassi, the so-called father of the birth control pill, tells a story that illustrates this foible well. His first scientific breakthrough was the chemical synthesis of cortisone—a naturally occurring corticosteroid with immunosuppressive actions used to treat inflammatory reactions. In 1952 he and many other high-powered chemists met at a prestigious Gordon Conference to share their findings in the cortisone field. This research was so "hot" that a professor from Harvard, either Robert Burns Woodward (a Nobel Prize winner) or his colleague Gilbert Stork (Djerassi cannot remember which), commented that anything with the words "synthesis of cortisone" was sure to be published. This observation led Djerassi, Woodward, Stork, and Lewis H. Sarett (a chemist who worked for the pharmaceutical giant Merck) to concoct a joke paper claiming a revolutionary new synthesis of cortisone from "neohamptogenin," a fictitious substance they had "discovered" in maple syrup from New Hampshire. They decided to see just how gullible their colleagues had become in the rush to be the first to find a simple and inexpensive synthesis of cortisone.

The fictitious cortisone paper was attributed to two nonexistent chemists and was replete with fake literature citations. At the crucial synthetic step—a step that would have revolutionized chemistry—they used the dodge that details (which they did not have) would be given in a subsequent paper. In other words, "take the proof of our assertion that this chemical reaction has been carried out on faith"—something no scientist should do. The trap was laid; the clues were placed; and on the last day of the conference, Woodward read this nonsensical communication to his audience of chemical experts.

Djerassi describes Woodward's deadpan demeanor as being so convincing that at one point a Swiss chemist actually stood up to claim priority for the nonexistent discovery. A repeat performance at Harvard a few days later was also successful. Not a single graduate student or research fellow doubted the reality of Woodward's presentation. Stork, one of the coauthors of this farce, was so irritated by the gullibility displayed by his colleagues that he attempted to point out the fraud to a group of the Harvard attendees—without success. Djerassi and the others were taken aback by this turn of events and did not submit their article for publication for fear that it might actually be taken as a serious contribution to the literature.[6]

The point of Djerassi's story is that authority—even wishful thinking—is just as powerful and prevalent in science and medicine as it is in any other sphere of human endeavor. In the case of AIDS, tremendous political pressure existed for American scientists to discover the cause and the cure of a disease first recognized in the United States. The topic was much hotter than cortisone synthesis ever was. Priority, patent rights for AIDS tests, national honor, and political and social exigencies all played their roles. As Djerassi's story suggests, scientists are not immune to these pressures. Add the full authority and financial backing of the National Institutes of Health, the Centers for Disease Control, and the Department of Health and Human Services to the already plausible assertion that AIDS was due to an infectious agent, and HIV becomes an all-too-believable solution. Scientists in 1992 cannot be expected to behave differently from scientists in 1952. A plausible story told by a powerful figure—particularly one who offers to pay you to buy into that story, as the NIH does through its grant system—is difficult to doubt.

An even more important element of the HIV/AIDS story may be physicians. Recent studies by Dr. Thomas Chalmers of Harvard University and Marlys and Charles Witte and Ann Kirwan of the University of Arizona, among others, have demonstrated that physicians are perhaps the most authority oriented of all professionals.[7] They are evaluated in medical school not on the basis of their critical thinking skills, their creativity, or their independence but their ability to learn quickly, to memorize well, to act prudently, and to be able to quote authority extensively. They want and are paid for having answers, not questions. Since the American Medical Association, following the lead of the NIH and the National Academy of Sciences, officially espouses the dogma that HIV = AIDS, every medical school and every biology and medicine textbook follows suit—and so do physicians. HIV is their answer—no questions asked. It is no wonder that the vast majority of physicians are satisfied that if HIV infection is being treated, AIDS is being treated.

The medical and research communities are the greatest asset in the fight against AIDS and, potentially, also the greatest threat to its success. There can be no breakthroughs without research, but breakthrough research is not possible when conformity is rewarded and skeptical inquiry punished. AIDS may continue to plague modern society, just as other preventable infections such as puerperal fever plagued our forebears, because of the close-mindedness of the very physicians whose job it is to diagnose, treat, and prevent these diseases. A century ago, they let patients die by denying that germs had anything to do with disease. Today they may be letting them die by insisting that the germ is everything. A happy medium must exist somewhere in between, but it will be reached only when informed skepticism and razor-sharp reasoning skills become as valuable assets to scientists and physicians as are advanced degrees, an infallible memory, or respect for authority.

Treatment Implications of Multifactorial or Cofactor Theories of AIDS

The point of this book has been to establish that our ignorance about AIDS is much greater than our knowledge. Thus, P. G. Smith, R. H. Morrow, and J. Chin of the World Health Organization Global Programme on AIDS and Special Programme for Research and Training in Tropical Diseases have recently identified the following as "important areas for further research" concerning AIDS in developing nations. One can clearly read their list, with minor modifications, as a programmatic statement of important areas for research in developed nations as well:

> Firstly, to determine what are the important interactions of HIV with the endemic tropical diseases. Secondly, to investigate whether concurrent HIV infections affect the efficacy of therapeutic or preventative measures against those diseases and, thirdly, to develop, where necessary, improved control strategies and tools. Epidemio-

logical studies are likely to feature prominently in the identification and quantification of the interactions. In considering the interaction of any two infections, questions of the following form are of concern: To what extent does the presence of one of the infections affect the susceptibility to the other and/or influence the natural history of that infection (e.g., by changing the spectrum of severity, clinical manifestations, response to treatment or infectivity to others). For example, is a person infected with HIV more susceptible to malaria? . . . Conversely, it is of interest to know whether there is an interactive effect of malaria on HIV infection. Are those with malaria more susceptible to infection with HIV?[1]

Smith and his colleagues go on to caution that the same sorts of interactions may occur between HIV and many other diseases and that "the likelihood of particular forms of interaction may also depend on host factors (e.g., age, pregnancy, malnutrition.)"[2] Until we know what these factors are for certain and can answer the associated questions concerning interactions for all of the permutations of all of the diseases found concurrently in AIDS patients in all risk groups throughout the world, we may be able neither to control nor adequately to treat AIDS.

One particular area is of great concern: whether we can arrest the progress of AIDS by eliminating exposure to cofactor agents associated with it. Multiple, concurrent infections are clearly at the top of the list. Whether HIV infection sets the stage for multiple, concurrent infections or results from the same set of factors that produce these multiple infections in the first place, the fact that these infections act synergistically has a variety of implications for the treatment and prophylaxis of AIDS. One is that protection against non-HIV infectious agents may be just as effective in preventing AIDS, or particular manifestations of AIDS, as protection against HIV. Such protection might consist of a variety of measures, including vaccination against specific infections or treatment of these infections with antibiotics. In this regard, a large number of experimental vaccines are on the horizon for disease agents associated with AIDS, including trypanosomiasis, malaria, varicella-zoster virus, cytomegalovirus, hepatitis A, and herpes simplex viruses.[3] Any of these licensed in the near future may be of great benefit to those at risk for AIDS. In the meantime, vaccination against hepatitis B virus and *Mycobacteria* may be beneficial. Indeed, physicians in Sweden have noticed that although the majority of AIDS patients in the United States become infected with *Mycobacterium avium* complex and tuberculosis is becoming an ever-increasing problem, only 10 percent of Swedish AIDS patients are so infected. One major difference between public health measures in Sweden and in the United States, Canada, and many other Western nations is that vaccination with bacille Calmette-Guérin (BCG)—a mycobacterial vaccine—is general in Sweden and extremely unusual in the United States and elsewhere. These Swedish physicians therefore wonder: "Does vaccination with *Bacillus Calmette-Guerin* protect against AIDS?"[4] We cannot afford to leave this question unanswered.

There is evidence that other prophylactic measures extend the lives of those with HIV infections and AIDS. Long-term AIDS survivor Michael Callen, in his book *Surviving AIDS,* lists a range of medications that are currently being tested, some with apparent success, as prophylactic measures against most of the major opportunistic diseases associated with AIDS: Bactrim and other sulfa drugs, and aerosol pentamidine, for *Pneumocystis carinii* pneumonia; pyrimethamine (Daraprim) with and without lucovorin (folinic acid) for toxoplasmosis; a series of antimycobacterial agents; fluconazole and Itraconazole for fungal infections; and so forth.[5] Gancyclovir, acyclovir, and foscarnet—drugs that are effective against herpes viruses such as cytomegalovirus and herpes simplex virus—have been found to increase longevity of AIDS patients in those who can tolerate the drugs.[6] These findings suggest that each of the infections associated with AIDS may provide a target for therapeutic or preventative measures that will effectively increase the quality of life and extend the life span of those people at risk for or already suffering from AIDS. Thus far, no major attempts by the pharmaceutical industry or national or international clinical trials have been made to approach AIDS from this preventative or therapeutic standpoint. If AIDS is truly multifactorial, however, it may be the only viable approach.

One caveat concerning long-term prophylaxis for AIDS is in order. As I have pointed out repeatedly, chronic use of antibiotics can lead to immune suppression. Pentamidine, trimethoprim, AZT (Zidovudine, Retrovir), DDI, and other antibiotics are known to cause suppression of red and white blood cell production if used chronically. There are, however, almost no long-term studies of the effects of chronic exposure to the vast majority of drugs that might be used prophylactically in AIDS.[7] If people at risk for AIDS are to be treated chronically with any medication, they must be monitored carefully for side effects and concomitant investigation of whether these antibiotics may depress cellular replication, cause liver or kidney damage, or deplete key vitamins, minerals, or other nutrients is essential. We do not want to be in the position of saying that we cured the patient but the treatment killed him.

Another little-explored approach to treating AIDS may prove useful. If AIDS is due either to HIV plus cofactors or to a variety of synergistic infections, then preventing any key infection may lessen the risk of AIDS substantially. It may prove easier to stop a mycoplasmal or cytomegalovirus infection than to stop HIV. Notably, Hamilton and his co-workers have demonstrated in mice that the synergistic effects of cytomegalovirus with *Pseudomonas, Staphylococcus,* or *Candida* are completely eliminated by prior vaccination of the mice with a cytomegalovirus vaccine.[8] A vast literature exists demonstrating a similar phenomenon for protection against autoimmune diseases. For example, experimental allergic encephalomyelitis, the model for demyelinization introduced in Chapter 5, is caused by a synergistic effect of myelin basic protein and a mycobacterial cofactor. Vaccination of animals with either of these antigens protects them against the autoimmunity induced by the combination.[9] Moreover, recent studies indicate that large amounts of either antigen inoculated into an animal suffering from encephalomyelitis will halt the clinical symptoms.[10] Results such as these proffer the hope that not only can the synergistic effects of co-infections be prevented by vaccination, but they can be reversed in AIDS patients if the specific triggering antigens can be identified. For example, if Montagnier and Lo are correct in identifying *Mycoplasmas* as necessary cofactors for AIDS, vaccination, antibiotic treatment, or autoimmune therapies may be easier to develop for these *Mycoplasmas* than for HIV, and they may be just as effective for preventing or treating AIDS.

There may be limits to the degree to which medical intervention is possible in AIDS, however. The synergism of multiple, concurrent infections makes it difficult to treat any individual infection effectively. Much smaller amounts of virus or bacterium—sometimes hundreds of thousands or millions of times less—are necessary to induce morbidity and mortality when co-infections are present than when they are not. Thus, any treatment must be unusually effective to be of any use to those who have AIDS or are at risk for it. Moreover, the type of treatment that might at first glance appear to be most beneficial to AIDS patients—that is, treating them with immunological stimulators, such as thymic hormones, vaccines, lymphokines, and so forth—may actually exacerbate the course of AIDS once it has been initiated, since both HIV and many of the other viruses associated with AIDS replicate much more efficiently in stimulated than in unstimulated cells.

One very odd possibility is also raised by alternative theories of AIDS, particularly by the theories that incorporate autoimmunity as a major event in the progress of the disease. Immunosuppressive drugs may actually benefit AIDS patients. As paradoxical as this may sound, there is limited evidence that it may have some validity. Corticosteroids are now in general use to supplement pentamidine and TMP-SMX prophylaxis for *Pneumocystis* pneumonia, and a number of trials of other immunosuppressive regimens have shown limited promise in AIDS patients.[11] The results make sense if the major cause of T cell death is due to lymphocytotoxic antibodies (that is, an autoimmune civil war). One way to tone down this civil war is to slow down the immune system itself. The danger is that in slowing down the immune system, the same sort of immune deficiency is created that you are trying to prevent. What is needed, and what does not yet exist to treat lymphocytotoxic autoimmunity, are drugs or vaccines that specifically turn off or eliminate the B cell attack on the T cells. If such therapies could be developed, either to downregulate antibody production in general or, even better, to inhibit the autoimmune reaction itself without affecting antibody production otherwise, they would most certainly promote the health of AIDS patients, if not cure them. Moreover, the development of such therapies would have applications to many autoimmune conditions besides those associated with AIDS. This is an area in which many breakthroughs await.

In the meantime, various aspects of medical practice must change to accommodate the possibility that HIV is not the sole agent responsible for AIDS. First, HIV seropositivity should not be used as an uncritical marker for imminent death from AIDS. HIV may be no more than a serious warning that a patient has multiple risks that need to be ferreted out and controlled or corrected. Among the clinical studies that need to become commonplace, if alternatives to the HIV-only theory are to be properly investigated, are tests for malnutrition in all its varied forms; a complete history of all drug use, whether it is intravenous or not; a complete medical history, including all chronic medications and repeated infections; an in-depth analysis of the types of sex a patient engages in, including all forms of anal sex, shared use of sex toys, and so forth; a complete history of sexually transmitted diseases; studies of what active viral, bacterial, fungal, and parasitic diseases a patient has; and tests for all possible autoimmune manifestations associated with AIDS, including the presence of lupuslike antibodies and lymphocytotoxic antibodies. None of these can be treated properly until they are known to be present.

Similarly, people at risk for AIDS, and those with AIDS, need to know that there are positive measures they can take that will drastically lower the probability of disease progression. Dr. C. Wayne Callaway, former director of the Mayo Clinic Nutrition Clinic has argued cogently in his book *Surviving with AIDS: A Comprehensive Program of Nutritional Co-Therapy* that malnutrition is preventable in those at risk for AIDS, and treatable in most cases. He has shown that nutritional therapy greatly improves the quality of life and survival of people with AIDS.[12] Drug use can be terminated with appropriate help. Sexual pleasures can be gained in ways that do not also carry the risk of repeated sexually transmitted diseases, potential damage to the anus and lower intestines, or immunologic exposure to semen. Common sense can lower the risk of contracting many viral, bacterial, fungal, and parasitic infections.

Indeed, several studies strongly suggest that a much broader approach to AIDS that includes specific remedies for malnutrition, elimination of all drug use, proper hygiene, safer sex measures, and behavioral modification can have profound effects on AIDS risk and the development of overt disease even among people who are already HIV seropositive. Without a doubt, the most far-reaching work in this area has been carried out by Dr. Maurizio Lucá Moretti of the InterAmerican Medical and Health Association, based in Boca Raton, Florida. He has been collaborating with his colleagues in Italy for several years on a study of 508 former intravenous drug abusers, all of whom were HIV seropositive. These men were voluntarily confined to a rehabilitation center, where their daily lives were under the management of the

house staff. Most of the patients were found to be severely malnourished on arrival at the rehabilitation center, 397 of them
chronically so. Their drug use was completely eliminated, nutritional status corrected, and even their sexual activity limited (the
rehabilitation center is a monastery in which the patients sleep
in small groups under supervision). None of these patients had a
sexually transmitted disease after entering the rehabilitation center. Of 139 HIV-seropositive individuals with an average daily
heroin use of more than five years' duration prior to entering the
center, none had developed infections associated with or any
symptoms of AIDS an average of 4.1 years after documented seroconversion.[13] Compared with all HIV-seropositive drug addicts in
the United States, this figure is phenomenal. Fully 32 percent of
HIV-seropositive drug addicts develop full-blown AIDS within
two years following seroconversion and over 50 percent by four
years.[14]

Similar, though not so striking, results were achieved in a study
of 297 asymptomatic HIV-positive intravenous drug abusers in
Switzerland who were studied for sixteen months. In the Swiss
study, only drug use was monitored. During that time, the risk of
developing overt AIDS was three times higher among the addicts
who continued to inject drugs than it was among those who
stopped their drug use. No attempt to measure or ameliorate malnutrition or to control infections was made.[15]

Dr. Elena Buimovici-Klein and Dr. Michael Lange and their colleagues have reported similar results working with gay men. They
recruited 100 gay men in New York City into their AIDS program
between November 1981 and March 1982. They carefully monitored viral load, immunological parameters, and exposure to allogeneic antigens (such as sperm) in each patient every six months.
They advised their patients to avoid all known immunosuppressive agents and gave them concrete suggestions about how to do
so. They concluded in January 1988 that their approach was successful: "Informing our study population of the results of their
viral and immunological examination contributed to a change in
their life styles; this may explain the fact that 63 percent of seropositive subjects are well and healthy, and that of the 42 seronegative subjects, only three have seroconverted and none has developed AIDS."[16] These results are particularly noteworthy given the
completely voluntary basis upon which these homosexual men

adopted, or did not adopt, measures to eliminate immunosuppressive risks. At the time this study began, the average period of latency between HIV infection and full-blown AIDS was less than three years for homosexual men. Therefore, 80 percent of the seropositives should have developed AIDS by 1988. In fact, only 37 percent did so. Also, up to 1986, the rate of spread of HIV was greater than 5 percent per year among the homosexual community so that at least two HIV seronegatives could have been expected to seroconvert each year in Buimovici-Klein's group, for a total of twelve or more people. They found only three. Moreover, these results were achieved without the benefit of any anti-HIV drugs or vaccines. In other words, the development of AIDS in high-risk groups, even among HIV-seropositive people, is clearly controllable (though perhaps not curable) by means of curtailing or eliminating immunosuppressive risks.

Other studies confirm that HIV need not be a death knell. Recall the existence of low-risk individuals who had apparently been infected by HIV, developed antibody, and then successfully eliminated the infection. Particularly memorable is the report from UCLA that among thirty-one HIV-seronegative homosexual men from whom HIV had been isolated at least once during 1989, thirty of these men showed no evidence of HIV infection in October 1991 after repeated attempts at HIV isolation, and all thirty were HIV-antibody negative. None of the men had been treated with antiretroviral drugs, but, notably, all of these men had altered their life-styles in the intervening two years from high-risk modes involving promiscuous anal intercourse to low-risk modes involving only a couple of partners and rigorous safer sex practices.[17] Once again, controlling exposure to cofactor infections and allogeneic antigens seems to have had a highly salutory effect on AIDS risk. As Luc Montagnier recently said in an interview, "HIV infection doesn't necessarily lead to AIDS. There are some people who could escape that. It may be a minority, but we can hope by treatments to increase this number. . . . If activation by microorganisms is important, I think [people testing positive for HIV] should reduce their risks of being exposed to such microbes, and have long-term antibiotic treatment."[18]

Results such as these from highly reputable AIDS physicians must cause us to reevaluate less orthodox approaches to AIDS as well. Some AIDS patients documented in Michael Callen's *Surviv-

ing AIDS have survived full-blown AIDS for a decade or more.[19] One significant factor these men share is that few have been treated according to the preferred regimens touted by mainstream AIDS investigators. Rather than taking AZT or DDI, many of these men have taken as few pharmacological agents as possible or turned to herbal "medicines" and wholistic health "cures." An emphasis on prophylaxis against further disease, maximal nutrition, elimination of drug- and sex-related risks, optimism in mental outlook, and other positive measures are common. Continued promiscuity and drug use are rare in this group.

Amazingly, however, no formal study of long-term survivors of AIDS, or of people with extremely long HIV latency periods, has ever been done—another striking lacuna in our knowledge that must be filled if we are to learn how to prevent and treat AIDS effectively. We must learn from those who are successful, whoever they are and whatever their mode of success. Attention must be paid to extraordinary long-term survival, control of T cell numbers, and regulation of HIV activity among HIV-positive people and those with AIDS reported from physicians working, for example, at the Biocenter in Princeton, New Jersey, the Golden Phoenix Center in Rimrock, Arizona, and Joyce Willoughby's practice in Los Angeles, where addiction treatment, nutritional supplementation, and herbal, mental health, and other (in some cases, crackpot) approaches are substituted for pharmacological treatments of AIDS.

While it is easy to scoff at these wholistic and counterculture approaches to treating AIDS, three facts must be considered. First, at least half of the current pharmacopoeia, including old standbys such as aspirin, digitalis, and the opiate painkillers, began as folk cures. They were often presented in formulations that contained a good deal of nonsense as well, but when the nonsense was stripped away, a nugget of scientifically verifiable sense remained.[20] We must consider the possibility that some nontraditional approaches to AIDS may, beneath their irrational veneers, also contain viable nutritional, herbal, and behavioral approaches to disease—perhaps even chemical leads to pharmacologically active compounds—that science needs to tease out.

Second, physicians have been loath to accept even well-documented treatment modalities such as hypnosis and acupuncture because the underlying phenomena have no rational explanation within our current theoretical framework. Alternative hypotheses to AIDS provide alternative frameworks for interpreting as valid some otherwise unexplainable treatments or "cures." For example, a wholistic approach to AIDS focusing on nutrition and behavior modification and emphasizing a positive mental image may bolster the immune system and simultaneously curtail exposure to drugs and infections, leading to improved health. This is a reasonable prediction of the cofactor and multifactorial theories of AIDS that differentiate them clearly from the HIV-only theory.

And finally, we must always remember when we evaluate novelties in science that the greatest medical breakthroughs, whether they have been the invention of the germ theory of disease, the development of antisepsis, the use of vaccination, or the discovery of the first antibiotics, were all met, initially, with total skepticism by the medical community. I would not be surprised if the most important innovators in AIDS research and treatment turn out to be peripheral members of the research and treatment communities.[21] We must remain aware of these historical verities as we continue to monitor the progress of AIDS research.

Rethinking Public Health Policy

Clearly any significant change in the theory of what causes AIDS will result in concomitant changes in public policy. Consider the possibilities.

Assume that HIV is not sufficient and may not be necessary to cause AIDS. The usual response to scientists such as Peter Duesberg, me, and other HIV critics is that we are "irresponsible" for raising such a possibility publicly. To question HIV, we are told, is tantamount to undermining all of the safer sex and clean needle messages that have thus far formed the basis for public policy on AIDS control. If people think that HIV does not cause AIDS, the argument goes, they will not take steps to protect themselves from it, and the death rate will soar.

Duesberg has recently fallen into this trap. Despite what I consider to be quite extensive evidence demonstrating that safer sex and clean needle programs have substantially lowered the risk of AIDS for those people who adopt the procedures rigorously, Duesberg now claims that since, in his view, AIDS = drugs, "safer sex" is pointless. Do not use drugs, Duesberg claims, and you can have intercourse anyway you want without getting AIDS. I cannot agree. Whether HIV requires cofactors or AIDS is a multifactorial, synergistic or autoimmune disease triggered by allogeneic insults such as semen or blood exposure with appropriate infections, safer sex is an essential ingredient in the recipe that will control AIDS. In my view, in fact, our current public policies do not go far enough to protect against AIDS.

To my knowledge, except for Duesberg, no scientific investigator who has challenged the HIV-only theory of AIDS believes that we should abandon safer sex practices or clean needle programs to control the spread of AIDS. On the contrary, it is an almost unanimous belief among critics of the HIV-only theory that such public policy measures are insufficient to control AIDS because the policies do not recognize the myriad non-HIV immunosuppressive agents that also put people at risk. I have written that

> the recognition that non-HIV immunosuppressive factors play a causative role in AIDS, far from undermining current public health attempts to control AIDS through "safe sex" and control of drug use, suggests that current HIV-based measures are not working because they are *too narrow.* For example, providing clean needles to addicts does not obviate the immunosuppressive effects of opiate abuse, its frequent concomitant malnutrition, or the prostitution which is often used to pay for the illicit drugs. Gay men need to know that even "safe sex" measures will not protect them from the immunosuppressive effects of recreational drugs, antibiotic abuse, rectal damage following anal intercourse, fisting, and anal masturbation, malnutrition following gay bowel syndrome, or from infections contracted by anal-oral [or oral-genital] contact. Heterosexuals need to realize that a combination of recreational drug abuse, poor eating habits, bulemia or anorexia, and promiscuous anal intercourse can be profoundly immunosuppressive whether they become infected with HIV or not. Physicians also need to become more aware of the immunosuppressive risks of some of the common procedures they use such as blood transfusions, factor therapies, and long-term antibiotic use.[1]

One could add to this list the problems that result from multiple, concurrent infections, no matter how they are acquired, and the additional risk that autoimmune diseases may pose.

I am far from alone in this position. Dr. Victor Lorian has tried to bring the risk posed by anal intercourse among heterosexuals, independent of HIV, to the attention of the medical community.[2] Joseph Sonnabend and his colleagues have often reiterated their position that all immunosuppressive risks add to the progressive decline in AIDS patients and must all be guarded against.[3] Dr. Maurizio Lucá Moretti has argued long and loud that the list of immunosuppressive agents that must be eliminated includes multiple infections, street drugs and pharmaceutical agents, anal intercourse, and malnutrition.[4] And Berkeley virologist Harry Rubin has responded to charges that Duesberg's questioning of the HIV theory is "irresponsible" with the argument, "In fact, [Duesberg's] view of HIV as an indicator for infection by other agents would logically call for even greater sanitary precautions. Indeed, his detractors might well be tarred with the same brush [irresponsibility] for discounting the role of immunosuppressive practices and for unwarranted reductionism of such a complex syndrome [to a single cause: HIV]."[5] What is at issue if HIV is not the sole necessary and sufficient cause of AIDS is not, then, the irrelevancy of current AIDS policies but the insufficiency of them.

Indeed, one of the most sobering scenarios imaginable is that public health policy will not recognize the existence of non-HIV immunosuppressive agents in AIDS risk groups. Consider the consequences of developing a vaccine against HIV if HIV turns out not to be the sole cause of AIDS. Presumably, people at risk for AIDS due to sexual or drug habits will go back to high-risk activities in the belief that they are now "safe." Not only may they not be safe from AIDS, they will certainly not be safe from the myriad of other immunosuppressive and medical problems that accompany high-risk forms of sex, drug use, multiple sexually transmitted diseases, and malnutrition. These issues transcend AIDS. I believe that it is irresponsible for the medical community and for government policymaking bodies to ignore the broad medical implications of non-HIV immunosuppressive factors associated with AIDS. These factors will continue to be present and to pose medical threats no matter what our response to HIV. People will die of preventable causes because of our exclusive focus on HIV.

Until a breakthrough in AIDS treatment or prophylaxis is forthcoming (if such a breakthrough is possible for such a complicated syndrome), we must rely on public health measures of the most

general kind. Some of these measures have already proved quite
effective, and their results are compatible with both HIV-only and
alternative theories of AIDS. For example, voluntary behavioral
changes undertaken by people at risk for AIDS have proved effective in controlling the spread and incidence of AIDS. There is no
doubt that so-called safer sex practices combined with the fear
of AIDS has had a profound and salutory effect on the health of
homosexual men in general. The incidence of new cases of HIV,
hepatitis B virus, gonorrhea, and syphilis, all other sexually transmitted diseases, and AIDS has decreased dramatically in the gay
community since 1985.[6] This is due, in part, to increased use of
condoms, more care in choosing partners, decreased promiscuity,
and an apparent decrease in the frequency of anal intercourse, fisting, and related forms of sex.[7] For example, in Amsterdam, the
yearly incidence of new HIV infections among homosexual men
rose from 3.0 percent in 1981 to 8.8 percent in 1984 and then fell
to 0.0 percent in 1987.[8] Homosexual communities in London, San
Francisco, and Denver, Colorado, had the same experience. In
London, the incidence of HIV infection also tailed off significantly
after 1985, along with all other sexually transmitted diseases, including hepatitis B virus and syphilis.[9] In San Francisco, the yearly
rate of new HIV infections dropped from a high of 6.0 percent in
1985 to 0.7 percent in 1987 and has continued to fall.[10] In Denver,
the incidence of HIV seropositivity reached a peak of 52 percent
of homosexual male residents in 1984, and the yearly number of
gonorrhea cases peaked at over 1,800 per year in the same group.
The incidence of HIV seropositivity has actually dropped yearly
since 1985, as have the number of gonorrhea cases. By 1989 only
about 45 percent of homosexual men were HIV seropositive, suggesting that the number of newly infected men is extremely small.
Just how small this number may be can be guessed at from gonorrhea cases, which fell a drastic 95 percent to a mere 90 cases per
year.[11]

   The importance of these many studies demonstrating that control of immunological risk factors can lead to quite effective control of AIDS is twofold. First, they demonstrate the continued validity of one of the oldest and most fundamental truths of medical
science: Public health measures are always more effective in controlling disease than are all the medicines in the world. Neither
vaccines nor medicines have led to the virtual elimination of ty-

phoid, cholera, typhus, or plague in the industrialized countries of the world. These required nothing more than the simple expedients of improved sanitation, sewage systems, and the control of pests. Few of us think twice about using a toilet and then washing our hands afterward. Cleanliness is a requirement for food preparers. And yet a mere two hundred years ago, the daily sanitary habits we take for granted did not exist. Disease, and fear of disease, shape our lives in ways that influence culture itself. AIDS will have the same salutory effect.

Second, if the basic thesis of this book is correct—that AIDS is not caused by a simple HIV infection but is a syndrome requiring multiple, concurrent causes of immune suppression—then the tremendous drop in the incidence of sexually transmitted diseases associated with safer sex means that the risk of immune suppression due to multiple infections and allogeneic exposure decreases drastically, the risk of autoimmunity decreases even more, and the probability of developing an HIV infection concurrent with an appropriate cofactor infection diminishes. In other words, the same measures that are meant to control the spread of HIV are necessary to control the spread of both allogeneic and infectious cofactors in AIDS. In consequence, I predict that the incidence of AIDS itself, as distinct from HIV seropositivity, will decrease much more quickly than the rate of HIV infection itself. More and more HIV-seropositive gay men will live longer and healthier lives. Indeed, the fact that the latency period between HIV infection and AIDS has been increasing yearly indicates just such an effect. This latency period should continue to increase, until many HIV-seropositive people are living healthy lives for several decades. Then, if not before, we will finally recognize that HIV does not equal AIDS.

There is, however, a socioeconomic aspect of AIDS we cannot ignore. One out of every 100 AIDS patients in the United States is currently incarcerated in the prison facilities of a single state: New York. Forty-seven percent of these inmates are Hispanic, 38 percent black, and 13 percent white. Poverty, drug abuse, and repeated criminal activity are prominent risk factors for nearly all of these inmates.[12] AIDS risk for blacks and Hispanics in general is many times higher than whites, in direct proportion with their socioeconomic status.[13] The same phenomenon has been seen in Canada with regard to increased incidence and mortality from

AIDS among native Indians, despite universally available health care. If we want to control AIDS, it is not vaccines, antiretroviral drugs, or other medical miracles that we need. We need to solve the social, economic, health education, and medical care problems that create the conditions that permit AIDS to develop in the first place.

New Directions in Biomedical Research

Medical breakthroughs are needed. The benefit of rethinking AIDS is that the alternative theories predict specific, new things that can be done. These predictions provide the crucial tests that distinguish between adequate and inadequate theories, and, in some cases, they may provide the surprises that make doing science worthwhile.

Some medical studies are necessary regardless of whether HIV is the sole cause of AIDS. Among these are such notable questions as: What is the long-term prognosis for patients with different pathological states requiring multiple blood transfusions? What types of transfusions (whole blood, plasma, granulocytes, and so forth) are most immunosuppressive and carry the greatest risk of long-term morbidity and mortality? What is the rate of death among HIV-negative hemophiliacs and drug abusers, and what diseases do they die of? To what extent does HIV, when controlled for amount of factor transfused, corticosteroid and painkiller used, and concurrent infections activated, increase morbidity and mortality among hemophiliacs? What is the prevalence of malnutrition among people at high risk for AIDS? What are the manifestations? How easily can these forms of malnutrition be prevented and with what effect on AIDS incidence? How many people at risk for AIDS use nonintravenous drugs, and what are the long-term effects of these drugs on the immune system, on susceptibility to infection, and on longevity? Can effective treatment of drug addiction, associated malnutrition, and infectious disease load prevent AIDS even among HIV-seropositive addicts as Lucá Moretti has found in Italy? Can similar approaches be adapted to HIV-seropositive gay men, hemophiliacs, and transfusion patients with any success? Questions such as these must be answered by controlled studies before any evaluation of the role of HIV in AIDS can be established in these risk groups or their immunological risks properly identified and ameliorated. Such glaring medical lacunae have no business existing, AIDS or no AIDS.

Other studies are necessary to differentiate the various AIDS hypotheses. Some are so simple that no one, apparently, has thought to try them (or they have been unable to obtain funding to do so).[1] For example, blood transfusions, clotting factor treatments, street drugs, pharmaceutical agents including anesthetics and antibiotics, semen components, malnutrition, and chronic or multiple concurrent infections have all been shown to cause immune deficiencies separately, yet no one has tested these factors in combinations similar to the ones that occur in AIDS patients. If the multifactorial, synergistic theory of AIDS is valid, then treating mice, rats, or rabbits with some combination of, say, multiple blood transfusions, cytomegalovirus infection, anesthetics and surgery, and opiate painkillers could well result in the immunological suppression typical of AIDS. Other combinations of factors might also have adverse effects mimicking AIDS, such as repeated exposure to allogeneic semen and sexually transmitted disease agents combined with chronic abuse of antibiotics and street drugs, or a combination of addictive drug use, malnutrition, antibiotic abuse, and repeated exposure to septicemic and suppurative infections. Since most healthy animals are known to be resistant to opportunistic infections such as cytomegalovirus, *Pneumocystis carinii,* and *Candida albicans* but can contract these infections when chemically immunosuppressed, the development of these opportunistic infections following treatments like those suggested above would suffice to model AIDS in the absence of HIV infection.[2]

Another set of experiments could test the possibility that AIDS is due to specific combinations of HIV with other infectious and noninfectious immunosuppressive agents.[3] These experiments are somewhat more difficult and expensive to carry out, since the only animals other than human beings that can be infected with HIV are chimpanzees. So far, chimpanzees infected with HIV have not developed AIDS, even eight years after infection. No one has apparently injected HIV into a malnourished chimpanzee addicted to opiate drugs or cocaine and subjected to repeated infections. Chimpanzees have not been inoculated with semen and subjected to multiple sexually transmitted diseases to produce the antisemen lymphocytotoxic autoantibodies that typify gay men with AIDS. No chimpanzee infected with HIV has been given repeated blood transfusions contaminated with hepatitis viruses, cytomegalovirus, and Epstein-Barr virus, exposed simultaneously to anesthetics, and then given painkillers to model the risks of transfusion patients. Most important, no chimpanzee has yet been co-infected with both HIV and Lo or Montagnier's *Mycoplasmas,* or even HIV with cytomegalovirus, HTLV-I, or any of the other viruses known to transactivate each other. In other words, almost nothing of our vast knowledge concerning the synergistic actions of HIV with other immunosuppressive factors that has been demonstrated in test tube studies or revealed by the epidemiology of AIDS has been tested in animal models of AIDS. It is time to do so.

Among the interesting experiments that are also in need of exploration are ones designed to create the autoimmunity that is typical of AIDS. Many specific leads now exist as to the possible agents responsible for this autoimmunity. Hoffmann and his colleagues have identified specific lymphocyte targets, some of which are present in transfused blood and semen. Sonnabend and others have identified specific protein targets on sperm that may induce autoimmunity against T cells. Hobbs and I have preliminary evidence that many so-called cofactors for AIDS, including cytomegalovirus, Epstein-Barr virus, and hepatitis viruses, may interact with HIV, *Mycoplasmas,* and *Mycobacteria* to produce such autoimmunity. These leads must also be followed up if autoimmunity in AIDS is to be understood.

Critics of the alternative theories of AIDS presented here may object to this diversion of resources. Why take away valuable time, money, and effort from studies of the *true* cause of AIDS, they may ask, when many of these studies may do nothing more than validate the HIV-only theory in the end? I believe that such studies are justified for several reasons. First, they are logically necessary for testing the HIV-only theory against its alternatives. Theories need testable controls just as much as experiments do. More important, all of the tests just suggested (and many others in addition) are needed from a purely medical point of view, totally divorced from considerations of AIDS. Hundreds of thousands of people, perhaps even millions worldwide, are regularly transfused, given anesthetics, operated upon, exposed to multiple immunosuppressive viruses, and then given opiate painkillers. Millions of people develop malaria, schistosomiasis, filariasis, and other parasitic infections every year and are then exposed to blood transfusions, other infectious agents, or a battery of antibiotics. Millions of people engage or have engaged in anal intercourse and may have antisemen antibodies that can cross-react with their T cells. We need to know whether they are more susceptible to disease or cancer than are other people. Hundreds of thousands of people contract, whether by chance or by the nature of their lifestyles, multiple, concurrent infections. We do not know what the immunological effects of these, and the many other combinations of immunosuppressive agents described in this book may be for anyone, let alone for AIDS patients. Millions of people develop autoimmune diseases each year, many of them identical to the autoimmune complications experienced by AIDS patients, and we know the causes of none of them. By carrying out the research program proposed here, we will gain much needed information not only about AIDS but about immunological risks and diseases that affect hundreds of millions of people who will never get AIDS. Everyone will benefit from the research, no matter what answers it yields.

What If the HIV-Only Theory Turns Out to Be Right?

Let me reiterate on the final pages of this book my most important message: Because a strong case can be made that HIV is not the sole necessary and sufficient cause of AIDS does not mean that HIV is not, in fact, the cause or a cause of AIDS. It does not mean that the past decade of research has been totally fruitless. HIV is part of AIDS; what proportion of AIDS is yet to be determined. All that we can be sure of at present is our ignorance. That ignorance requires the palliative of new research, new exploration, new experiment, new tests. This book has laid out the outline of what needs to be done. I ask nothing more than that we retain a critical, questioning attitude toward all theories of AIDS until such a time as all of the anomalies, paradoxes, contradictions, and outstanding questions concerning HIV, cofactors, and AIDS are fully resolved.

The only path to truth is to continue questioning—questioning even things that are taken to be undeniable facts. I call this the

"faith of a heretic" after a book of the same title by philosopher
Walter A. Kauffmann.[1] The faith of the heretic is not an easy phi-
losophy to live by. It requires that belief in the results of science
be reached only after the route to those results has been critically
retraced and every effort has been made to find flaws in it. Few
people have the time, patience, or personality for such science
searching. The few who do are readily discouraged by the vast
majority whose belief in their own results and in the results of
other scientists is less skeptical. The point of this skepticism is
not, at least for me, to overthrow HIV but to reinvestigate the
assumptions underlying AIDS research and AIDS policy in order
to determine whether any important observations and experiments have been overlooked. It is totally irrelevant to me what the
outcome of this rethinking and researching is, as long as it is done
well and done properly.

I could not care less whether AIDS is caused by HIV, by HIV
with appropriate cofactors, or by non-HIV causes. I do care that
we investigate all of these possibilities and reject all but one of
them. Until research is done on the alternatives, we do not know,
and we cannot know for certain, the cause or causes of AIDS.
Assurance in science comes only through elaborating as many possible explanations as can be imagined for a phenomenon and eliminating all that can possibly be eliminated.[2] That process has not
been employed in AIDS research. The door on alternative possibilities was prematurely closed before anyone had a chance to see
how many rooms there were, let alone explore their treasures. We
must research them now or risk basing our AIDS policies, prophylactic measures, and medical treatments on incomplete and ineffective knowledge.

I fear that the notion of elaborating possibilities and eliminating
all but one makes the science of AIDS sound like a game. In some
senses it is. Scientists are game players and puzzle solvers, in the
most professional sense. The game, in this case, is not a mere intel-
lectual challenge played in laboratories for the enjoyment or en-
lightenment of the investigators. Human lives depend on the outcome of AIDS research. If we have adopted the wrong strategy in
this most mortal of games, if we do not understand all of the rules
nature has devised for AIDS, then people will continue to die, perhaps needlessly. To the extent that this is so, the scientific and

medical communities will be at fault in failing to recognize their own prejudices and ignorance.

I have put my scientific reputation on the line in this book in order to make certain that we accept nothing about AIDS uncritically. If appropriate studies eventually confirm the HIV-only theory of AIDS and I end up looking the fool for having questioned a dogma that additional tests validate, so be it. It is better to lose my scientific reputation than anyone's life. I suspect, however, that both my reputation and people's lives will be better served by the challenge I offer.

```
┌─────────────────────────────────────────┐
│  ╔═══════════════════════════════════╗   │
│  ║              NOTES                 ║   │
│  ╚═══════════════════════════════════╝   │
└─────────────────────────────────────────┘
```

A key to the abbreviations used for journals cited here may be found in the Index Medicus, *which is available in libraries having a medical collection.*

CHAPTER 1 ANOMALIES: OBSERVATIONS ABOUT AIDS THAT DO NOT FIT THE PICTURE

Anomaly 1: HIV and AIDS Are Older Than We Think: The Prehistory of AIDS, 1872–1979

1. Ellis FA. 1934. Multiple idiopathic hemorrhagic sarcoma of Kaposi. Report of a case in an American Negro. *Arch Dermatol Syphilol* 30:706–708.

2. Collins MD, Fisher H. 1950. A case of generalized hemangiosarcomatosis erroneously considered as generalized tuberculosis. *Am Rev Tuberculosis* 61:257–262.

3. Wyatt JP, Hemsath FA, Soash MD. 1951. Disseminated cytomegalic inclusion disease in an adult with primary refractory anemia and transfusional siderosis. *Am J Clin Patholol* 21:50–55.

4. Wyatt JP, Simon T, Trumbull ML, Evans M. 1953. Cytomegalic inclusion pneumonitis in the adult. *Am J Clin Patholol* 23:353–362.

5. Williams G, Stretton TB, Leonard JC. 1960. Cytomegalic inclusion disease and *Pneumocystis carinii* infection in an adult. *Lancet* ii:951–955; Williams G, Stretton TB, Leonard JC. 1983. AIDS in 1959? *Lancet* ii:1136.

6. Lyons HA, Vinijchaikul K, Hennigar GR. 1961. *Pneumocystis carinii* pneumonia unassociated with other disease. *Arch Intern Med* 108:929–936; Hennigar GR, Vinijchaikul K, Roque AL, Lyons HA. 1961. *Pneumocystis carinii* pneumonia in an adult: Report of a case. *Am J Clin Pathol* 35:353–364, 1961.

7. Witte MH, Witte CL, Minnich LL, Finley PR, Drake WL Jr. 1984. AIDS in 1968. *JAMA* 251:2657; Elvin-Lewis M, Witte M, Witte C, et al. 1973. Systemic chlamydial infection associated with generalized lymphedema and lymphangiosarcoma. *Lymphol* 6:113–121.

8. Lindboe CF, Froland SS, Wefring DW, et al. 1986. Autopsy findings in three family members with a presumably acquired immunodeficiency syndrome of unknown etiology. *Acta Pathol Microbiol Immunol Scand A* 94:117–123.

9. Gottlieb GJ, Ragaz A, Rogel JV, et al. 1981. A preliminary communication on extensively disseminated Kaposi's sarcoma in young homosexual men. *Am J Dermatopathol* 3:111–114.

10. Centers for Disease Control. 1981. Follow-up on Kaposi's sarcoma and *Pneumocystis* pneumonia. *MMWR* 30 (33):409–410.

11. Brownstein MH, Shapiro L, Skolnik P. 1973. Kaposi's sarcoma in community practice. *Arch Dermatol* 107:137–138.

12. Oettle AG. 1962. Geographical and racial differences in the frequency of Kaposi's sarcoma as evidence of environmental or genetic causes. *Unio Internationalis Contra Cancrum, Acta* 18:330–363.

13. Nicastri AD, Hutter RVP, Collins HS. 1965. *Pneumocystis carinii* pneumonia in an adult. *NY State J Med*:2149–2154.

14. Young LS. 1984. Preface. In: Young LS, ed. 1984. *Pneumocystis carinii Pneumonia. Pathogenesis. Diagnosis. Treatment.* New York: Marcel Dekker, vii.

15. Young. 1984. *Pneumocystis carinii Pneumonia*, vii–viii.

16. Young. 1984. *Pneumocystis carinii Pneumonia*, 154.

17. Luddy RE, Champion LAA, Schwartz AO. 1977. *Pneumocystis carinii* pneumonia with pneumatocoele formation. *Am J Dis Child* 131:470.

18. Levenson SM, Warren RD, Richman SD, Johnston GS, Chabner BA. 1976. Abnormal pulmonary gallium accumulation in *P carinii* pneumonia. *Radiology* 119:395; Turbiner EH, Yeh SDJ, Rosen PP, Bains MS, Benna RS. 1978. Abnormal gallium scintigraphy in *Pneumocystis carinii* pneumonia with a normal chest radiograph. *Radiology* 127:437; Dee P, Winn W, McKee K. 1979. *Pneumocystis carinii* infection of the lung: Radiologic and pathologic correlation. *Am J Radiol* 132:741; Sirotzky L, Memoli V, Roberts JL, Lewis EJ. 1978. Recurrent *Pneumocystis* pneumonia with normal chest roentgenograms. *JAMA* 240:1513; Zakowski RC, Gottlieb MS, Groopman J. 1984. Acquired immunodeficiency syndrome (AIDS), Kaposi's sarcoma, and *Pneumocystis carinii* pneumonia. In: Young. 1984. *Pneumocystis carinii Pneumonia*, 195–226.

19. Centers for Disease Control. 1989. Update: Acquired immunodeficiency syndrome—United States, 1981–1988. *MMWR* 38:229.

20. Hamperl H. 1956. Pneumocystis infection and cytomegaly of the lungs in the newborn and adult. *Am J Pathol* 32:1–6.

21. Hamperl. 1956. *Am J Pathol* 32:1–6; Nichols PW. 1982. Letter to the editor. *New Engl J Med* 306:934–935.

22. Corbitt G, Bailey AS, Williams G. 1990. HIV infection in Manchester, 1959. *Lancet* i:51; Froland SS, Jenum P, Lindboe CF, et al. 1988. HIV-1 infection in Norwegian family before 1970. *Lancet* i:1344–1345; Garry RF, Witte MH, Gottlieb AA, et al. 1988. Documentation of an AIDS virus infection in the United States in 1968. *JAMA* 260:2085–2087.

23. Moore JD, Cone EJ, Alexander SS Jr. 1985. HTLV-III seropositivity in 1971–72: Parenteral drug abusers—a case of false positives or evidence of viral exposure? *New Engl J Med* 314:1387–1388.

24. Huminer D, Rosenfeld JB, Pitlik SD. 1987. AIDS in the pre-AIDS era. *Rev Infect Dis* 9:1102–1108; Huminer D, Pitlik SD. 1988. Further evidence for the existence of AIDS in the pre-AIDS era. *Rev Infect Dis* 10:1061.

25. Katner HP, Pankey GA. 1987. Evidence for a Euro-American origin of human immunodeficiency virus (HIV). *J Natl Med Assoc* 79:1068–1072.

26. Breimer LH. 1984. Did Moritz Kaposi describe AIDS in 1872? *Clio Medica* 19:156–159; Root-Bernstein RS. 1990. AIDS and Kaposi's sarcoma pre-1979. *Lancet* i:969.

27. Root-Bernstein RS. 1990. Non-HIV immunosuppressive factors in AIDS: A multifactorial, synergistic theory of AIDS aetiology. *Res Immunol* 141:815–838.

28. Van der Meer G, Brug SL. 1942. Infection à Pneumocystis chez l'homme et chez les animaux. *Ann Soc belge de méd trop* 22:301–307; McMillan GC. 1947. Fatal inclusion-disease pneumonitis in adult. *Am J Pathol* 23:995–1003.

29. Symmers WSC. 1960. Generalized cytomegalic inclusion-body disease associated with pneumocystis pneumonia in adults: Report of three cases. *J Clin Pathol* 13:1–21. Morris L, Distenfeld A, Amorosi E, Karpatkin S. 1982. Autoimmune thrombocytopenic purpura in homosexual men. *Ann Intern Med* 96:714–717; Stricker RB. 1990. Pathogenesis and treatment of HIV-related immune thrombocytopenia. In: Volberding P, Jacobson MA, eds. 1990. *AIDS Clin Rev 1990* New York: Marcel Dekker, 97–114.

30. Muller G. 1960. Interstitielle plasmacellulare Pneumonie nach Corticosteroidbehandlung. *Frankfurter Zeitschrift Pathologie* 70:657–675.

31. Sheldon WH. 1962. Pulmonary *Pneumocystis carinii* infection. *J Pediatr* 61:780; Pavlica F. 1962. The first observation of congenital pneumocystic pneumonia in a fully developed stillborn child. *Ann Paediatr* 198:177–184; Huneycutt HC, Anderson WR, Hendry WS. 1964. *Pneumocystis carinii* pneumonia. *Am J Clin Patholol* 41:411–418; Rao M, Steiner P, Victoria MS, James P, et al. 1977. *Pneumocystis carinii* pneumonia. Occurrence in a healthy American infant. *JAMA* 238:2301–2302. Radman JC. 1973 *Pneumocystic carinii* pneumonia in an adopted Vietnamese infant. *JAMA* 230:1561–1563; Anonymous. 1974. Rare type of pneumonia in Vietnamese patients. *JAMA* 230:13; Centers for Disease Control. 1976 *Pneumocystis carinii* pneumonia in Vietnamese orphans. *MMWR* 25:15; Giebink GS, Sholler L, Keenan TP, Franciosi RA, Quie FG. 1976. *Pneumocystis carinii* in two Vietnamese refugee infants. *Pediatrics* 58:115; Hughes WT, Feldman S, Akur RJA, Verzosa MS, Hustu HO, Simon JV. 1974. Protein-calorie malnutrition—a host determinant for *Pneumocystis carinii* infection. *Am J Dis Child* 128:44.

32. Williams G, Stretton TB, Leonard JC. 1960. Cytomegalic inclusion disease and *Pneumocystis carinii* infection in an adult. *Lancet* ii:951–955.

33. Anderson CD, Barrie HJ. 1960. Fatal *Pneumocystis* pneumonia in an adult. *Am J Clin Patholol* 34:365.

34. Hennigar GR, Vinijchaikul K, Roque AL, Lyons HA. 1961. *Pneumocystis carinii* pneumonia in an adult. Report of a case. *Am J Clin Patholol* 35:353–364; Lyons HA, Vinijchaikul K, Hennigar GR. 1961. *Pneumocystis carinii* pneumonia unassociated with other disease. *Arch Intern Med* 108:929–936. Kaftori JK, Bassan H, Gellei B, Griffel B. 1962. *Pneumocystis carinii* pneumonia in the adult. Report of a primary case. *Arch Intern Med* 109:114–122, 438–446; Watanabe JM, Chinchinian H, Weitz C, McIlvanine S. 1965. *Pneumocystis carinii* pneumonia in a family. *JAMA* 193:119; LeClair RA. 1969. Descriptive epidemiology of interstitial pneumocystic pneumonia. *Am Rev Respir Dis* 99:542–547; Brunet JB, Bouvet E, Leibowitch J, et al. 1983. Acquired immunodeficiency syndrome in France. *Lancet* i:700–701. See also references in note 31.

35. Burke BA, Good RA. 1973. *Pneumocystis carinii* infection. *Medicine* 52:23–50; Robbins JB. 1967. *Pneumocystis carinii* pneumonitis: A review. *Pediatr Res* 1:131–158; Robbins JB. 1968. Immunological and clinico-pathological aspects of *Pneumocystis carinii* pneumonitis. In: Bergsma D, Good RA, eds. 1968. *Immunologic*

Deficiency Diseases in Man. New York: National Foundation, 219–242.

36. Robinson JJ. 1961. Two cases of pneumocystis. Observation in 203 adult autopsies. *Arch Pathol* 71:156–159.

37. Esterly JA. 1968. *Pneumocystis carinii* in lungs of adults at autopsy. *Amer Rev Respir Dis* 97:935–937.

38. Norman L, Kagan IG. 1973. Some observations on the serology of *Pneumocystis carinii* infections in the United States. *Infect Immunol* 8:317.

39. Meuwissen JHET. 1976. Infections with *Pneumocystis carinii*. *Natl Cancer Inst Monogr* 43:133–136.

40. Centers for Disease Control. 1987. Revision of the CDC surveillance definition for acquired immunodeficiency syndrome. *MMWR* 36 Suppl: 1S–15S.

41. Zimmerman SL, Frutchey L, Gibbes JH. 1947. Meningitis due to Candida (Monilia) albicans with recovery. *JAMA* 135:145–147.

42. Woods JW, Manning IH Jr, Patterson SCN. 1951. Monilial infections complicating the therapeutic use of antibiotics. *JAMA* 145:207–211.

43. Andren L, Theander G. 1956. Roentgenographic appearances of oesophageal moniliasis. *Acta Radiologica [Stock]* 46:571–574; Eban RE, Symers DA. 1959. Oesophageal moniliasis—radiological appearances. *J Faculty Radiol* 10:164–165.

44. Louria DB, Stiff DP, Bennett B. 1962. Disseminated moniliasis in the adult. *Medicine* 41:307–337; case JM, 337.

45. Sanders E, Levinthal C, Donner MW. 1962. Monilial esophagitis in a patient with hemoglobin SC disease. *Ann Intern Med* 57:650–654.

46. Higgs JM, Wells RS. 1972. Chronic mucocutaneous candidiasis: associated abnormalities of iron metabolism. *Br J Dermatol* 86 (Suppl 8):88–102; Fletcher J, Mather J, Lewis MJ, Whiting G. 1975. Mouth lesions in iron-deficient anemia: Relationship to *Candida albicans* in saliva and to impairment of lymphocyte transformation. *J Infect Dis* 131:44–50.

47. Grieve NWT. 1964. Monilial oesophagitis. *Br J Radiol* 37:551–554; Holt JM. 1968. Candida infection of the oesophagus. *Gut* 9:227–231; Guyer PB, Brunton FJ, Rooke HWP. 1971. Candidiasis of the oesophagus. *Br J Radiol* 44:131–136.

48. Brown JW, McKee WM. 1972. Acute monilial esophagitis occurring without underlying disease in a young male. *Digestive Dis* 17:85–88.

49. Sheft DJ, Shrago G. 1970. Esophageal moniliasis. *JAMA*

213:1859–1862; Mann NS, Caplash VK. 1975. Monilial esophagitis. *South Med J* 68:479–480.

50. Kodsi BE, Wickremesinghe PC, Kozinn PJ, Iswara K, Goldberg PK. 1976. Candida esophagitis. A prospective study of 27 cases. *Gastroenterol* 71:715–719.

51. Wells RS, Higgs JM, MacDonald A, Valdimarsson H, Holt PJL. 1972. Familial chronic muco-cutaneous candidiasis. *J Med Genetics* 9:302–310; Orringer MB, Sloan H. 1978. Monilial esophagitis: An increasingly frequent cause of esophageal stenosis? *Ann Thoracic Surg* 26:364–374; Scott BB, Jenkins D. 1982. Gastro-oesophageal candidiasis. *Gut* 23:137–139.

52. Bogen E, Kessel J. 1937. Monilial meningitis. *Arch Pathol* 23:909; reviewed in Louria DB, Stiff DP, Bennett B. 1962. Disseminated moniliasis in the adult. *Medicine* 41:309, Table 1; Odds FC. 1979. *Candida and Candidosis.* Leicester: Leicester University Press, ch. 19.

53. Joachim H, Polayes SH. 1940. Subacute endocarditis and systemic mycosis (Monilia). *JAMA* 115:205–208; Polayes SH, Emmons CW. 1941. Final report on the identification of the organism of the previously reported case of subacute endocarditis and systemic mycosis (Monilia). *JAMA* 117:1533–1534.

54. Wikler A, Williams EG, Douglass ED, Emmons CW, Dunn RC. 1942. Mycotic endocarditis. *JAMA* 119:333–336.

55. Odds. 1979. *Candida*, Tables 41 and 49; Factor S et al. 1979. Sudden death in a narcotic addict four months following aortic valve replacement. *Am Heart J* 89:233; Louria DB, Stiff DP, Bennett B. *Medicine* 41:307–337, 1962; Louria DB, Hensle T, Rose J. 1967. The major medical complications of heroin addiction. *Ann Intern Med* 67:1–22; Sapira JD. 1968. The narcotic addict as a medical patient. *Am J Med* 45:555–588; Elliot JH et al. 1979. Mycotic endophthalmitis in drug abusers. *Am J Ophthalmol* 88:66–72; Aguilar GL, et al. 1979. Candida endophthalmitis after intravenous drug abuse. *Arch Ophthalmol* 97:96–100.

56. Orringer and Sloan. 1978. *Ann Thoracic Surg* 26:364–374.

Anomaly 2: HIV Is Neither Necessary Nor Sufficient to Cause AIDS

1. Burkett E. 1990. HIV. Not guilty? *Tropic (Miami Herald),* 23 Dec, 12–17.

2. Grmek MD, 1990. RC Maulitz, J Duffin, trans. *History of AIDS: Emergence and Origin of a Modern Pandemic.* Princeton NJ: Princeton University Press, 60–70.

3. Grmek. 1990. *History of AIDS*, 71–82; Crewdsen J. Science under the microscope. *Chicago Tribune*, 19 Nov 1989, sec. 5; Crewdsen J. 1991. Gallo case. *Science* 255, 10–11; Cohen J. 1991. Gallo case. *Science* 254:946–947; Anonymous. 1992. Verdicts are in on the Gallo probe. *Science* 256:735–739.

4. Grmek. 1990. *History of AIDS*, 60–70.

5. Lemaitre M, Guétard D, Hénin Y, Montagnier L, Zerial A. 1990. Protective activity of tetracycline3 analogs against the cytopathic effect of the human immunodeficiency virus in CEM cells. *Res Virol* 141:5.

6. Lusso P, Ensoli B, Markham PD, et al. 1989. Productive dual infection of human CD4+ T lymphocytes by HIV-1 and HHV-6. *Nature* 337:370–373; Lusso P, Veronese FM, Ensoli B, et al. 1990. Expanded HIV-1 cellular tropism by phenotypic mixing with murine endogenous retroviruses. *Science* 247:848–852.

7. Gallo RC. 1989. My life stalking AIDS. *Discover*, Oct 1989:30–36.

8. Lo S-C, Dawson MS, Wong DM, et al. 1989. Identification of *Mycoplasma incognitus* infection in patients with AIDS: An immunohistochemical, *in situ* hybridization and ultrastructural study. *Amer J Trop Med Hyg* 41:601–616; Lo S-C, Hayes MM, Wang RY-H, Pierce PF, Kotani H, Shih JW-K. 1991. Newly discovered mycoplasma isolated from patients infected with HIV. *Lancet* 338:1415–1418.

9. Burkett. 1990. *Tropic (Miami Herald)*, 23 Dec, 12–17.

10. Wright K. 1990. Mycoplasmas in the AIDS spotlight. *Science* 248:682–683.

11. Swift D. 1990. Does HIV cause AIDS? *Medical Post* (Toronto), 18 Dec 1990, 10.

12. Afrasiabi R, Mitsuyasu R, Nishanian P, et al. 1986. Characterization of a distinct subgroup of high-risk persons with Kaposi's sarcoma and good prognosis who present with normal T4 cell number and T4:T8 ratio and negative HTLV-III/LAV serologic test results. *Am J Med* 81:969–973; Friedman-Kein AE, Saltzman BR, Cao Y, et al. 1990. Kaposi's sarcoma in HIV-negative homosexual men. *Lancet* i:1344–1345; Garcia-Muret MP, Soriano V, Pujol RM, Hewlett I, Clotet B, De Moragas JM. 1990. AIDS and Kaposi's sarcoma in HIV-negative bisexual man. *Lancet* i:969–970; Castro A, Pedreira J, Soriano V, et al. 1992. Kaposi's sarcoma and disseminated tuberculosis in HIV-negative individual. *Lancet* i:4 April; Veyssier-Belot C, Couderc L, Desgranges C. 1990. Kaposi's sarcoma and HTLV-I infection. *Lancet* 336:575; Ho D, Moudgil

T, Robin H, Alam M, Wallace B, Mizrachi Y. 1989. HIV-1 in seronegative patient with visceral Kaposi's sarcoma and hypogammaglobulinemia. *Am J Trop Med* 86:349–351; Rappersberger K, Tschaachler E, Zonzits E, et al. 1990. Endemic Kaposi's sarcoma in HIV-1 seronegative persons: Demonstration of retrovirus-like particles in cutaneous lesions. *J Invest Dermatol* 95:371–381; Bowden F, McPhee D, Deacon N, et al. 1991. Antibodies to gp41 and nef in otherwise HIV-negative homosexual men with Kaposi's sarcoma. *Lancet* 337:1313–1314; Huang YQ, Li JJ, Rush MG, et al. 1992. HPV-16-related DNA sequences in Kaposi's sarcoma. *Lancet* 339:515–518.

13. Mannucci PM, Quattrone P, Mattturri L. 1986. Kaposi's sarcoma without human immunodeficiency virus antibody in a hemophiliac. *Ann Intern Med* 105:466.

14. Janssen RS, Saykin AJ, Kaplan JE, et al. 1988. Neurological complications of human immunodeficiency virus infection in patients with lymphadenopathy syndrome. *Ann Neurol* 23:49–55.

15. Pankhurst C, Peakman M. 1989. Reduced CD4+ T cells and severe oral candidiasis in absence of HIV infection. *Lancet* i:672; Gatenby PA. 1989. Reduced CD4+ T cells and candidiasis in absence of HIV infection. *Lancet* i:1027; Daus H, Schwarze G, Radtke H. 1989. Reduced CD4+ count, infections and immune thrombocytopenia without HIV infection. *Lancet* ii: 559–560.

16. Jacobs JL, Libby DM, Winters RA, et al. 1991. A cluster of *Pneumocystis carinii* pneumonia in adults without predisposing illnesses. *New Engl J Med* 324:246–250. Gupta S, et al. 1992. *Proc Natl Acad Sci USA* (15 August).

17. Root-Bernstein RS. 1990. Do we know the cause(s) of AIDS? *Persp Biol Med* 33:480–500; Root-Bernstein RS. 1990. Non-HIV immunosuppressive factors in AIDS: A multifactorial, synergistic theory of AIDS etiology. *Res Immunol* 141:815–838.

18. Associated Press. 1992. New AIDS-like illness comes to light at conference. *Lansing St J*, 22 July, 3A; Fackelmann KA. 1992. The baffling case of an AIDS-like malady. *Sci News* 142:70; Altman LK. 1992. Two experts questioning report about possible new AIDS virus. *NY Times*, 2 Aug, pt 1, p 1; Fumento M, 1992. A complicated disease won't have simple answers. *LA Times*, 28 July, B7; Heimoff S, 1992. Test ideas with science, not scorn. *LA Times*, 28 July, B7; Cohen J. 'Mystery' virus meets the skeptics. *Science* 257:1032–1034.

19. Okello DO, Sewankambo N, Goodgame R, et al. 1990. Absence of bacteremia with Mycobacterium avium-intracellulare in Ugandan patients with AIDS. *J Infect Dis* 162:208–210.

20. Schmutzhard E, Fuchs D, Hengster P, et al. 1989. Retroviral infections (HIV-1, HIV-2, and HTLV-I) in rural northwestern Tanzania. Clinical findings, epidemiology, and association with infections common in Africa. *Am J Epidemiol* 130:309–318.

21. Widi-Wirski R, Berkley S, Downing R, et al. 1988. Evaluation of the WHO clinical case definition for AIDS in Uganda. *JAMA* 260:3286–3289.

22. Centers for Disease Control. 1989. Update: Acquired immunodeficiency syndrome—United States, 1981–1988. *MMWR* 38:229–236.

Anomaly 3: Where Is the HIV Anyway?

1. Holmes KK, Johnson DW, Trostle HJ. 1970. An estimate of the risk of men acquiring gonorrhoea by sexual contact with infected females. *Am J Epidemiol* 108:136–144; Wright DJM, Daunt O. 1973. How infectious is gonorrhea? *Lancet* i:208; Platt R, Rice PA, McCormack WM. 1983. Risk of acquiring gonorrhea and prevalence of abnormal adrenal findings among women recently exposed to gonorrhea. *JAMA* 250:3205–3209; Hooper RR, Reynolds GH, Jones OG, et al. 1978. Cohort study of venereal disease: 1. The risk of gonorrhea transmission from infected women to men. *Am J Epidemiol* 108:136–144; Centers for Disease Control. 1979. *STD Fact Sheet* U.S. Dept. of Health, Education, and Welfare publication 79–8195.

2. Peterman TA, Stoneburner RL, Allen JR, Jaffe HW, Curran JW. 1988. Risk of human immunodeficiency virus transmission from heterosexual adults with transfusion-associated infections. *JAMA* 259:55–59; Padian N, Wiley J, Winkelstein W. 1987. Male to female transmission of human immunodeficiency virus: Current results, infectivity rates, and San Francisco population seroprevalence estimates. Presented at the Third International Conference on AIDS, Washington, DC, 4 June, 1987; Padian NS, Shiboski SC, Jewell NP. 1990. The effect of number of exposures on the risk of heterosexual HIV transmission. *J Infect Dis* 161:883–887.

3. Anonymous. 1989. Lovers, liars and other strangers. *Newsweek*, 27 Feb, 61; Chin P, Knapp D. 1989. The price of betrayal. *People Weekly* 31:180–183, Hudson R, Davidson S. 1986. *Rock Hudson: His Story*. New York: William Morrow.

4. Anonymous. 1991. Magic's wife tests negative again. *Lansing State J*, 19 Dec, 3A; Associated Press. 1992. It's a "magical" baby! *Lansing State J*, 5 June, 1A.

5. See note 2.

6. Holmes KK, Johnson DW, Trostle JH. 1970. An estimate of the risk of men acquring gonorrhea by sexual contact with infected females. *Am J Epidemiol* 91:170–174. Holmes KK, Mardh P-A, eds. *International Perspectives on Neglected Sexually Transmitted Diseases* Washington DC: Hemisphere Publishing Co.; Wright DJM, Daunt O. 1973. How infectious is gonorrhoea? *Lancet* i:208; Kingsley LA, Rinaldo CR Jr, Lyter DW, Valdiserri RD, Belle SH, Ho M. 1990. Sexual transmission efficiency of hepatitis B virus and human immunodeficiency virus among homosexual men. *JAMA* 264:230–234.

7. Zagury D, Bernard J, Leibowitch J, et al. 1984. HTLV-III in cells cultured from semen of two patients with AIDS. *Science* 226:449–451.

8. Ho DD, Schooley RT, Rota TR, et al. 1984. HTLV-III in the semen and blood of a healthy homosexual man. *Science* 226:451–453.

9. Wolff H, Anderson DJ. 1988. Potential human immunodeficiency virus-host cells in human semen. *AIDS Res Human Retroviruses* 4:1–2. Wolff H, Mayer K, Seage G, Politch J, Horsburgh CR, Anderson D. 1992. A comparison of HIV-1 antibody classes, titers, and specificities in paired semen and blood samples from HIV-1 seropositive men. *AIDS* 5:65–69; Anderson DJ, O'Brien TR, Politch JA, et al. 1992. Effects of disease stage and zidovudine therapy on the detection of human immunodeficiency virus type 1 in semen. *JAMA* 267:2769–2774.

10. Levy J. 1988. The transmission of AIDS: the case of the infected cell. *JAMA* 259:3037–3038.

11. Zagury et al. 1984. *Science* 226:449–451.

12. Levy. 1988. *JAMA* 259:3037–3038.

13. Levy. 1988. *JAMA* 259:3037–3038.

14. Krieger JN, Coombs RW, Collier AC, et al. 1991. Recovery of human immunodeficiency virus type 1 from semen: Minimal impact of stage of infection and current antiviral chemotherapy. *J Infect Dis* 163:386–388;

15. Levy. 1988. *JAMA* 259:3037–3038.

16. Kolata G. 1988. Discovery on AIDS virus raises a possibility of safe fatherhood. *NY Times* 17 Oct, pt 1, p 1.

17. Van Voorhis BJ, Martinez A, Mayer K, Anderson DJ. 1991. Detection of human immunodeficiency virus type 1 in semen from seropositive men using culture and polymerase chain reaction deoxyribonucleic acid amplification techniques. *Fertil Steril* 55:588–594.

18. Buimovici-Klein E, Lange M, Ong KR, Grieco MH, Cooper LZ. 1988. Virus isolation and immune studies in a cohort of homosexual men. *J Med Virol* 25:371–385; Lo SC, Hayes MM, Wang RY, Pierce PF, Kotani H, Shih JW. 1991. Newly discovered mycoplasma isolated from patients infected with HIV. *Lancet* 338:1415–1418.

19. Skolnik PR, Kosloff BR, Bechtel LJ, et al. 1989. Absence of infectious HIV-1 in the urine of seropositive viremic subjects. *J Infect Dis* 160:1056–1060; Kawashima H, Bandyopadhyay S, Rutstein R, Plotkin SA. 1991. Excretion of human immunodeficiency type 1 in the throat but not in the urine by infected children. *J Pediatr* 118:80–82.

20. Nadal D, Hunziker UA, Shupbach J, et al. 1989. Immunological evaluation in the early diagnosis of prenatal or perinatal HIV infection. *Arch Dis Child* 64:662–669; European Collaborative Study. 1988. Mother-to-child transmission of HIV infection. *Lancet* ii:1039–1043; Italian Multicentre Study. 1988. Epidemiology, clinical features, and prognostic factors of paediatric HIV infection. *Lancet* ii:1043–1046; Blance S, Rouzioux C, Moscato MG, et al. 1989. A prospective study of infants born to women seropositive for human immunodeficiency virus type 1. *New Engl J Med* 320:1643–1648; Johnson JP, Nair P, Hines SE, et al. 1989. Natural history and serologic diagnosis of infants born to human immunodeficiency virus–infected women. *Am J Dis Child* 143:1147–1153; Wolinsky SM, Wike CM, Korber TM, et al. 1992. Selective transmission of HIV type 1 variants from mothers to infants. *Science* 255:1134–1137.

21. Levy. 1988. *JAMA* 259:3037–3038.

22. Shaw G. 1990. Artificially inseminated woman gets AIDS, sues. *Medical Post*, 18 Dec, 5.

23. Gaines J. 1990. A scandal of artifical insemination. *NY Times Mag Part 2*, 7 Oct, 22–29.

24. Chiasson MA, Stoneburner, RL, Joseph SC. 1990. Human immunodeficiency virus transmission through artificial insemination. *J AIDS* 3:69–72. Gaines. 1990. *NY Times Mag Part 2*, 7 Oct, 28.

25. Centers for Disease Control. 1990. HIV-1 infection and artifical insemination with processed semen. *MMWR* 39:249–251.

26. Centers for Disease Control. 1990. *MMWR* 39:249–251.

27. Stewart GJ, Tyler JPP, Cunningham AL, et al. 1985. Transmission of human T-cell lymphotropic virus type III (HTLV-III) by artificial insemination by donor. *Lancet* ii:581–584; Joan Shenton, Meditel Productions, London, England: personal communication.

Anomaly 4: HIV Exposure Without Seroconversion or AIDS

1. Peterman TA, Stoneburner RL, Allen JR, Jaffe HW, Curran JW. 1988. Risk of human immunodeficiency virus transmission from heterosexual adults with transfusion-associated infections. *JAMA* 259:55–58.

2. Centers for Disease Control. 1991. HIV/AIDS Surveillance Report, July:1–18.

3. Nahmias S. 1989. *J Sex Res* 26:15; *Lansing State J* 1 Feb 1989, 3A.

4. See notes 5–12.

5. Desjarlais D, Friedman S, Novick D, et al. 1988. HIV-1 infection among intravenous drug users in Manhattan, New York City, from 1987–1988. *JAMA* 261:1008–1012; Division of Substance Abuse Services, New York State. 1985. Statewide comprehensive five-year plan 1984–85 through 1988–89. Second Annual Update, Oct 1985.

6. Seidlin M, Krasinski K, Bebenroth D, Itri V, Paolino AM, Valentine F. 1988. Prevalence of HIV infection in New York call girls. *J AIDS* 1:150–154.

7. Wallace J. 1989. Case presentations of AIDS in the United States. In: Ma P, Armstrong D, eds. *AIDS and Infections of Homosexual Men*. 2d ed. Boston: Butterworths, 285–295.

8. Rosenberg MJ, Weiner JM. 1988. Prostitutes and AIDS: A health department priority? *Am J Public Health* 78:418–423.

9. Coutinho RA, van der Helm T. 1986. [No indications of LAV/HTLVIII in non-drug-using prostitutes in Amsterdam] *Nederlands Tijdscrift voor Geneeskunde* 130:509.

10. Luthy R, Ledergerber B, Tauber M, Siegnthaler W. 1987. Prevalence of HIV antibodies among prostitutes in Zurich, Switzerland. *Klinische Wochenschrift* 65:287–288.

11. Kopp W, Dangl-Erlach E. 1986. [HTLV-III monitoring in prostitutes in Vienna.] *Wiener Klinische Wochenschrift* 98:695–698.

12. Brenky-Fandeux D, Fribourg-Blanc A. 1985. HTLV-III antibody in prostitutes. *Lancet* ii:1424; Day S, Ward H, Harris JRW. 1988. Prostitute women and public health. Br Med J 297:1585; Hyams KC, Escamilla J, Papadimos TJ, et al. 1988. HIV infection in a non-drug abusing population. Scand J Infect Dis 21:353.

13. Wallace. 1989. In: Ma and Armstrong, eds. 1989. *AIDS and Infections*, 285–295.

14. Shiokawa Y. 1988. [Depressed cellular immunity in prostitutes and male homosexuals predisposes to HIV.] *Jap J Bacteriol* 43:559–564.

15. Wallace. 1989. In: Ma and Armstrong, eds. 1989. *AIDS and Infections*, 285–295.

16. Piot P, Laga M. 1988. Prostitutes: A high risk group for HIV infection? *Sozial- und Präventativmedicin* 33:336–339.

17. Cameron DW, Simonsen JN, D'Costa LJ, et al. 1989. Female to male transmission of human immunodeficiency virus type 1: risk factors for seroconversion in men. *Lancet* ii:403–407; Plant ML, Plant MA, Peck DF, Setters J. 1989. The sex industry, alcohol and illicit drugs: Implications for the spread of HIV infection. *Br J Addiction* 84:53–59.

18. Burke DS, Brundage JF, Goldenbaum M, et al. 1990. Human immunodeficiency virus infections in teenagers. *JAMA* 263:2074–2077; Garland FC, Mayers DL, Hickey TM, et al. 1989. Incidence of human immunodeficiency virus seroconversion in US Navy and Marine Corps personnel, 1986–1988. *JAMA* 262:3161–3165.

19. Centers for Disease Control. 1989. Summary of Surveillance Data. *MMWR* 38, No. S-4:36–38, Figures 7–11; CDC, personal communication.

20. Stewart GJ, Tyler JPP, Cunningham AL, et al. 1985. Transmission of human T-cell lymphotropic virus type III (HTLV-III) by artificial insemination by donor. *Lancet* ii:581–584.

21. Padian N, Winkelstein W, unpublished data reported in Levy J. 1988. The transmission of AIDS: The case of the infected cell. *JAMA* 259:3037–3038.

22. Padian et al. 1988. *JAMA* 259:3037.

23. Padian NS, Shiboski SC, Jewell NP. 1990. The effect of number of exposures on the risk of heterosexual HIV transmission. *J Infect Dis* 161:883–887.

24. Kim HC, Raska K 3d, Clemow L, et al. 1988. Human immunodeficiency virus infection in sexually active wives of infected hemophilic men. *Am J Med* 85:472–476.

25. Marcus R. 1988. The CDC Cooperative Needlestick Surveillance Group. Surveillance of health care workers exposed to blood from patients infected with human immunodeficiency virus. *New Engl J Med* 319:1118–1123; CDC, HIV/AIDS Surveillance Report, July 1991.

26. Hagen MD, Meyer KB, Pauker SG. 1988. Routine preoperative screening for HIV. Does the risk to the surgeon outweigh the risk to the patient? *JAMA* 259:1357–1360.

27. Henry K, Campbell S, Jackson B, et al. 1990. Longterm follow-up of health care workers with worksite exposure to human immunodeficiency virus. *JAMA* 263:1765–1766.

28. Anonymous. 1990. Doctors with HIV. *Lancet* 336:432.

29. Centers for Disease Control. 1991. Update: Transmission of HIV infection during an invasive dental procedure—Florida. *MMWR* 40:21–33, 18 Jan; Associated Press. 1992. AIDS sufferer wants tests for health-care workers. *Lansing State J*, 29 Jan, 3B.

30. Associated Press. 1992. *Lansing State J*, 29 Jan, 3B.

31. Associated Press. 1992. Friend: AIDS dentist did it on purpose. *Lansing State Journal*, 6 Nov, 3A.

32. Palca J. 1992. The case of the Florida dentist. *Science* 255:392–394.

33. Golembiewski C. 1991. AIDS. A healthy fear? *Lansing State J*, 4 Nov, 1A–5A; Walsh E. 1991. Dentist's AIDS death shakes Illinois town. *Washington Post*, 20 July, A1–A10.

34. Menitove JE. 1989. Status of recipients of blood from donors subsequently found to have antibody to HIV. *New Engl J Med* 315:1095–1096.

35. Ward JW, Bush TJ, Perkins HA, et al. 1989. The natural history of transfusion-associated infection with human immunodeficiency virus. *New Engl J Med* 321:947–952.

36. Ludlum CA, Tucker J, Steel CM, et al. 1985. Human T-lymphotropic virus type III (HTLV-III) infection in seronegative hemophiliacs after transfusions of factor VIII. *Lancet* ii:233–236.

37. Gazengel C, Courouce AM, Torchet MF, et al. 1984. Use of HBV vaccine in hemophiliacs. *Scand J Haematal* Suppl 40, 33:323–328.

38. Tabor E. 1984. Review of the transmission of hepatitis by clotting factor concentrates. *Scand J Haematol* Suppl 40, 33:303–308.

Anomaly 5: Sometimes HIV Disappears

1. Dufoort G, Couroucé A-M, Ancell-Park R, Bletry O. 1988. No clinical signs 14 years after HIV-2 transmission via blood transfusion. *Lancet* ii:510.

2. Cuthbert RJG, Ludlum CA, Tucker J, et al. 1990. Five year prospective study of HIV infection in the Edinburgh haemophiliac cohort. *Br Med J* 301:956–961.

3. Burger H, Weiser B, Robinson WS, et al. 1985. Transient antibody to lymphadenopathy-associated virus/human T-lymphotropic virus type III and T-lymphocytic abnormalities in the wife of a man who developed the acquired immunodeficiency syndrome. *Ann Intern Med* 103:545–547.

4. Sonnex C. 1986. Transient antibody to human immunodeficiency virus [Letter]. *Ann Intern Med* 105:464–465.

5. Kuritsky JN, Rastogi SC, Faich GA, et al. 1986. Results of nation-wide screening of blood and plasma for antibodies to HTLV-III. *Transfusion* 26: 205–207; Marwick C. 1985. Blood banks give HTLV-III testing positive appraisal after 5 months. *JAMA* 254:1681–1683; Ward JW, Grindon AJ, Feorino PM, et al. 1986. Laboratory and epidemiologic evaluation of an enzyme immunoassay for antibodies to HTLV-III. *JAMA* 256:357–361; Burke DS, Brundage JF, Redfield RR, et al. 1988. Measurement of the false positive range in a screening program for human immunodeficiency virus infecitons. *New Engl J Med* 319:961–964; Papadopulos-Eleopolus E, Turner VJ, Papadimetriou JM. 1993. Is a positive western blot proof of HIV infection? *Bio/Technology*, in press.

6. Barnes DM. 1988. Losing AIDS antibodies. *Science* 240:1407; Horsburgh CR Jr, Ou CY, Jason J, et al. 1990. Concordance of polymerase chain reaction with human immunodeficiency virus antibody detection. *J Infect Dis* 162:542–545.

7. Perrin LH, Zubler R, Hirschel B, et al. 1988. Négativation d'une sérologie positive pour le virus d'immunodeficience humaine (VIH). A propos de deux observations. *Schweizische Medecinische Wochenschrift* 118/45:1641–1644.

8. Fribourg-Blanc A. 1988. Deux observations d'annulation spontanee d'une sero-positivité HIV. *Médecine et maladie infectieuse* 18:216–218.

9. Jaffe HW, Feorino PM, Darrow WW, et al. 1985. Persistent infection with human T-lymphotropic virus type III/lymphadenopathy-associated virus in apparently healthy homosexual men. *Ann Intern Med* 102:627.

10. Imagawa D, Detels R. 1991. HIV-1 in seronegative homosexual men. *New Engl J Med* 325:1250–1251; Farzadegan H, Polis MA, Wolinsky SM, et al. 1988. Loss of human immunodeficiency virus type I (HIV-1) antibodies with evidence of viral infection in asymptomatic homosexual men. *Ann Intern Med* 108:785–790.

11. Gibbons J, Cory JM, Hewlett IK, Epstein JS, Eyster ME. 1990. Silent infections with human immunodeficiency virus type 1 are highly unlikely in multitransfused seronegative hemophiliacs. *Blood* 76:1924–1926.

12. Blanche S, Rouzioux C, Moscato M-LG, et al. 1989. A prospective study of infants born to women seropositive for human immunodeficiency virus type 1. *New Engl J Med* 320:1643–1648; Greenberg AE, Nguyen-Dinh DP, Mann JM, et al. 1988. The association between malaria, blood transfusions, and HIV seropositivity in a pediatric population in Kinshasa, Zaire. *JAMA* 259:545–

549; Gaetano C, Scano G, Carbonari M, et al. 1987. Delayed and defective anti-HIV IgM response in infants. *Lancet* i:631; Aiuti F, Luzi G, Mezzaroma I, Scano G, Papetti C. 1987. Delayed appearance of HIV infection in children. *Lancet* ii:858.

13. Barnes. 1988. Losing HIV antibodies. *Science* 240:1407.
14. Imagawa, Detels. 1991. *New Engl J Med* 325:1250–1251.
15. Evans AS. 1989. Does HIV cause AIDS? An historical perspective. *J AIDS* 2:107–113.
16. Brown P. 1992. AIDS researchers focus on immune response. *New Scientist*, 8 August, 16.
17. Burkett. 1990. *Tropic (Miami Herald)]*, 15; CDC, personal communcation; Warth JA. 1989. The risk of developing AIDS in hemophiliac and homosexual men. *JAMA* 262:3129–3130.
18. Callen M. 1990. *Surviving AIDS*. New York: Harper Collins, 5, Ch. 1.

CHAPTER 2   THE ROLE OF HIV IN AIDS

The Evolving Definition of AIDS

1. Centers for Disease Control. 1982. Update on acquired immune deficiency syndrome (AIDS)—United States. *MMWR* 31 (37):507–508.
2. Associated Press. 1987. Patients infected with AIDS after kidney transplants. *Lansing State J*, 29 Sep, 3A; Dummer JS, Erb S, Breinig MK, et al. 1989. Infection with human immunodeficiency virus in the Pittsburgh transplant population. A study of 583 donors and 1043 recipients, 1981–1986. *Transplantation* 47:134–140.
3. Kjellstrand CM, Hylander B, Collins AC. 1990. Mortality on dialysis—on the influence of early start, patient characteristic, and transplantation and acceptance rates. *Am J Kidney Dis* 15:483–490; Held PJ, Brunner F, Odaka M, Garcia JR, Port FK, Gaylin DS. 1990. Five-year survival for end-stage renal disease patients in the United States, Europe, and Japan, 1982–1987. *Am J Kidney Dis* 15:451–457.
4. Salt A, Sutehall G, Sargaison M, et al. 1990. Viral and toxoplasma gondii infections in children after liver transplantation. *J Clin Patholol* 43:63–67; Singh N, Dummer JS, Kusne S, et al. 1988. Infections with cytomegalovirus and other herpes viruses in 121 liver transplant recipients: Transmission by donated organ and the effect of OKT3 antibodies. *J Infect Dis* 158:124–131.
5. Centers for Disease Control. 1981. *Pneumocystis* pneumonia—Los Angeles. *MMWR* 30 (21):250–252.

6. Selik RM, Haverkos HW, Curran JW. 1984. Acquired immune deficiency syndrome (AIDS) trends in the United States, 1978–1982. *Am J Med* 76:493–500.

7. Centers for Disease Control. 1985. Revision of the case definition of acquired immunodeficiency syndrome for national reporting—United States. *MMWR* 34 (25):373–375.

8. Centers for Disease Control. 1985. *MMWR* 34(25):373–375.

9. Selik et al. 1984. *Am J Med* 76:493–500.

10. Centers for Disease Control. 1989. Update: Acquired immunodeficiency syndrome—United States, 1981–1988. *MMWR* 38:229–236.

11. Centers for Disease Control. 1987. Revision of the CDC surveillance case definition for acquired immunodeficiency syndrome. *MMWR* 36 (Suppl 1S):3S–15S; Imrey HH. 1988. AIDS case definition. *Science* 240:1263.

12. Centers for Disease Control. 1991. Extension of public comment period for revision of HIV infection classification system and expansion of AIDS surveillance case definition. *MMWR* 40:891.

13. Cimons M. 1991. Federal government to expand definition of AIDS. *LA Times*, 9 Aug, A37.

14. Cimons. 1991. *LA Times*, 9 Aug, A37; Associated Press. 1992. Center holds off on AIDS definition. *Lansing State J*, 5 Jan.

15. Eckholm E. 1991. Facts of life. More than inspiration is needed to fight AIDS. *NY Times*, 1 August, sec. 4, p1.

16. Grmek M. 1990. RC Maulitz, J Duffin, trans. *History of AIDS*. Princeton: Princeton University Press, 109.

17. Centers for Disease Control. 1992. Mortality patterns—United States, 1989. *MMWR* 41:121–125.

18. Callen M. 1990. *Surviving AIDS*. New York: Harper Collins.

The Case for HIV as the Cause of AIDS

1. Blattner W, Gallo RC, Temin HM. 1988. HIV causes AIDS. *Science* 242:514–517.

2. Duesberg PH. 1987. Retroviruses as carcinogens and pathogens: Expectations and reality. *Cancer Res* 47:1199–1220; Duesberg PH. 1988a. HIV is not the cause of AIDS. *Science* 241:514–516; Duesberg PH. 1988b. Human immunodeficiency virus and acquired immunodeficiency syndrome: Correlation but not causation. *Proc Natl Acad Sci USA* 86:755–764; Duesberg PH. 1989. Responding to "The AIDS debate." *Naturwissenschaften* 77:97–102.

3. Weber J. 1988. AIDS and the 'guilty' virus. *New Sci* 5 May, 32–33.

4. Weber. 1988. *New Sci* 5 May, 32–33.
5. Madden DL, Tzan NR, Roman GC, et al. 1987. HIV and HTLV-I antibody studies. Pregnant women in the 1960s, patients with AIDS, homosexuals and individuals with tropical spastic paralysis. *Yale J Biol Med* 60:569–574.
6. Biggar RJ, Buskell-Bales Z, Yakshe PN, Caussy D, Gridley G, Seeff, L. 1990. Antibody to human retroviruses among drug users in three east coast American cities, 1972–1976. *J Infect Dis* 163:57–63; Goedert JJ, Kessler CM, Aledort LM, et al. 1989. A prospective study of human immunodeficiency virus type 1 infection and the development of AIDS in subjects with hemophilia. *New Engl J Med* 321:1141–1148; Evatt BL, Gomperts ED, McDougal JS, Ramsey RB. 1985. Coincidental appearance of LAV/HTLV-III antibodies in hemophiliacs and the onset of the AIDS epidemic. *New Engl J Med* 312:483–486.
7. Lazzarin A, Crocchiolo P, Galli M, et al. 1986. Milan as possible starting point of LAV/HTLV-III epidemics among Italian drug addicts. *Boll 1st Sieroter Milanese* 66:9–13.
8. Madhok R, Melbye M, Lowe GDO, et al. 1985. HTLV-III antibody in sequential plasma samples: from haemophiliacs 1974–1984. *Lancet* i:524–525.
9. Auerbach DM, Bennet JV, Brachman PS, et al. 1982. Report of the Centers for Disease Control Task Force on Kaposi's Sarcoma and Opportunistic Infections: Epidemiological aspects of the current outbreak of Kaposi's sarcoma and opportunistic infections. *New Engl J Med* 306:248–252.
10. Wormser GP, Krupp LB, Hanrahan JP, et al. 1983. Acquired immunodeficiency syndrome in male prisoners. *Ann Intern Med* 98:297–303; Jaffe H, Choi K, Thomas PA, et al. 1984. National case-control study of Kaposi's sarcoma and *Pneumocystis carinii* pneumonia in homosexual men: Part 1, Epidemiologic results. *Ann Intern Med* 99:145–151.
11. Mason JM, Evatt BL, Chorba TL, Ramsey RB. 1984. Acquired immunodeficiency syndrome (AIDS) in hemophiliacs. *Scand J Haematol* Suppl 40, 33:349–356.
12. Selik RM, Haverkos HW, Curran JW. 1984. Acquired immune deficiency syndrome (AIDS) trends in the United States 1978–1982. *Am J Med* 76:493–500.
13. Eigen M. 1989. The AIDS debate. *Naturwissenschaften* 76:341–350.
14. Blattner et al. 1988. *Science* 242:514–517.
15. Weiss RA, Jaffe HW. 1990. Duesberg, HIV and AIDS. *Nature* 345:659–670.

16. Auerbach DM, Darrow WW, Jaffe HW, Curran JW. 1984. Cluster of cases of the acquired immune deficiency syndrome. Patients linked by sexual contact. *Am J Med* 76:487–492; Centers for Disease Control. 1982. A cluster of Kaposi's sarcoma and *Pneumocystis carinii* pneumonia among homosexual male residents of Los Angeles and Orange counties, California. *MMWR* 31:305–307; Biggar RJ, Melbye M, Ebbesen P, et al. 1984. Low T-lymphocyte ratios in homosexual men. Epidemiological evidence for a transmissible agent. *JAMA* 251:1441–1446.

17. DeVita R, ed. 1985. *AIDS: Etiology, Diagnosis, Treatment, and Prevention*. Philadelphia: Lippincott, 68; Jonas Salk, personal communication.

18. Weiss, Jaffe. 1990. *Nature* 345:659–670.

19. Weiss, Jaffe, 1990. *Nature* 345:659–670.

20. Ginsberg HS. 1988. Scientific forum on AIDS: A summary. Does HIV cause AIDS? *J AIDS* 2:165–172.

21. Dalgleish AG, Beverly PCL, Clapham PR, et al. 1985. The CD4 (T4) antigen is an essential component of the receptor for the AIDS retrovirus. *Nature* 312:763–767; Klatzmann D, Champagne E, Chamaret S, et al. 1985. T-lymphocyte T4 molecule behaves as the receptor for human retrovirus LAV. *Nature* 312:767–769.

Problems with the Evidence that HIV Causes AIDS

1. Lazzarin A, Crocchiolo P, Galli M, et al. 1986. Milan as possible starting point of LAV/HTLV-III epidemics among Italian drug addicts. *Boll 1st Sieroter Milanese* 66:9–13.

2. Grmek MD. 1990. In: Maulitz RC, Duffin J, trans. *History of AIDS*. Princeton: Princeton University Press, 125.

3. Moore JD, Cone EJ, Alexander SS Jr. 1985. HTLV-III seropositivity in 1971–72: Parenteral drug abusers—a case of false positives or evidence of viral exposure? *New Engl J Med* 314:1387–1388.

4. Walzer PD, Perl DP, Krogstad DJ, Rawson PG, Schultz MG. 1974. *Pneumocystis carinii* pneumonia in the United States: Epidemiologic, diagnostic, and clinical features. *Ann Intern Med* 80:83–93. Also in: Robbins JB, DeVita VT Jr, Dutz W, eds. 1976. *Symposium on Pneumocystis carinii Infection*. Bethesda, US Dep Health, Education and Welfare, National Institute of Health, National Cancer Institute Monograph, No. 43, DHEW Publication No. NIH 76:930.

5. Young LS. 1984. Preface. In: Young LS, ed. 1984. *Pneumocystis carinii Pneumonia: Pathogenesis. Diagnosis. Treatment*. New York: Marcel Dekker, vi–xiii.

6. Simon HB, Guerry D, Breslow A, Kirkpatrick CH. 1973. Opportunistic pathogens in the immunologically hyperresponsive host. *Am J Med* 55:856–864.

7. Jaffe HW, Choi K, Thomas PA, et al. 1983. National case-control study of Kaposi's sarcoma and *Pneumocystis carinii* pneumonia in homosexual men: Part 1, Epidemiologic results. *Ann Intern Med* 99:145–151; Selik RM, Haverkos HW, Curran JW. 1984. Acquired immune deficiency syndrome (AIDS) trends in the United States 1978–1982. *Am J Med* 76:493–500; Young. 1984. In: Young, ed. 1984. *Pneumocystis carinii Pneumonia*, vi–xiii; Mason JM, Evatt BL, Chorba TL, Ramsey RB. 1984. Acquired immunodeficiency syndrome (AIDS) in hemophiliacs. *Scand J Haematol* Suppl 40, 33:349–356.

8. Rao M, Steiner P, Victoria MS, et al. 1977. *Pneumocystis carinii* pneumonia. Occurrence in a healthy American infant. *JAMA* 238:2301–2302; Furio MM, Wordell CJ. 1985. Treatment of infectious complications of acquired immunodeficiency syndrome. *Clin Pharmacol* 4:539–554.

9. Zakowski RC, Gottlieb MS, Groopman J. 1984. Acquired immunodeficiency syndrome (AIDS), Kaposi's sarcoma, and *Pneumocystis carinii* pneumonia. In: Young. 1984. *Pneumocystis carinii Pneumonia*, 208–209.

10. Gilkey F. 1982. Opportunistic infections and Kaposi's sarcoma in homosexual men. *New Engl J Med* 306:935–936.

11. Jensen OM, Mouridsen HT, Petersen HS, et al. 1982. Kaposi's sarcoma in homosexual men: Is it a new disease? *Lancet* i:1027.

12. Leonidas JL, Hyppolite N. 1983. Haiti and the acquired immunodeficiency syndrome. *Ann Intern Med* 98:1020–1021.

13. Auerbach DM, Bennet JV, Brachman PS, et al. 1982. Report of the Centers for Disease Control Task Force on Kaposi's Sarcoma and Opportunistic Infections: Epidemiological aspects of the current outbreak of Kaposi's sarcoma and opportunistic infections. *New Engl J Med* 306:248–252.

14. Centers for Disease Control. 1981. *MMWR* 30:5 June, 250–252; Safai B, Good RA. 1981. Kaposi's sarcoma: A review and recent developments. *Clin Bull* 10:62–69.

15. Biggar RJ, Horm J, Fraumeni JF Jr, Greene MH, Goedert JJ. 1984. Incidence of Kaposi's sarcoma and mycosis fungoides in the United States including Puerto Rico, 1973–1981. *J Natl Cancer Inst* 73:89–94.

16. Clemmesen J. 1982. Kaposi's sarcoma in homosexual men: Is it a new disease? *Lancet* ii:51–52; Brownstein MH, Shapiro L, Skolnik P. 1973. Kaposi's sarcoma in community practice. *Arch Dermatol*

107:137–138; Biggar RJ, Nasca PC, Burnett WS. 1988. AIDS-related Kaposi's sarcoma in New York City in 1977. *New Engl J Med* 318:252.

17. Biggar et al. 1984. *J Nat Cancer Inst* 73:89–94.

18. Henig RM. 1983. AIDS: A new disease's deadly odyssey. *NY Times Mag*, 6 Feb, 32.

19. Aronson DL. 1983. Pneumonia deaths in haemophiliacs. *Lancet* ii:1023.

20. Rizza CR, Spooner RJD. 1983. Treatment of haemophilia and related disorders in Britain and Northern Ireland during 1976–80: report on behalf of the directors of haemophilia centres in the United Kingdom. *Br Med J* 286:929–933.

21. Melbye M, Froebel KS, Madhok R, et al. 1984. HTLV-III seropositivity in European haemophiliacs exposed to factor VIII concentrate imported from the USA. *Lancet* ii:1444–1445; Mason JM, Evatt BL, Chorba TL, Ramsey RB. 1984. Acquired immunodeficiency syndrome (AIDS) in hemophiliacs. *Scand J Haematol* Suppl 40, 33:349–356.

22. Brunet JB, Bouvet E, Leibowitch J, et al. 1983. Acquired immunodeficiency syndrome in France. *Lancet* i:700–701.

23. Blattner W, Gallo RC, Temin HM. 1988. HIV causes AIDS. *Science* 242:514–517.

24. Weiss RA, Jaffe HW. 1990. Duesberg, HIV and AIDS. *Nature* 345:659–670.

25. Peterman TA, Stoneburner RL, Allen JR, et al. 1988. Risk of human immunodeficiency virus transmission from heterosexual adults with tranfusion-associated infections. *JAMA* 259:55–58.

26. Blattner et al. 1988. *Science* 242:514–517.

27. Blattner et al. 1988. *Science* 242:514–517.

28. CDC. 1989. *MMWR* 38:229–236.

29. CDC. 1991. HIV/AIDS Surveillance. July:1–18.

30. Kreiss JK, Kasper CK, Fahey JL, et al. 1984. Nontransmission of T-cell subset abnormalities from hemophiliacs to their spouses. *JAMA* 251:1450–1454.

31. CDC. 1991. HIV/AIDS Surveillance. July:1–18.

32. Auerbach DM, Darrow WW, Jaffe HW, Curran JW. 1984. Cluster of cases of the acquired immune deficiency syndrome. Patients linked by sexual contact. *Am J Med* 76:487–492.

33. Rubin H. 1988. Etiology of AIDS. *Science* 241:1389–1390.

34. Stewart GT. 1990. Uncertainties about AIDS and HIV. *Lancet* i:1325; Jonas S. 1988. AIDS: An alternative scenario. *Congressional Record [USA]* 15 Sep.

35. Peterman TA, Stoneburner RL, Allen JR, Jafe HW, Curran JW.

1988. Risk of human immunodeficiency virus transmission from heterosexual adults with transfusion-associated infections. *JAMA* 259:55–59.

36. Eigen M. 1989. The AIDS debate. *Naturwissenschaften* 76:349–350.

37. Griffen BE. 1989. Burden of proof. *Nature* 338:670.

38. Clark SJ, Saag MS, Decker WD, et al. 1991. High titers of cytopathic virus in plasma of patients with symptomatic primary HIV-1 infection. *New Engl J Med* 324:954–960; Daar ES, Moudgil T, Meyer RD, Ho DD. 1991. Transient high levels of viremia in patients with primary human immunodeficiency virus type 1 infection. *New Engl J Med* 324:961–964; Busch MP, Amad ZE, Sheppard HW, Ascher MS. 1991. Primary HIV-1 infection. *New Engl J Med* 325:733.

39. Auerbach et al. 1984. *Am J Med* 76:487; Burkett E. (1990). HIV: Not guilty? *Tropic (Miami Herald)*, 23 Dec, 12–17; CDC, personal communication.

40. Burkett. 1990. *Tropic*, 23 Dec, 12–17.

41. Nowak MA, Anderson RM, McLean AR, Wolfs TFW, Goudsmit J, May RM. 1991. Antigenic diversity thresholds and the development of AIDS. *Science* 254:963–969.

42. E.g., Booth W. 1988. A rebel without a cause of AIDS. *Science* 239:1485–1488; Moore JP, Blanc DF. 1991. Immunological incompetence in AIDS. *AIDS* 5:455–456; Ward R. 1988. Mainstream scientists confront unorthodox view of AIDS. *Nature* 332:574.

43. Hooper C. 1991. AIDS brain research broadens in search of mechanisms. *J NIH Res* 3:17–19.

44. Marx J. 1991. Clue found to T cell loss in AIDS. *Science* 254:798–800.

45. Nowak et al. 1991. *Science* 254:963–969.

46. McLean AR, Nowak MA. 1992. Models of interactions between HIV and other pathogens. *J Theor Biol* 155:69–86.

47. Sarngadharan MG, Markham PD. 1987. The role of human T-lymphotropic retroviruses in leukemia and AIDS. In: Wormser GP, ed. 1987. *AIDS—Acquired Immune Deficiency Syndrome—and Other Manifestations of HIV Infection*. Park Ridge, NJ: Noyes Publications, 197.

48. Barre-Sinoussi F, Chermann J-C, Rey F, et al. 1983. Isolation of a T lymphotropic retrovirus from a patient at risk for acquired immune deficiency syndrome (AIDS). *Science* 220:868–871; Popovic M, Sarngadharan MG, Reed E, et al. 1984. Detection, isolation

and continuous production of cytopathic human T lymphotropic retrovirus (HTLV-III) from patients with AIDS and pre-AIDS. *Science* 224:497–500.

49. Popovic et al. 1984. *Science* 224:497–500; Gallo RC. 1987. The AIDS virus. *Sci Am* 256:45–56.

50. Sarngadharan and Markham. 1987. In: Wormser, ed. 1987. *AIDS*, 197–198.

51. Zagury D, Bernard J, Leonard R, et al. 1986. Long-term cultures of HTLV-III-infected T cells: A model of cytopathology of T-cell depletion in AIDS. *Science* 231:850–853; Sarngadharan and Markham. 1987. In: Wormser, ed. 1987. *AIDS*, 197–198.

Koch's Postulates, Etiological Criteria, and AIDS

1. Levy J. 1988. The transmission of AIDS: The case of the infected cell. *JAMA* 259:3037–3038.

2. Kion TA, Hoffmann GW. 1991. Anti-HIV and anti-anti-MHC antibodies in alloimmune and autoimmune mice. *Science* 253:1138.

3. Sayers MH, Beatty PG, Hanson JA. 1986. HLA antibodies are a cause of false positive reactions in screening enzyme immunoassays for antibodies to HTLV-III. *Transfusion* 26:113–115; Blanton M, Balakrishnan K, Dumaswala O, et al. 1987. HLA antibodies in blood donors with reactive screening tests for antibody to the immunodeficiency virus. *Transfusion* 27:118–119; Shi LR, Schaude R, Wei D, et al. 1991. 33 of 197 murine monoclonal antibodies against total HIV-1 protein and against NEF protein and against overlapping synthetic NEF lipopeptides presenting a distinct pattern of cross-reactivity with normal human lymphoid tissue. In: Abstracts VIII International Conference on AIDS, Florence, II:150; Mathé G. 1992. Is the AIDS virus responsible for the disease? *Biomed Pharmacother* 446:1–2.

4. Ginsberg HS. 1988. Scientific forum on AIDS: A summary. Does HIV cause AIDS? *J AIDS* 2:165–172.

5. Ginsberg. 1988. *J AIDS* 2:165–172.

6. E.g., Weber J. 1988. AIDS and the 'guilty' virus. *New Sci* 5 May: 32–33; Weiss RA, Jaffe HW. 1990. Duesberg, HIV and AIDS. *Nature* 345:659–670; Blattner W, Gallo RC, Temin HM. 1988. HIV causes AIDS. *Science* 242:514–517.

7. Evans AS. 1976. Causation and disease: The Henle-Koch postulates revisited. *Yale J Biol Med* 49:175–195; Evans AS. 1978. Causation and disease: A chronological journey. *Am J Epidemiol* 108:249–258.

8.   Rivers TM. 1937. Viruses and Koch's postulates. *J Bacteriol* 33:1.

9.   Heubner RJ. 1957. The virologist's dilemma. *Ann NY Acad Sci* 67:430.

10. DeVita R, ed. 1985. *AIDS: Etiology, Diagnosis, Treatment, and Prevention.* Philadelphia: Lippincott, 31–32.

11. DeVita, ed. 1985. *AIDS: Etiology,* 35 note.

12. Ginsberg. 1988. *J AIDS* 2:165–172.

13. Evans AS. 1989. Does HIV cause AIDS? An historical perspective. *J AIDS* 2:107–113.

14. Duesberg 1989. Does HIV cause AIDS? *J AIDS* 2:514–515.

15. Duesberg PH. 1988. HIV is not the cause of AIDS. *Science* 241:514–517.

16. Evans AS. 1989. Does HIV cause AIDS? Author's reply. *J AIDS* 2:515–517.

Multifactorial, Synergistic Disease Models of AIDS

1.   Modified from various writings and correspondence with Dr. Leslie Klevay, Human Nutrition Research Center, Grand Forks, ND.

2.   Evans AS. 1976. Causation and disease: The Henle-Koch postulates revisited. *Yale J Biol Med* 49:175–195; Evans AS. 1982. The clinical illness promotion factor: A third ingredient. *Yale J Biol Med* 55:193–199; Evans AS. 1989. Does HIV cause AIDS? An historical perspective. *J AIDS* 2:107–113.

3.   Evans AS. 1989. Does HIV cause AIDS? Author's Reply. *J AIDS* 2:515–517.

4.   Koch R. 1884. *Über die cholerabacterien.* Berlin: G. Richmer.

5.   Witte MH, Stuntz M, Witte CL, Way D. 1989. AIDS, Kaposi's sarcoma, and the gay population. The germ or the terrain? *Int J Dermatol* 28:585–586.

6.   Dubos R, Escande J-P. 1979. *Quest: Reflections on Medicine, Science, and Humanity.* New York: Harcourt Brace Jovanovich, 45.

7.   Degre M, Glasgow LA. 1968. Synergistic effect in viral-bacterial infection. I. Combined infection of the respiratory tract in mice with parainfluenza virus and *Haemophilus influenzae. J Infect Dis* 118:449–462.

8.   Shope E. 1931. Swine influenza, I. Experimental transmission and pathology. *J Exp Med* 54:349–359; Lewis PA, Shope RE. 1931. Swine influenza. II. A hemophilic bacillus from the respiratory tract of infected swine. *J Exp Med* 54:361–372; Shope RE. 1931. Swine influenza, III. Filtration experiments and etiology. *J Exp Med* 54:373–385.

9. Shope. 1931. *J Exp Med* 54:373–385. Degre and Glasgow. 1968. *J Infect Dis* 118:449–462.

10. Gorbach SI, Bartlett JG. 1974. Anaerobic infections (third of three parts). *New Engl J Med* 290:1289–1294.

11. Dudding BA, Wagner SC, Zeller JA, Gmelich JT, French GR, Top FH Jr. 1972. Fatal pneumonia associated with adenovirus type 7 in three military trainees. *New Engl J Med* 286:1289–1292.

12. Gorbach, Bartlett. 1974. *New Engl J Med* 290:1289–1294.

13. Witebsky E, Rose NR, Terplan K, Pain JR, Egan RW. 1957. Chronic thyroiditis and autoimmunization. *JAMA* 164:1439–1447; Root-Bernstein RS. 1991. Self, nonself, and the paradoxes of autoimmunity. In: Tauber AI, ed. 1991. *Organism and the Development of Self*. Needham, MA: Kluwer Academic, 159–209.

14. Root-Bernstein. 1991. In: Tauber, ed. 1991. *Organism and Self*, 159–209.

15. Root-Bernstein RS. 1990. Do we know the cause(s) of AIDS? *Persp Biol Med*, 480–500; Root-Bernstein RS. 1990. Non-HIV immunosuppressive agents in AIDS: A multifactorial, synergistic theory of AIDS aetiology. *Res Immunol* 141:815–838; Sonnabend JA. 1984. The etiology of AIDS. *AIDS Res* 1:1–15; Sonnabend JA. 1989. AIDS: An explanation for its occurrence among homosexual men. In: Ma P, Armstrong D, eds. 1989. *AIDS and Infections of Homosexual Men*. 2d ed. Boston: Butterworths, 449–470; Papadopoulos-Eleopulos E. 1988. Reappraisal of AIDS. Is the oxidation induced by the risk factors the primary cause? *Med Hypotheses* 25:151–162.

CHAPTER 3 IMMUNOSUPPRESSION AND AIDS

Noninfectious Immunosuppressive Agents Associated with AIDS

1. Allen JR. 1984. Epidemiology: United States. In: Ebbesen P, Biggar RJ, Melbye M, eds. 1984. *AIDS: A Basic Guide for Clinicians*. Munksgaard: W. B. Saunders, 16.

2. Center for Disease Control. 1987. Revision of the CDC surveillance case definition for acquired immunodeficiency syndrome. *MMWR* 36 (Suppl 1S): 3S–15S.

3. Sonnabend JA, Saadoun S. 1984. The acquired immunodeficiency syndrome: A discussion of etiologic hypotheses. *AIDS Res* 1:107–120.

4. Floersheim GL. 1978. [Immunosuppression as a concomitant ef-

fect.] *Schwiz Med Wochenschr* 108:1449–1460; Chandra RK. 1983. Preface. In: Chandra RK, ed. 1983. *Primary and Secondary Immunodeficiency Disorders* Edinburgh: Churchill Livingstone.

5. Sonnabend JA, Witkin SS, Putilo DT. 1983. Acquired immunodeficiency syndrome: Opportunistic infections and malignancies in male homosexuals. *JAMA* 249:2370–2373; Sonnabend JA, Witkin SS, Purtilo DT. 1984. A multifactorial model for the development of AIDS in homosexual men. *NY Acad Sci* 437:177; Sonnabend and Saadoun. 1984. *AIDS Res* 1:107–120; Sonnabend JA. 1984. The etiology of AIDS. *AIDS Res* 1:1–15; Sonnabend JA. 1989. AIDS; An explanation for its occurence among homosexual men. In: Ma P, Armstrong D, eds. 1989. [1984] *AIDS and Infections of Homosexual Men*. 2nd rev ed, Boston: Butterworths, 449–470.

Immunosuppression due to Semen Components

1. Metalnikoff S. 1900. Etudes sur la spermatoxine. *Ann Inst Pasteur* 14:577–603.
2. Favilli G. 1931. The effect of testicle extract on red blood cells in vitro. *J Exp Med* 54:197–206.
3. Stites DP, Erikson RP. 1975. Suppressive effect of seminal plasma on lymphocyte activation. *Nature* 253:727–729; Prakash C, Coutinho A, Moller G. 1976. Inhibition of *in vitro* immune responses by a fraction from seminal plasma. *Scand J Immunol* 5:77–85.
4. Marcus ZH, Freisheim JH, Houk JL, Herman JH, Hess EV. 1978. *In vitro* studies in reproductive immunology. 1. Suppression of cell-mediated immune response by human spermatozoa and fractions isolated from human seminal plasma. *Clin Immunol Immunopathol* 9:318–326; Marcus ZH, Hess EV, Herman JH, Troiano P, Freisheim J. 1979. *In vitro* studies in reproductive immunology. 2. Demonstration of the inhibitory effect of male genital tract constituents on PHA-stimulated mitogenesis and E-rosette formation of human lymphocytes. *J Reprod Immunol* 1:97–107; Franken DR, Slabber CF. 1981. Reproductive immunology: The inhibitory effect of human seminal plasma from normozoospermic men on *in vitro* lymphocyte cultures. *Andrologia* 13:537–540; Erikson KL. 1984. Seminal plasma inhibits lymphocyte response to T-dependent and independent antigens *in vitro*. *Immunology* 52:721–726.
5. Richards JM, Bedford JM, Witkin SS. 1983. Immune response to allogeneic insemination via the rectum in the rabbit. *Fed Proc* 42:1334.
6. Hurtenbach U, Shearer GM. 1982. Germ cell–induced immune

suppression in mice. Effect of inoculation of syngeneic spermatozoa on cell-mediated immune responses. *J Exp Med* 155:1719.

7. Mathur S, Goust J-M, Williamson HO, et al. 1981. Cross-reactivity of sperm and T-lymphocyte antigens. *Am J Reprod Immunol* 1:113–118; Witkin SS, Sonnabend J. 1983. Immune responses to spermatozoa in homosexual men. *Fertil Steril* 39:337–342; Mavligit GM, Talpaz M, Hsia FT, et al. 1984. Chronic immune stimulation by sperm alloantigens: Support for the hypothesis that spermatazoa induce immune dysregulation in homosexual males. *JAMA* 251:237–241; James K, Hargreave TB. 1984. Immunosuppression by seminal plasma and its possible clinical significance. *Immunol Today* 5:357.

8. Shearer GM, Rabson A. 1984. Semen and AIDS. *Nature* 308:230; Shearer GM. 1984. Immune suppression and recognition of class II (Ia) antigens: A possible factor in the etiology of acquired immune deficiency syndrome. In: Friedman-Kien AD, Laubenstein LJ, eds. 1984. *AIDS: The Epidemic of Kaposi's Sarcoma and Opportunistic Infections*. New York: Masson.

9. Williamson JD. 1984. Semen polyamines in AIDS pathogenesis. *Nature* 310:103; Schopf RE, Schramm P, Benes P, Morsches B. 1984. Seminal plasma-induced suppression of the respiratory burst of polymophonuclear leukocytes and monocytes. *Andrologia* 16:124–128.

10. Marcus ZH, Lunenfeld B, Weissenberg R, Lewin LM. 1987. Immunosuppressant material in human seminal fluid: Inhibition of blast transformation and NK activity by seminal fluid from patients of a male infertility clinic. *Gynecol Obstet Invest* 23:54–59.

11. Marcus et al. 1987. *Gynecol Obstet Invest* 23:54–59.

12. James K, Harvey J, Bradbury AW, Hargreave TB, Cullen RT. 1983. The effect of seminal plasma on macrophage function—a possible contributory factor in sexually transmitted disease. *AIDS Res* 1:45–57.

13. Lord EM, Sensabaugh GF, Stites DP. 1977. Immunosuppressive activity of human seminal plasma I. Inhibition of *in vitro* lymphocyte activation. *J Immunol* 118:1704–1711; Anderson DJ, Tarter TH. 1982. Immunosuppressive effects of mouse seminal plasma components *in vivo* and *in vitro*. *J Immunol* 128:535–539.

14. James et al. 1983. *AIDS Res* 1:45–57; Turner MJ, White JO, Soutter WP. 1990. Human seminal plasma inhibits the lymphocyte response to infection with Epstein-Barr virus. *Gynecol Oncol* 37:60–65.

15. Witkin, Sonnabend. 1983. *Fertil Steril* 39:337–342; Mavligit et al. 1984. *JAMA* 251:237–241.

16. Darrow WW, Echenberg DF, Jaffe HW, et al. 1987. Risk factors for human immunodeficiency virus (HIV) infections in homosexual men. *Am J Public Health* 77:479–483.

17. Amerongen HM, Weltzin R, Farnet CM, Michetti P, Haseltine WA, Neutra MR. 1991. Transepithelial transport of HIV-1 by intestinal M cells: A mechanism for transmission of AIDS. *J AIDS* 4:760–765.

18. Seeff LB, Wright EC, Zimmerman HG, et al. 1978. Type B hepatitis after needle-stick exposure: Prevention with hepatitis B immune globulin. Final report of the Veterans Administration Cooperative Study. *Ann Intern Med* 88:285–293; Alter MJ, Francis DP. 1989. Hepatitis B virus transmission between homosexual men: A model for AIDS. In: Ma P, Armstrong D., eds. *AIDS and Infections of Homosexual Men* Boston: Butterworths, 99–108.

19. Shearer and Rabson. 1984. *Nature* 308:230; Mavligit et al. 1984. *JAMA* 251:237–241.

Immunosuppression due to Addictive and Recreational Drugs

1. Quinn TC, Glasser D, Cannon R, et al. 1988. Human immunodeficiency virus infection among patients attending clinics for sexually transmitted diseases. *New Engl J Med* 318:197.

2. National Commission on AIDS. 1991. *Report: The Twin Epidemics of Substance Use and HIV.* Washington DC: National Commission on AIDS.

3. Moss AR. 1987. AIDS and intravenous drug use: The real heterosexual epidemic. *Br Med J* 294:389.

4. Pillai RM, Watson RR. 1990. In vitro immunotoxicology and immunopharmacology: Studies on drugs of abuse. *Toxicol Lett* 53:269–283.

5. Ellinwood EH, Gawin FH. 1988. What we don't know about cocaine. *AAAS Observer* 2 Sep, 5; Ellinwood EH, Gawin FH. 1988. Cocaine. *New Engl J Med* 318:1173–1176.

6. Watson RR. 1990. Preface. *Drugs of Abuse and Immune Function.* Boca Raton: CRC Press.

7. Faith RE. 1983. Effects of cocaine exposure on immune functions. *Toxicol* 3:56–60; Fuchs D, Dierich MP, Hausen A, et al. 1987. Activated T cells in addition to LAV/HTLV-III infection: A necessary precondition for development of AIDS. *Cancer Detection and Prevention*, Suppl, 1:583.

8. Van Dyke C, Stesin A, Jones R, Chuntharapai A, Seaman W. 1986. Cocaine increases natural killer cell activity. *J Clin Invest* 77:1387–1390; Havas HF, Dellaria M, Schiffman G, Geller EB,

Adler MW. 1987. Effect of cocaine on the immune response and host resistance in BALB/c mice. *Int Arch Allergy Appl Immunol* 83:377; Bagasra O, Forman L. 1989. Functional analysis of lymphocyte subpopulations in experimental cocaine abuse. I. Dose-dependent activation of lymphocyte subsets. *Clin Exp Immunol* 77:289–293; Watson ES, Murphy JC, El-Sohly HN, El-Sohly MA, Turner CE. 1983. Effects of the administration of coca alkaloids on the primary immune responses of mice: Interaction with delta-9-tetrahydrocannabinol and ethanol. *Toxicol Appl Pharmacol* 71:1–13.

9. Bagasra, Forman. 1988. *Clin Exp Immunol* 77:289; Klein TW, Newton CA, Friedman H. 1988. Suppression of human and mouse lymphocyte proliferation by cocaine. In: Bridge TP et al., eds. 1988. *Psychological, Neuropsychiatric, and Substance Abuse Aspects of AIDS.* New York: Raven Press, 139–143.

10. Klein et al. 1988. In: Bridge et al., eds. 1988. *Aspects of AIDS,* 139–143.

11. Fuchs et al. 1987. *Cancer Detection and Prevention,* Suppl, 1:583.

12. Peterson PK, Gekker G, Chao CC, Schut R, Molitor TW, Balfour HH Jr. 1991. Cocaine potentiates HIV-1 replication in human peripheral blood mononuclear cell cocultures. Involvement of transforming growth factor-beta. *J Immunol* 146:81–84.

13. Donohoe RM, Falek AA. 1988. Neuroimmunomodulation by opiates and other drugs of abuse: Relationship to HIV infection and AIDS. In: Bridge et al., eds. 1988. *Aspects of AIDS,* 145–158.

14. Cantacuzene J. 1898. Nouvelles recherches sur le mode de destruction des vibrions dans l'organisme. *Ann Instit Pasteur J Microbiologie* 12:273–300.

15. Rubin G. 1904. The influence of alcohol, ether, and chloroform on natural immunity in its relation to leucocytosis and phagocytosis. *J Infect Dis* 1:425–444; Snel G. 1903. Immunität und Narkose. *Berl Klin Wochenschr* 40:212.

16. Reynolds R. 1910. The influence of narcotics on phagocytosis. *Lancet* i:569.

17. Arkin A. 1913. The influence of strychnin, caffein, chloral, antipyrin, cholesterol, and lactic acid on phagocytosis. *J Infect Dis* 12:408–424.

18. Atchard C, Bernard H, Gagneux C. 1909. Action de la morphine sur les propriétés leucocytaires. Leuco-diagnostic du morphinisme. *Bull Mem Soc Med Hosp Paris* 28:958.

19. Helpern, Rho, 1966; Geller SA, Stimmel B. 1973. Diagnostic confusion from lymphatic lesions in heroin addicts. *Ann Intern Med* 78:703.

20. Cushman P, Grieco MH. 1973. Hyperglobulinemia associated with narcotic addiction. *Am J Med* 54:320; Kreek MJ, Dodes L, Kane S, Knobler J, Martin R. 1972. Long-term methadone maintenance therapy: Effects on liver function. *Ann Intern Med* 77:598–602; Kreek MJ. 1973. Medical safety and side effects of methadone in tolerant individuals. *JAMA* 223:665–668; Sapira JD. 1968. The narcotic addict as a medical patient. *Am J Med* 45:555–588.

21. Brown, SM, Stimmel B, Taub RN, Kochwa S, Rosenfield RE. 1974. Immunologic dysfunction in heroin addicts. *Arch Intern Med* 134:1001–1006.

22. Hussey HH, Katz S. 1950. Infections resulting from narcotic addiction. *Am J Med* 9:186; Louria DB, Hensle T, Rose J. 1967. The major complications of heroin addiction. *Ann Intern Med* 67:1; Briggs JH, McKerron CG, Sounani RL, Taylor TJE, Andrews H. 1967. Severe systemic infections complicating "main line" heroin addiction. *Lancet* ii:1227; Cherubin CE. 1967. The medical sequelae of narcotic addiction. *Ann Intern Med* 67:23; Sapira, 1968. *Am J Med* 45:555–588. Cherubin CE. 1971. Infectious disease problems of narcotic addicts. *Arch Intern Med* 128:309; Dreyer NP, Fields BN. 1973. Heroin-associated infective endocarditis. A report of 28 cases. *Ann Intern Med* 78:699; Lang M, Salaki JS, Middleton J, et al. 1978. Infective endocarditis in heroin addicts: Epidemiological observations and some unusual cases. *Am Heart J* 96:144; Harris PD, Garret D. 1972. Susceptibility of addicts to infection and neoplasia. *New Engl J Med* 287:310; Brown et al. 1974. *Arch Intern Med* 134:1001–1006; Reichman LB, Felton CP, Edsall JR. 1979. Drug dependence, a possible new risk factor for tuberculosis disease. *Arch Intern Med* 139:337; Sadeghi A, Behmard S, Vesselinovitch SD. 1979. Opium: A potential urinary bladder carcinogen in man. *Cancer* 43:2315; Kornfeld H, Van de Stouwe RA, Lange M, et al. 1982. T-lymphocyte sub-populations in homosexual men. *New Engl J Med* 307:729–731.

23. Kerman RH, Floyd M, Van Buren CT, Kahan BD. 1980. Improved allograft survival of strong immune responder-high risk recipients with adjuvant antithymocyte globulin therapy. *Transplantation* 30:450–454.

24. McDonough RJ, Madden JJ, Falek A, et al. 1980. Alterations of T and null lymphocyte frequency in the peripheral blood of human opiate addicts: *In vivo* evidence for opiate receptor sites on T-lymphocytes. *J Immunol* 125:2539.

25. Falek A, Madden JJ, Shafer DA, Donohoe RM. 1986. Individual

differences in opiate-induced alterations at the cytogenetic, DNA repair, and immunologic levels: Opportunity for genetic assessment. *Natl Inst Drug Abuse Res Monogr Ser* 66:11–24; Donahoe RM, Nicholson JK, Madden JJ, et al. 1986. Coordinate and independent effects of heroin, cocaine, and alcohol abuse on T-cell E-rosette formation and antigenic marker expression. *Clin Immunol Immunopathol* 41:254–264; Donahoe RM. 1988. Opiates as immunocompromising drugs: The evidence and possible mechanisms. *NIDA Monograph Series* 90:105; Madden JJ, Donahoe RM. 1990. Opiate binding to cells of the immune system. In: Watson RR, ed. 1990. *Drugs of Abuse and Immune Function*. Boca Raton, FL: CRC Press, 213–228. Bernard A, Gelin C, Raynal B, Pham D, Goss C, Boumsell L. 1982. Phenomenon of human T cells rosetting with sheep erythrocytes analyzed with monoclonal antibody. *J Exp Med* 155:1317.

26. Donohoe RM, Bueso-Ramos C, Donohoe F, et al. 1987. Mechanistic implications of the findings that opiates and other drugs of abuse moderate T-cell surface receptors and antigenic markers. *Ann NY Acad Sci* 496:711–721.

27. Layon J, Idris A, Warzynski M, et al. 1984. Altered T-lymphocyte subsets in hospitalized intravenous drug abusers. *Arch Intern Med* 144:1376–1380; Maravelias CP, Coutselinis AS. 1984. Suppressive effects of morphine on human blood lymphocytes: An *in vitro* study. *IRCS Med Sci* 12:106; Peterson PK, Molitor TW, Chao CC, Sharp B. 1990. Opiates and cell-mediated immunity. In: Watson RR, ed. 1990. *Drugs of Abuse and Immune Function*. Boca Raton, Fl: CRC Press, 1–14; Weber RJ, Pert A. 1989. The periaqueductal gray matter mediates opiate-induced immunosuppression. *Science* 245:188–190; Shavit Y, DePaulis A, Martin FC, et al. 1986. Involvement of brain opiate receptors in the immune-suppressive effect of morphine. *Proc Natl Acad Sci (USA)* 83:7114–7117; Wybran J, Appelboom T, Famaey J-P, Govaerts A. 1979. Suggestive evidence for receptors for morphine and methionine-enkephalin on normal human blood T lymphocytes. *J Immunol* 123(1):1068–1070; Sibinga NES, Goldstein A. 1988. Opioid peptides and opiod receptors in cells of the immune system. *Ann Rev Immunol* 6:219.

28. Arora PK, Fride E, Petitto J, Waggie K, Skolnick P. 1990. Morphine-induced immune alterations in vivo. *Cell Immunol* 126:343–353.

29. Tubaro E, Borelli G, Croce C, Cavallo G, Santiangeli C. 1983. Effect of morphine on resistance to infection. *J Infect Dis* 148:656; Sein GM, Leung CMK, Dai S, Ogle CW. 1986. The immunores-

ponsiveness of morphine—dependent mice. *IRCS Med Sci* 14:742; Tubaro E, Santiangelli C, Belogi L, et al. 1987. Methadone vs morphine: Comparison of their effect on phagocytic functions. *Int J Immunopharmacol* 9:79.

30. Simon RH, Arbo TE. 1986. Morphine increases metastatic tumor growth. *Brain Res Bull* 16:363; Friedman H, Klein T, Specter S, et al. 1988. Drugs of abuse and virus susceptibility. In: Bridge TP et al., eds. 1988. *Psychological, Neuropsychiatric, and Substance Abuse Aspects of AIDS*. New York: Raven Press, 125–137.

31. Weisman M, Lerner N, Vogel WH, Schnoll S. 1973. Quality of street heroin. *New Engl J Med* 289:698–689; Vogel WH. 1971. Toxicity of drugs and "street" drugs: Medical and legal problems. *Contemp Drug Probl* 1:34.

32. Squinto SP, Mondal D, Block AL, Prakash O. 1990. Morphine-induced transactivation of HIV-1 LTR in human neuroblastoma cells. *AIDS Res Human Retroviruses* 6:1163–1168.

33. Rogers DF, Barnes PJ. 1989. Opioid inhibition of neurally mediated mucus secretion in human bronchi. *Lancet* i:930–931.

34. Kristal AR. 1986. The impact of the acquired immunodeficiency syndrome on patterns of premature death in New York City. *JAMA* 255:2306–2308; Bailey JA, Brown LS. 1989. Serologic investigations in a New York City cohort of parenteral drug users. *J Natl Med Assococ* 82:405–408.

35. Haverkos HW. 1988. Kaposi's sarcoma and nitrite inhalants. In: Bridge et al., eds. 1988. *Aspects of AIDS*, 165–172; Marmor M, Friedman-Kien AE, Laubenstein L, et al. 1982. Risk factors for Kaposi's sarcoma in homosexual men. *Lancet* i:1083–1087.

36. Lynch DW, Moorman WJ, Burg JR, et al. 1985. Subchronic inhalation toxicity of isobutyl nitrite in BALB/c mice. I. Systemic toxicity. II. Immunotoxicity studies. *J Toxicol Environ Health* 15:823–833, 835–846.

37. Osterloh J, Olson K. 1986. Toxicities of alkyl nitrites. *Ann Intern Med* 104:727.

38. Lotzova E, Savary CA, Hersh EM, et al. 1984. Depression of murine natural killer cell cytotoxicity by isobutyl nitrite. *Cancer Immunol Immunother* 17:130–134.

39. Koch R. 1884. *Über die cholerabacterien*. Berlin: G. Richmer.

40. Abbot AC. 1896. The influence of acute alcoholism on the normal vital resistance of rabbits to infection. *J Exp Med* 1:447; Rubin. 1904. *J Infect Dis* 1:425–444.

41. Abbot, G. 1896. *J Exp Med* 1:447; Labbe M. 1902. L'alcool et la résistance de l'organisme aux maladies. *Presse Med* 2:786; Capps

JA, Coleman GH. 1923. Influence of alcohol on prognosis of pneumonia in Cook County hospital. *JAMA* 80:750; Pickrell KL. 1938. The effect of alcoholic intoxication and ether anesthesia on resistance to pneumococcal infection. *Bull Johns Hopkins Hosp* 63:238. Adams AG, Jordan C. 1984. Infections in the alcoholic. *Med Clin North Am* 68:179–200; Bagasra O, Kajdacsy-Balla A. 1988. Ethanol-induced immunomodulations in chronic alcoholism. *Clin Immunol Newslett* 9:9; Bagasra O, Howeedy A, Dorio R, Kajdacsy-Balla A. 1987. Functional analysis of T-cell subsets in chronic experimental alcoholism. *Immunology* 61:63; Berenyi MR, Straus B, Cruz D. 1974. *In vitro* and *in vivo* studies of cellular immunity in alcoholic cirrhosis. *Am J Dig Dis* 19:199–205; Brayton RG, Stokes PE, Schwartz MS, et al. 1970. Effect of alcohol and various diseases on leukocyte mobilization, phagocytosis and intracellular bacterial killing. *New Engl J Med* 282:123–128; Ericsson CD, Kohl S, Pickering LK, et al. 1980. Mechanisms of host defense in well nourished patients with chronic alcoholism. *Alcoholism Clin Exp Res* 4:261–265; Gluckman SJ, Dvorak VC, MacGregor RR. 1977. Host defenses during prolonged alcohol consumption in a controlled environment. *Arch Intern Med* 136:1539–1543; Jerrells TR, Marietta CA, Bone G, Weight FF, Eckardt MJ. 1988. Ethanol-associated immunosuppression. In: Bridge TP et al., eds. *Psychological, Neuropsychiatric, and Substance Abuse Aspects of AIDS.* New York: Raven Press, pp. 173–185; Louria DB. 1963. Susceptibility to infection during experimental alcohol intoxication. *Trans Assoc Am Physicians* 76:102–112.

42. Descotes J. 1988. *Immunotoxicology of Drugs and Chemicals.* Amsterdam: Elsevier, 407.

43. Bayer BM, Flores CM. 1990. Effects of morphine on lymphocyte function: Possible mechanisms of interaction. In: Watson RR, ed. 1990. *Drugs of Abuse and Immune Function.* Boca Raton, FL: CRC Press, 151–174.

Immunosuppression due to Anesthesia and Surgery

1. Graham EA. 1911. The influence of ether and ether anesthesia on bacteriolysis, agglutination, and phagocytosis. *J Infect Dis* 8:147–175.

2. Rubin G. 1904. The influence of alcohol, ether, and chloroform on natural immunity in its relation to leucocytosis and phagocytosis. *J Infect Dis* 1:425–444.

3. Howard RJ, Simmons RL. 1974. Acquired immunologic deficiencies after trauma and surgical procedures. *Surg Gynecol Obstet* 139:771–782.

4. Wingard DW, Lang R, Humphrey LJ. 1967. Effect of anesthesia on immunity. *J Surg Res* 7:430; Humphrey LJ, Wingard DW, Lang R. 1969a. Effect of halothane on spleen cells; *in vitro* studies of reversibility of immunosuppression. *Surgery* 65:939–945; Humphrey LJ, Wingard DW, Lang R. 1969b. The effect of surgery and anesthesia on the immunologic responsiveness of the rat. *Surgery* 65:946–953; Park SK, Brody JI, Wallace HA, et al. 1971. Immunosuppressive effect of surgery. *Lancet* i:53–55; Bruce DL. 1972. Halothane inhibition of phytohaemagglutinin induced transformation of lymphocytes. *Anesthesiology* 36:201; Slawikowski GJM. 1960. Tumour development in adrenalectomised rats given inoculations of aged tumour cells after surgical stress. *Cancer Res* 20:316; Frederickson EL, Humphrey LJ. 1971. An *in vitro* study of the effect of halothane on the production of antibodies by white blood cells. In: Fink BR, ed. 1971. *Cellular Biology and Toxicity of Anesthetics*. Baltimore: Williams and Wilkins, 261–279; reviewed in Descotes J. 1988. *Immunotoxicology of Drugs and Chemicals*. Amsterdam: Elsevier, 156–165; Kehlet H, Wandall JH, Hjortso NC. 1982. Influence of anesthesia and surgery on immunocompetence. *Regional Anesth* 7:568; Cascorbi HF. 1981. Effect of anesthetics on the immune system. *Int Anesth Clin* 19:69; Nalda MA, Garcia M. 1982. Un concepta a incluir en anestesiologia y reanimacion: su interaction immunologica. *Rev Espagnola Anest Rean* 29:333; Bruce DL, Wingard DW. 1971. Anesthesia and the immune response. *Anesthesiology* 34:271; Cullen BF, Van Belle G. 1975. Lymphocyte transformation and changes in leukocyte counts. *Anesthesiology* 43:563; Hamid J, Bancewicz J, Brown R, Ward C, Irwing MH, Ford WL. 1984. The significance of changes in blood lymphocyte populations following surgical operations. *Clin Exp Immunol* 56:49; Salo M. 1978. Effect of anaesthesia and open-heart surgery on lymphocyte responses to phytohemagglutinin and concanavalin-A. *Acta Anaesth Scand* 22:471; Boileau S, Lancelot A, Laxenaire MC, Moneret-Vautrin DA. 1981. Modifications de la réactivité cutanée à la phyohemagglutinine entrainées par l'anesthésie générale. *Ann Anesth Franc* 3:231; Espanol T, Todd GB, Soothill JF. 1974. The effect of anaesthesia on the lymphocyte responses to phyohaemagglutinin. *Clin Exp Immunol* 18:73.

5. Saba TM. 1970. Opsonic depletion after surgery. *Nature* 228:781.

6. Slade MS, Simmons RL, Unis E, Greenberg LJ. 1975. Immunodepression after major surgery in normal patients. *Surgery* 78:363–372.

7. Espanol et al. 1974. *Clin Exp Immunol* 18:73; Riddle PR, Berenbaum MC. 1967. Postoperative depression of the lymphocyte response to phyohemagglutinin. *Lancet* i:746.

8. Tsuda T, Kahan BD. 1983. The effects of anesthesia on the immune response. In: Chandra RK, ed. 1983. *Primary and Secondary Immunodeficiency Disorders*. Edinburgh: Churchill Livingston, 253–262; Stevenson GWQ, Hall SC, Rudnick S, Seleny FL. Stevenson HC. 1990. The effect of anesthetic agents on the human immune response. *Anesthesiology* 72:542–552.

9. Hole A, Dakke O. 1984. T lymphocytes and the subpopulation of T helper and T suppressor cells measured by monoclonal antibodies (T11, T4 and T8) in relation to surgery under epidural and general anaesthesia. *Acta Anaesth Scand* 28:296–301.

10. Bruce DL, Eide KA, Linde HW, Eckenhoff JE. 1968. Causes of death among anesthesiologists: A 20 year study. *Anesthesiology* 29:565; Schumann D. 1990. Nitrous oxide anaesthesia: Risks to health personnel. *Int Nurs Rev* 37:214–217.

11. Oladimeji M, Gumishaw AD, Patterson KG, Goldstone AH. 1982. Effect of surgery on monocyte function. *Br J Surg* 69:145; Wang BS, Wolfe JHN, Wu AVO, Heacock EH, Mannick JA. 1980. Immunosuppression associated with surgical trauma. *Surg Forum* 31:61; Slade et al. 1975. *Surgery* 78:363–372; Vermesse G, Camus D, Wattre P, Capron A, Gautier-Benoit C. 1978. Modifications immunitaires dans les suite opératoires immédiates. *Nouv Presse Med* 7:529; Riddle, Berenbaum. 1967. *Lancet* i:746.

12. Gieraerts R, Narvalgund A, Vaes L, Soetens M, Chang J-L, Jahr J. 1987. Increased incidence of itching and herpes simplex in patients given epidural morphine after cesarean section. *Anesth Analg* 66:1321; Crone LL, Conly JM, Klark KM, et al. 1988. Recurrent herpes simplex virus labialis and the use of epidural morphine in obstetric patients. *Anesth Analg* 67:318.

Immunosuppression due to Pharmaceutical Agents

1. Stevenson HC, Fauci AS. 1983. Immunosuppression secondary to pharmacologic agents. In: Chandra RK, ed. 1983. *Primary and Secondary Immunodeficiency Disorders*. Edinburgh: Churchill Livingstone, 232–341.

2. Stevenson, Fauci. 1983. In Chandra, ed. 1983. *Immunodeficiency Disorders*, 232–341.

3. Reviewed in Descotes J. 1988. *Immunotoxicology of Drugs and Chemicals*. Amsterdam: Elsevier, 19–101.
4. Munster AM, Loadholdt CB, Leary AG, Barnes MA. 1977. The effect of antibiotics on cell-mediated immunity. *Surgery* 81:692–696; Tarnawski A, Batko B. 1973. Antibiotics and immune processes. *Lancet* i:674.
5. Roszkowski W et al. 1985. Antibiotics and immunomodulation: Effects of cefotaxime, anikacin, mezlocillin, piperacillin, and clindamycin. *Med Microbiol Immunol (Berlin)* 173:279–289.
6. Tarnawski, Batko. 1973. *Lancet* i:674; Munster et al. 1977. *Surgery* 81:692–696.
7. Reviewed in Descotes. 1988. *Immunotoxicology*, 107–111.
8. Grey D, Hamilton-Miller JMT. 1977. Trimethoprim-resistant bacteria: Cross-resistance patterns. *Microbios* 19:45–54.
9. Reviewed in Descotes. 1988. *Immunotoxicology*, 112–128.
10. Neumann HH. 1982. Use of steroid creams as possible cause of immunosuppression in homosexuals. *New Engl J Med* 306:935.
11. Goldman D. 1956. The major complications of treatment of psychotic states with chlorpromazine and reserpine and their management. *Psychiatry Res Rep Am Psychiat Assoc* 4:79; Wardell DW. 1957. Untoward reactions to tranquilizing drugs. *Am J Psychiatry* 113:745; Kane FJ Jr. 1963a. Oral moniliasis following the use of thorazine. *Am J Psychiatry* 120:187–188; Kane FJ Jr. 1963b. Severe oral moniliasis complicating chlorpromazine therapy. *Am J Psychiatry* 119:890; Kane FJ Jr, Anderson WB. 1964. A fourth occurrence of oral moniliasis during tranquilizer therapy. *Am J Psychiatry* 120:1199–2000; Pollack T, Buck IF, Kalnins L. 1964. An oral syndrome complicating psychopharmacotherapy: Study II. *Am J Psychiatry* 121:384–386; Peyron M, Gorceix A. 1972. Role possible des substances neuroleptiques dans les atteintes bronchopulmonaires du sujet agé. *J Europ Toxicol* 5:2; Giroud M, Guffat JJ, Le Bris H, Pucheu H, Rouviere JP. 1978. Insuffisance respiratoire aigue et traitement psychiatrique au long court. *Lyon Med* 239:251; Aizepy P, Durocher A, Gay R, et al. 1979. Accidents médicamenteaux graves chez l'adulte. *Nouv Presse Med* 8:1315; Hilpert F, Ricome JL, Aizepy P. 1980. Insuffisances respiratoires aigues durant les traitements au long cort par les neuroleptiques. *Nouv Presse Med* 9:2897.
12. Nahas GG, Desoize B, Leger C. 1979. Effects of psychotropic drugs on DNA synthesis in cultured lymphocytes. *Proc Soc Exp Biol Med* 160:344; Peyron, Gorceix. 1972. *J Europ Toxicol* 5:2.
13. Reviewed in Descotes. 1988. *Immunotoxicology*, 130–147.
14. Descotes. 1988. *Immunotoxicology*, 139–141.

Immunosuppression due to Malnutrition

1. McMillan GC. 1947. Fatal inclusion-disease pneumonitis in adult. *Am J Pathol* 23:995–1003.

2. Gajdusek DC. 1976. *Pneumocystis carinii* as the cause of human disease: Historical perspective and magnitude of the problem. Introductory remarks. In: Robbins JB, DeVita VT Jr, Dutz W, eds. 1976. *Symposium on Pneumocystis carinii Infection.* National Cancer Institute Monograph 43, 1–11.

3. Roueché B. 1983. A matter of taste. *Sat Evening Post* 255:66.

4. Brody J. 1981. *Jane Brody's Nutrition Book.* New York: Norton.

5. Chandra RK. 1983a. Introduction. In: Chandra RK, ed. 1983. *Primary and Secondary Immunodeficiency Disorders.* Edinburgh: Churchill Livingstone, 4.

6. Chandra RK. 1983b. Malnutrition. In: Chandra RK, ed. 1983. *Immunodeficiency Disorders*, 187–203; Chandra RK, Gupta S, Singh H. 1982. Inducer and suppressor T-cell subsets in protein-energy malnutrition: Analysis by monoclonal antibodies. *Nutr Res* 2:21–26.

7. Dowd PS, Heatley RV. 1984. The influence of undernutrition on immunity. *Clin Sci* 66:241–248.

8. Coovadia HM, Parent MA, Loening WE, et al. 1974. An evaluation of factors associated with the depression of immunity in malnutrition and in measles. *Am J Clin Nutr* 27:665; Beisel WR, Edelman R, Nauss K, Suskind RM. 1981. Single-nutrient effects on immunological functions. *JAMA* 245:53–58.

9. Robbins SL. 1974. *Pathological Basis of Disease.* Philadelphia: Saunders, 718.

10. Newberne PM. 1977. Effect of folic acid B, choline and methionine on immunocompetence and cell-mediated immunity. In: Suskind RM, ed. 1977. *Malnutrition and the Immune Response.* New York: Raven Press, 333–336.

11. Chandra RK, Newberne RM. 1977. In: Chandra, Newberne, eds. 1977. *Mechanisms of Interaction.* New York: Plenum, 152–171; Beisel et al. 1981. *JAMA* 245:53–58.

12. Chandra. 1983b. In: Chandra, ed. 1983. *Immunodeficiency Disorders*, 187–203.

13. Tasman-Jones C. 1980. Zinc deficiency states. *Adv in Intern Med* 26:97–114; Tarnawski A, Batko B. 1973. Antibiotics and immune processes. *Lancet* i:674; Munster AM, Loadholdt CB, Leary AG, Barnes MA. 1977. The effect of antibiotics on cell-mediated immunity. *Surgery* 81:692–696; Allen JI, Kay NE, McClain EJ. 1981. Severe zinc deficiency in humans: Association with revers-

ible T-lymphocyte dysfunction. *Ann Intern Med* 95:154–157; Chandra RK, Au B. 1980. Single nutrient deficiency and cell-mediated immune responses. I. Zinc. *Am J Clin Nutr* 33:736–738; Chesters JK, Will M. 1981. Measurement of zinc flux through plasma in normal and endotoxin-stressed pigs and the effects of Zn supplementation during stress. *Br J Nutr* 46:119–130; Beisel WR. 1982. Single nutrients and immunity. *Am J Clin Nutr* 35:449–451; Fernandes G, Nair M, Onoe K, Tanaka T, Floyd R, Good RA. 1979. Impairment of cell-mediated immunity functions by dietary zinc deficiency in mice. *Proc Natl Acad Sci USA* 76:457–461; Fraker PJ, Haas SM, Luecke RW. 1977. Effect of zinc deficiency on the immune response of the young adult A/J mouse. *J Nutr* 107:1889–1893.

14. Chandra. 1983b. In: Chandra, ed. 1983. *Immunodeficiency Disorders*, 187–203; Gross RJ, Newberne PM. 1980. Role of nutrition in immunological function. *Physiol Rev* 60:188–297.

15. Beisel et al. 1981. *JAMA* 245:53–58; Chandra. 1983b. In: Chandra, ed. 1983. *Immunodeficiency Disorders*, 187–203.

16. Boyne R, Arthur JR. 1986. The response of selenium-deficient mice to *Candida albicans* infection. *J Nutr* 116:816–822; Boyne R, Arthur JR, Wilson AB. 1986. An in vivo and in vitro study of selenium deficiency and infection in rats. *J Comp Pathol* 96:379–386; Reffett JK, Spears JW, Brown TT. 1988. Effect of dietary selenium on the primary and secondary immune response in calves challenged with infectious bovine rhinotracheitis virus. *J Nutr* 118:229–234.

17. Hughes WT, Price RA, Sisko F, et al. 1974. Protein-calorie malnutrition. A host determinant for *Pneumocystis carinii* infection. *Am J Dis Child* 128:44–52.

Immunosuppression Due to Blood Transfusions and Clotting Factors

1. Browning JD, More I, Boyd JF. 1980. *J Clin Pathol* 33:11–18.
2. Opelz G, Terasaki PI. 1976. Prolongation effect of blood transfusions on kidney graft survival. *Transplantation* 22:380–382.
3. Salvatierra O Jr, Vincenti F, Amend W, et al. 1980. Deliberate donor-specific blood transfusions prior to living related renal transplantation. *Ann Surg* 192:543–552.
4. Salvatierra et al. 1980. *Ann Surg* 192:543–552.
5. Salvatierra et al. 1980. *Ann Surg* 192:543–552.
6. Fischer E, Lenhard V, Seifert P, et al. 1980. Blood transfusion-induced suppression of cellular immunity in man. *Human Immu-*

nol 3:187–194; Lenhard V, Maassen G, Grosse-Wilde H, et al. 1983. Effect of blood transfusions on immunoregulatory mononuclear cells in prospective transplant recipients. *Transplant Proc* 15:1011–1015; Kessler CM, Shulof RS, Goldstein A, et al. 1983. Abnormal T lymphocyte subpopulations associated with transfusions of blood-derived products. *Lancet* i:991; Kaplan J, Sarnaik S. Gitlin J, Lusher J. 1984. Diminished helper/suppressor lymphocyte ratios and natural killer activity in recipients of repeated blood transfusions. *Blood* 64:308–310; Hibberd AD. 1984. The immunosuppressive effect of blood transfusion. *New Zealand Med J* 97:158–159; Klatzmann D et al. 1984. Suppression of lymphocyte reactivity by blood transfusions in uremic patients. III. Regulation of cell-mediate lympholysis. *Transplantation* 38:222–226; Waymack JP, Robb E, Alexander JW. 1987. Effect of transfusion on immune function in a traumatized animal model. *Arch Surg* 122:935–939.

7.   Francis DMA, Shenton BK. 1981. Blood transfusion and tumour growth: Evidence from laboratory animals. *Lancet* 2:871; Burrows L, Tartter P. 1982. Effect of blood transfusions on colonic malignancy recurrence rate. *Lancet* 2:662.

8.   Cumming PD, Wallace EL, Schoor JB, Dodd RY. 1989. Exposure of patients to human immunodeficiency virus through the transfusion of blood components that test antibody-negative. *New Engl J Med* 321:941–946; Ward JW, Bush TJ, Perkins HA, et al. 1989. The natural history of transfusion-associated infection with human immunodeficiency virus. *New Engl J Med* 321:947–952.

9.   Adler SP. 1983. Transfusion-associated cytomegalovirus infections. *Rev Infect Dis* 5:977–993; Wilhelm JA, Matter L, Schopfer K. 1986. The risk of transmitting cytomegalovirus to patients receiving blood transfusions. *J Infect Dis* 154:169–172; Benecke JS, Tegtmeier GE, Alter HJ, Luetkemeyer RB, Solmon R, Bayer WL. 1984. Relation of titers of antibodies to CMV in blood donors to the transmission of cytomegalovirus infection. *J Infect Dis* 150:883–888.

10. Wilhelm et al. 1986. *J Infect Dis* 154:169–172.

11. Luthardt T, Siebert H, Lösel I, Quevedo M, Todt R. 1971. Cytomegalovirus infections in infants with blood exchange transfusions after birth. *Klin Wochenschr* 49:81–86.

12. Ades E, Hinson A, Morgan S. 1980. Immunological studies in sickle-cell disease. *Clin Immunol Immunopathol* 17:459–566; Morgan J, Waring NP, Daul CB, Ohene-Fempong K, DeShazo

RD. 1985. Persistent lymphadenopathy associated with hyper-transfusion in sickle-cell disease. *J Allergy Clin Immunol* 76:869–875.

13. Passaleva A, Massai G, Morfini M, Longo G, Rossi Ferrini PL, Ricci M. 1983. Circulating immune complexes in haemophilia and von Willebrand's disease. *Scand J Haematol* 31:466–474.

14. Beddall AC, Al-Rubei K, Williams MD, Hill FGH. 1985. Lympho-cyte subset ratios and factor VIII usage in hemophilia. *Arch Dis Child* 60:530; Hultin MB, Dattwyler RJ, Lipton RA. 1989. Con-trolled prospective study of factor IX concentrate therapy and im-munodeficiency. *Am J Hematology* 31:71–72; Mahir WS, Millard RE, Booth JC, Flute PT. 1988. Functional studies of cell-mediated immunity in haemophilia and other bleeding disorders. *Br J Haematol* 69:367–370; Pollack S, Atias D, Yaffe G, Katz R, Schechter Y, Tatarsky I. 1985. Impaired immune function in hemophilia patients treated exclusively with cryoprecipitate: Rela-tion to duration of treatment. *Am J Hematol* 20:1; Jason J, Hol-man RC, Evatt BL. 1990. Relationship of partially purified factor concentrates to immune tests and AIDS. The Hemophilia-AIDS collaborative Study Group. *Am J Hematol* 34:262–269.

15. Madhok R, Gracie JA, Smith J, et al. 1990. Capacity to produce interleukin 2 is impaired in haemophilia in the absence and pres-ence of HIV 1 infection. *Br J Haematol* 76: 70–74; Madhok R, Smith J, Jenkins A, Lowe GD. 1991. T cell sensitization to factor VIII in haemophilia A? *Br J Haematol* 79:235–238; Schultz JC, Shahidi NT. 1990. Influence of fibronectin-containing blood prod-ucts on lymphocyte reactivity. *Transfusion* 30:799–807; Hay CR, McEvoy P, Duggan-Keen M. 1990. Inhibition of lymphocyte IL2-receptor expression by factor VIII concentrate: a possible cause of immunosuppression in haemophiliacs. *Br J Haematol* 75:278–281.

Immunosuppression Associated with Age

1. Kay MMB. 1983. Immunodeficiency in old age. In: Chandra RK, ed. 1983. *Primary and Secondary Immunodeficiency Disorders.* Edinburgh: Churchill Livingstone, 165–186.

2. Miller ME, Stiehm ER. 1983. Immunology and resistance to infec-tion. In: Remington JS, Klein JO, eds. 1983. *Infectious Diseases and the Fetus and Newborn Infant.* 2d ed. Philadelphia: Saunders, 27–68.

3. Miller, Stiehm. 1983. In: Remington and Klein, eds. 1983. *Infec-tious Diseases and the Fetus*, 27–68.

CHAPTER 4 MULTIPLE, CONCURRENT INFECTIONS AND AIDS

Antigenic Overload

1. Navarro C, Hagstrom JWC. 1982. Opportunistic infection and Kaposi's sarcoma in homosexual men. *Lancet* i:933.
2. Clark M, Gonnell M. 1981. Disease that plagues gays. *Newsweek*, 21 Dec.
3. Grmek MD. 1990. Maulitz RC, Duffin J, trans. *History of AIDS*. Princeton: Princeton University Press, 16.
4. Nowak MA, Anderson RM, McLean AR, Wolfs TFW, Goudsmit J, May RM. 1991. Antigenic diversity thresholds and the development of AIDS. *Science* 254:963–969; McLean AR, Nowak MA. 1992. Models of interactions between HIV and other pathogens. *J Theor Biol* 155:69–86.
5. Sarin PS, Gallo RC. 1984. Role of viruses in the etiology [of AIDS]. In: Ebbesen P, Biggar RJ, Melbye M, eds. 1984. *AIDS: A Basic Guide for Clinicians*. Philadelphia: Saunders, 173–180.

The Diseases Associated with AIDS

1. Rogers MF, Morens DM, Steward JA, et al. 1983. National case-control study of Kaposi's sarcoma and *Pneumocystis carinii* pneumonia in homosexual men: Part 2, Laboratory results. *Ann Intern Med* 99:151–158.
2. Carney WP, Rubin RH, Hoffman RA, Hansen WP, Healey K, Hirsch MS. 1981. Analysis of T lymphocyte subsets in cytomegalovirus mononucleosis. *J Immunol* 126:2114–2116; Specter S, Bendinelli M, Friedman H. 1989. *Virus-Induced Immunosuppression*. New York: Plenum.
3. Demmler GJ, O'Neil GW, O'Neil JH, Spector SA, Brady MT, Yow MD. 1986. Transmission of cytomegalovirus from husband to wife. *J Infect Dis* 154:545–546.
4. Lusso P, Ensoli B, Markham PD et al. 1989. Productive dual infection of human CD4 + T lymphocytes by HIV-1 and HHV-6. *Nature* 337:370–373; Lusso P, Veronese FM, Ensoli B, et al. 1990. Expanded HIV-1 cellular tropism by phenotypic mixing with murine endogenous retroviruses. *Science* 247:848–852.
5. Levine PH. 1984. The epidemiology of Epstein-Barr virus–associated malignancies. In: Levine PH, Ablashi DV, Kottaridis SD, eds. 1984. *Epstein-Barr Virus and Associated Diseases*. Boston: Martinus Nijhoff Publishing, 81–89; Rinaldo CR Jr, Kingsley, LA, Lyter DW, et al. Association of HTLV-III with Epstein-Barr virus

infection and abnormalities of T lymphocytes in homosexual men. *J Infect Dis* 154:556–561.

6. Levine. 1984. In: Levine et al., eds. 1984. *Epstein-Barr Virus*, 81–89.

7. Hodgson HJF, Wands JR, Isselbacher KJ. 1978. Alteration in suppressor cell activity in chronic active hepatitis. *Proc Natl Acad Sci USA* 75:1549–1553; Klingenstein RJ, Savarese AM, Dienstag JL, Rubin RH, Bahn AK. 1981. Immunoregulatory T cell subsets in acute and chronic hepatitis. *Hepatology* 1:523; Thomas HC, Brown D, Routhier G, et al. 1982. Inducer and suppressor T-cells in hepatitis B virus-induced liver disease. *Hepatology* 2:202–204; Barnaba V, Musca A, Cordova C, et al. 1983. Relationship between T cell subsets and suppressor cell activity in chronic hepatitis B virus (HBV) infection. *Clin Exp Immunol* 53:281–288; Pape GR, Rieber EP, Eisenberg J, et al. 1983. Involvement of the cytotoxic/suppressor T-cell subset in liver diseases. *Gastroenterology* 85:657–662; Dienstag JL. 1985. Immunopathogenesis of acute and chronic hepatitis B. In Gerety J, ed. 1985. *Hepatitis B*. Orlando, FL: Academic Press, 221–245; Nouri-Aria KT, Alexander GJM, Portmann B, Vergani D, Eddleston ALWF, Williams R. 1986. *In vitro* study of IgG production and concanavalin A induced suppressor cell function in acute and chronic hepatitis B virus infection. *Clin Exp Immunol* 64:50–58; Alexander GJ, Mondelli MM, Naumov NV, et al. 1986. Functional characterization of peripheral blood lymphocytes in chronic HBsAg carriers. *Clin Exp Immunol* 63:498–507; Pasquinelli C, Laure F, Chatenoud L. 1986. Hepatitis B virus DNA in mononuclear blood cells. A frequent event in hepatitis B surface antigen-positive and negative patients with acute and chronic liver disease. *J Hepatol* 3:95–103; Laure F, Chatenoud C, Pasquinelli C, et al. 1987. Frequent lymphocyte infection by hepatitis B virus in haemophiliacs. *Br J Haematol* 65:181–185; Noonan CA, Yoffe B, Mansell PW, Mielnick JL, Hollinger FB. 1986. Extrachromosomal sequences of hepatitis B virus DNA in peripheral blood mononuclear cells of acquired immune deficiency syndrome patients. *Proc Natl Acad Sci USA* 83:5698–5702.

8. Sarin PS, Gallo RC. 1984. Role of viruses in the etiology [of AIDS]. In: Ebbesen P, Biggar RJ, Melbye M, eds. 1984. *AIDS: A Basic Guide for Clinicians*. Philadelphia: Saunders, 173–180; Lusso P, Veronese FM, Ensoli B, et al. 1990. Expanded HIV-1 cellular tropism by phenotypic mixing with murine endogenous retroviruses. *Science* 247:848–852.

9. Lo S-C, Tsi S, Benish JR, Shih JW-K, Wear DJ, Wong DM. 1991. Enhancement of HIV-1 cytocidal effects in CD4+ lymphocytes by the AIDS-associated mycoplasma. *Science* 251:1074–1076.

10. Odds, FC. 1979. *Candida and Candidosis.* Leicester: Leicester University Press.
11. Witkin SS. 1985. Defective immune responses in patients with recurrent Candidiasis. *Infect Med* May/June: 129–132; Tuck R et al. 1979. Defective Candida killing in childhood malnutrition. *Arch Dis Child* 54:445–447.
12. Sega E, Schwartz J, Altboum Z, Vardinon N, Eylan E. 1977. Suppressive action of *Candida albicans* on the immune response in guinea pigs. *Microbios* 19:79–87; Giger DK et al. 1978. Experimental murine candidiaisis: Pathological and immune responses in T-cell depleted mice. *Infect Immunol* 21:729.
13. Witkin. 1985. *Infect Med*, May–June: 129–132.
14. Terry RJ, Hudson KM. 1983. Secondary immunodeficiencies in protozoan and helminth infections. In: Chandra RK, ed. 1983. *Primary and Secondary Immunodeficiency Disorders.* Edinburgh: Churchill Livingstone, 219–231; Virella G, Fudenberg HH. 1982. Secondary immunodeficiencies. In: Twomey JJ, ed. 1982. *The Pathophysiology of Human Immunologic Disorders.* Baltimore: Urban and Schwarzenberg, 91–124.
15. Reviewed in Terry, Hudson. 1983. In: Chandra, ed. 1983. *Immunodeficiency Disorders*, 219–231; Virella, Fudenberg. 1982. In: Twomey, ed. 1982. *Human Immunologic Disorders*, 91–124; Maleckar JR et al. 1984. Suppression of mouse lymphocyte responses to mitogens *in vitro* by Trypanosoma cruzi. *Int J Parasitol* 14:45–52.
16. Reviewed in Terry, Hudson. 1983. In: Chandra, ed. 1983. *Immunodeficiency Disorders*, 219–231.
17. Reviewed in Virella, Fudenberg. 1982. In: Twomey, ed. 1982. *Human Immunologic Disorders*, 91–124.
18. Reviewed in Virella, Fudenberg. 1982. In: Twomey, ed. 1982. *Human Immunologic Disorders*, 91–124; Ali SK, Puri S, Chandra S, Chandra RK. 1983. Miscellaneous disorders. In: Chandra, ed. 1983. *Immunodeficiency Disorders*, 280–289.
19. Williams RC Jr, Koster FT, Kilpatrick KA. 1983. Alterations in lymphocyte cell surface markers during various human infections. *Am J Med* 75:807–816.

The Incidence of Multiple, Concurrent Infections in Various Risk Groups Compared with American Heterosexuals

1. Clair PK, Embil JA, Fahey J. 1990. A seroepidemiologic study of cytomegalovirus infection in a Canadian recruit population. *Milit Med* 155:189–192; Buimovici-Klein E, Lange M, Ong, KR,

Grieco MH, Cooper LZ. 1988. Virus isolation and immune studies in a cohort of homosexual men. *J Med Virol* 25:371–385; Nahmias AJ, Lee FK, Beckman-Nahmias S. 1990. Sero-epidemiological and -sociological patterns of herpes simplex virus infection in the world. *Scand J Infect Dis* Suppl 69:19–36; Koo D. 1990. Chronic fatigue syndrome. A critical appraisal of the role of Epstein-Barr virus. *West J Med* 150:590–596.

2. Drew WL, Mintz L, Miner RC, Sands M, Ketterer B. 1981. Prevalence of cytomegalovirus infection in homosexual men. *J Infect Dis* 143:188–192; Buimovici-Klein et al. 1988. *J Med Virol* 25:371–385.

3. Bastien MR, Smith JG Jr. 1989. Prevention of hepatitis B. *Arch Dermatol* 125:212–215.

4. Bowell P, Mayne K, Puckett A, Entwistle C, Selkon J. 1989. Serological screening tests for syphilis in pregnancy: results of a five year study (1983–87) in the Oxford region. *J Clin Patholol* 42:1281–1284.

5. White NH, Yow MD, Demmler GJ, et al. 1989. Prevalence of cytomegalovirus antibody in subjects between the ages of 6 and 22 years. *J Infect Dis* 159:1013–1017.

6. Collier AC, Handsfield HH, Roberts PL, et al. 1990. Cytomegalovirus infection in women attending a sexually transmitted disease clinic. *J Infect Dis* 162:46–51.

7. Canavaggio M, Leckie G, Allain JP, et al. 1990. The prevalence of antibody to HTLV-I/II in United States plasma donors and in United States and French hemophiliacs. *Transfusion* 30:780–782; Taylor PE, Stevens CE, Pindyck J, Schrode J, Steaffens JW, Lee H. 1990. Human T-cell lymphotropic virus in volunteer blood donors. *Transfusion* 30:783–786.

8. Weber DJ, Rutala WA, Samsa GP, Sarubbi FA Jr, King LC. 1989. Epidemiological study of tuberculosis in North Carolina, 1966–1986. *South Med J* 82:1204–1214.

9. Nahmias AJ, Roizman B. 1973. Infection with herpes-simplex viruses 1 and 2 (third of three parts). *New Engl J Med* 289:781–789; Nahmias et al. 1990. *Scand J Infect Dis* Suppl 69:19–36.

10. Enck RE, Betts RF, Brown MR, Miller G. 1979. Viral serology (hepatitis B virus, cytomegalovirus, Epstein-Barr virus) and abnormal liver function tests in transfused patients with hereditary hemorrhagic diseases. *Transfusion* 19:32–38; Jackson JB, Erice A, Englund JA, Edson JR, Balfour HH Jr. 1988. Prevalence of cytomegalovirus antibody in hemophiliacs and homosexuals infected with human immunodeficiency virus type 1. *Transfusion* 28:187–189; Nordenfelt E. 1990. Epidemiology of hepatitis delta

virus and non-A, non-B hepatitis. *Scand J Infect Dis* Suppl 69:49–53.

11. Lairmore MD, Jason JM, Hartley TM, Khabbaz RF de B, Evatt BL. 1989. Absence of human T-cell lymphotropic virus type I coinfection in human immunodeficiency virus–infected hemophilic men. *Blood* 74:2596–2599; Canavaggio et al. 1990. *Transfusion* 30:780–782.

12. Cortes E, Detels R, Aboulafia D, et al. 1989. HIV-1, HIV-2, and HTLV-I infection in high-risk groups in Brazil. *New Engl J Med* 320:953–958.

13. Khabbaz RF, Hartel D, Lairmore M, et al. 1991. Human T lymphotropic virus type II (HTLV-II) infection in a cohort of New York intravenous drug users: An old infection? *J Infect Dis* 163:252–256; Biggar RJ, Buskell-Bales Z, Yakshe PN, Caussy D, Gridley G, Seeff L. 1991. Antibody to human retroviruses among drug users in three East Coast American cities, 1972–1976. *J Infect Dis* 163:57–63; Lee HH, Weiss SH, Brown LS, et al. 1990. Patterns of HIV-1 and HTLV-I/II in intravenous drug abusers from the middle Atlantic and central regions of the USA. *J Infect Dis* 162:347–352.

14. Taylor et al. 1990. *Transfusion* 30:783–786; Vranck R, Coenjaert A, Muylle L. 1990. A seroepidemiological survey of HTLV-I/HTLV-II in selected Belgian populations. *AIDS Res Human Retroviruses* 6:827–830.

15. Widell A, Hansson BG, Moestrup T, Nordenfelt E. 1983. Increased occurrence of hepatitis A with cyclic outbreaks among drug addicts in a Swedish community. *Infection* 11:198–200.

16. Bailey JA, Brown LS Jr. 1990. Serologic investigations in a New York City cohort of parenteral drug users. *J Natl Med Assoc* 82:405–408.

17. Selwyn PA, Hartel D, Lewis VA, et al. 1989. A prospective study of the risk of tuberculosis among intravenous drug users with human immunodeficiency virus infection. *New Engl J Med* 320:545–550.

18. Rolfs RT, Goldberg M, Sharrar RG. 1990. Risk factors for syphilis: Cocaine use and prostitution. *Am J Public Health* 80:853–857.

19. Fernandez-Cruz E, Fernandez A, Gutierrez C, et al. 1988. Progressive cellular immune impairment leading to development of AIDS: Two-year prospective study of HIV infection in drug addicts. *J Exp Immunol* 72:190–195.

20. Jackson et al. 1988. *Transfusion* 28:187–189.

21. Buimovici-Klein et al. 1988. *J Med Virol* 25:371–385; Drew WL, Mintz L, Miner RC, Sands M, Ketterer B. 1981. Prevalence of

cytomegalovirus infection in homosexual men. *J Infect Dis* 143:188–192.

22. Drew WL, Mintz L. 1989. Cytomegalovirus infection in healthy and immune deficient homosexual men. In: Ma P, Armstrong D, eds. *AIDS and Infections of Homosexual Men*. Boston: Butterworths, 119–130.

23. Drew et al. 1981. *J Infect Dis* 143:188–192; Buimovici-Klein et al. 1988. *J Med Virol* 25:371–385.

24. Rogers MF, Morens DM, Steward JA, et al. 1983. National case-control study of Kaposi's sarcoma and *Pneumocystis carinii* pneumonia in homosexual men: Part 2, Laboratory results. *Ann Intern Med* 99:151–158; Bartholemew C, Saxinger WC, Clark JW, et al. 1987. Transmission of HTLV-1 and HIV among homosexual men in Trinidad. *JAMA* 257:2604–2608.

25. Alter MJ, Francis DP. 1989. Hepatitis B virus transmission between homosexual men: A model for AIDS. In: Ma, Armstrong, eds. *AIDS*, 99–108.

26. Schreeder MT, Thompson SE, Hadler SC, et al. 1982. Hepatitis B in homosexual men: Prevalence of infection and factors related to transmission. *J Infect Dis* 146:7–15.

27. Coutinho RA, Albrecht-van Lent P, Lelie N, Nagelkerke N, Kuipers H, Rijsdijk T. 1983. Prevalence and incidence of hepatitis A among male homosexuals. *Br Med J* 287:1743–1745.

28. Nahmias et al. 1990. *Scand J Infect Dis* Suppl 69:19–36.

29. Alter MJ, Hadler SC, Margolis HS, et al. 1990. The changing epidemiology of hepatitis B in the United States. Need for alternative vaccination strategies. *JAMA* 263:1218–1222; Black FT. 1983. *Ureaplasma urealyticum* and *Mycoplasma hominis*. In: Holmes KK, Mardh P-A, eds. *International Perspectives on Neglected Sexually Transmitted Diseases*. Washington, DC: Hemisphere Publishing Co., 37–54; Enck RE, Betts RF, Brown MR, Miller G. 1979. Viral serology (hepatitis B virus, cytomegalovirus, Epstein-Barr virus) and abnormal liver function tests in transfused patients with hereditary hermorrhagic diseases. *Transfusion* 19:32–38; Fernandez-Cruz E, Fernandez A, Gutierrez C, et al. 1988. Progressive cellular immune impairment leading to development of AIDS: Two-year prospective study of HIV infection in drug addicts. *J Exp Immunol* 72:190–195; Hardman PK, Gier RE, Tegtmeier G. 1990. The incidence and prevalence of hepatitis B surface antibody in a dental school population. *Oral Surg Oral Med Oral Pathol* 69:399–402; Hart GJ, Woodward N, Johnson AM, Tighe J, Parry JV, Adler MW. 1991. Prevalence of HIV, hepatitis B and associated risk behaviours in clients of a needle-exchange in central Lon-

don. *AIDS* 5:543–547; Haverkos HW, Edelman R, Killen JY, Gottlieb MS. 1986. Herpes zoster and the classification of HTLV-III/LAV-related diseases. *J Infect Dis* 154:190; Hessol NA, Lifson AR, O'Malley PM, Doll LS, Jaffe HW, Rutherford GW. 1989. Prevalence, incidence, and progression of human immunodeficiency virus infection in homosexual and bisexual men in hepatitis B vaccine trials, 1978–1988. *Am J Epidemiol* 130:1167–1175; Holmberg SD, Stewart JA, Gerber AR, et al. 1988. Prior herpes simplex virus type 2 infection as a risk factor for HIV infection. *JAMA* 259:1048–1050; Hosker HS, Lindsay DC, Game F, Codd AA, Dale G, Record CO. 1988. Incidence and types of acute viral hepatitis in Newcastle upon Tyne. *Postgrad Med J.* 64:854–855; Hyams KC, Palinkas LA, Burr RG. 1989. Viral hepatitis in the US Navy, 1975–1984. *Am J Epidemiol* 130:319–326; Kallenius G, Hoffner SE, Svenson SB. 1989. Does vaccination with bacille Calmette-Guérin protect against AIDS? *Rev Infect Dis* 11:349–351; Krueger GRF. 1984. Pathology of Epstein-Barr virus (EBV)–associated disease (the lymphatic system). In: Levine PH, Ablashi DV, Pearson GR, Kottaridis SD, eds. 1984. *Epstein-Barr Virus and Associated Diseases.* Boston: Martinus Nijhoff Publishing, 106–131; Lang DJ, Kovacs AAS, Zaia JA, et al. 1989. Seroepidemiologic studies of cytomegalovirus and Epstein-Barr virus infections in relation to human immunodeficiency virus type 1 infection in selected recipient populations. *J Acq Immune Def Syn* 2:540–549; Ma P, Armstrong D. 1989. *AIDS and Infections of Homosexual Men.* 2d ed. Boston: Butterworths, passim; McAdam JM, Brickner PW, Scharer LL, Crocco JA, Duff AE. 1990. The spectrum of tuberculosis in a New York City men's shelter clinic (1982–1988). *Chest* 97:798–805; Meyer RD, Moudgil T, Detels R, Phair JP, Hirsch MS, Ho DD. 1990. Seroprevalence of human T cell leukemia viruses in selected populations of homosexual men. *J Infect Dis* 162:1370–1372; Miles MR, Olsen L, Rogers A. 1977. Recurrent vaginal Candidiasis. *JAMA* 238:1836–1839; Morlet A, DArke S, Guinan JJ, Wolk J, Gold J. 1990. Intravenous drug users who present to the Albion Street (AIDS) Centre for diagnosis and management of human immunodeficiency virus infection. *Med J Aust* 152:78–80; Mosley JW. 1975. The epidemiology of viral hepatitis: An overview. *Am J Med Sci* 270:253–270; Norkkrans G. 1990. Epidemiology of hepatitis B virus (HBV) infections with particular regard to current routes of transmission and development of cirrhosis and malignancy. *Scand J Infect Dis* 69:43–47; De Paoli P, Reitano M, Battistin S, et al. 1986. Immunologic abnormalities in intravenous drug abusers and relationship to the prolonged generalized lymphadenopathy syndrome in Italy. *Clin Exp Immunol* 64:451–456; Pereira LH, Embil JA, Haase DA, Manley KM. 1990. Cytomegalovirus infection among women attending a sexually transmitted disease clinic: Association with clinical symptoms and other sexually transmitted diseases. *Am J Epidemiol* 131:683–692; Quinnan GV Jr, Masur H, Rook AH, et al. 1984. Herpes virus infections in the acquired immune deficiency syndrome. *JAMA* 252:72–77; Rahman MA, Kingsley LA, Breinig MK, et al. 1989. Enhanced antibody responses to Epstein-Barr virus in HIV-infected homosexual men. *J Inf Dis* 159:472–479; Remington JS, Desmonts G. 1983. Toxoplasmosis. In: *Infectious Diseases of the Fetus and Newborn Infant* 2d ed. Philadelphia: Saunders, 144–263; Rinaldo CR Jr, Kingsley, LA, Lyter DW, et al. 1986. Association of HTLV-III with Epstein-Barr virus infection and abnormalities of T lymphocytes in homosexual men. *J Inf Dis* 154:556–561; Schachter J. 1983. *Clamydia tracomatis*. In: Holmes KK, March P-A, eds. *International Perspectives on Neglected Sexually Transmitted Diseases*. Washington, DC: Hemisphere Publishing Co., 7–36; Selwyn PA, Hartel D, Lewis VA, et al. 1989. A prospective study of the risk of tuberculosis among intravenous drug users with human immunodeficiency virus infection. *N Engl J Med* 320:545–550; Sixbey JW, Sirley P, Chesney PJ, Buntin DM, Resnick L. 1989. Detection of a second widespread strain of Epstein-Barr virus. *Lancet* ii:761–765; Sculley TB, Cross SM, Borrow P, Cooper DA. 1988. Prevalence of antibodies to Epstein-Barr virus nuclear antigen 2B in persons infected with the human immunodeficiency virus. *J Infect Dis* 158:186–192; Zimmerman HL, Potterat JJ, Dukes RL, et al. 1990. Epidemiological differences between chlamydia and gonorrhea. *Am J Public Health* 80:1338–1342.

30.   Sumaya CV, Boswell, RN, Enck Y, et al. 1986. Enhanced serological and virological findings of Epstein-Barr virus in patients with AIDS and AIDS-related complex. *J Infect Dis* 154:864–870.

31.   Fernandez-Cruz et al. 1988. *J Exp Immunol* 72:190–195.

32.   Specter S, Bendinelli M, Friedman H. 1989. *Virus-Induced Immunosuppression*. New York: Plenum, 15.

Synergism and Viral Transactivation as Causes of Immune Suppression

1.   Taussig MJ. 1975. Antigenic competition. In: Hobart MJ, McConnell I, eds. 1975. *The Immune System*. Oxford: Blackwell, 165.

2.   Huang S-W, Lattos DB, Nelson DB, Reeb K, Hong R. 1973.

Antibody-associated lymphotoxin in acute infection. *J Clin Invest* 52:1033–1040; Huang S-W, Hong R. 1973. Lymphopenia and multiple viral infections. *JAMA* 225:1120–1121.

3. MacDonald KL, Osterholm MT, Hedberg CW, et al. 1987. Toxic shock syndrome. A newly recognized complication of influenza and influenzalike illness. *JAMA* 257:1053–1063; Science News. 1987. New mean team: Flu and toxic shock. *Sci News* 14 Mar: 169.

4. Carlson E. 1983. Effect of strain of *Staphylococcus aureus* on synergism with *Candida albicans* resulting in mouse mortality and morbidity. *Infect Immun* 42:285–292.

5. Hamilton JR, Overall JC, Glasgow LA. 1976. Synergistic effect on mortality in mice with murine cytomegalovirus and *Pseudomonas aeruginosa*, *Staphylococcus aureus*, or *Candida albicans* infections. *Infect Immun* 14:982–989.

6. Bale JF Jr, Kern ER, Overall JC Jr, Glasgow LA. 1982. Enhanced susceptibility of mice infected with murine cytomegalovirus to intranasal challenge with *Escherichia coli*: Pathogenesis and altered inflammatory response. *J Infect Dis* 145:525–531.

7. Oill PA, Fiala M, Shofferman J, Byfield PA, Guze LB. 1977. Cytomegalovirus mononucleosis in a healthy adult. Association with hepatitis, secondary Epstein-Barr virus antibody response and immunosuppression. *Am J Med* 62:413–417.

8. Enck RE, Betts RF, Brown MR, Miller G. 1979. Viral serology (hepatitis B virus, cytomegalovirus, Epstein-Barr virus) and abnormal liver function tests in transfused patients with hereditary hermorrhagic diseases. *Transfusion* 19:32–38.

9. Wu JC, Lee SD, Yeh PF, et al. 1990. Isoniazid-rifampin-induced hepatitis in hepatitis B carriers. *Gastroenterology* 98:502–504; Kossaifi T, Dupon M, Le Bail G, Lacut Y, Balabaud C, Bioulac-Sage P. 1990. Perisinusoidal cell hypertrophy in a patient with acquired immunodeficiency syndrome. *Arch Pathol Lab Med* 114:876–879; Root-Bernstein. 1993. Idiotype-antiidiotype interactions between hepatitis B virus, Mycobacteria, and CD13 (monocyte) antibodies. Submitted.

10. Gendelman HE, Phelps W, Feigenbaum L, et al. 1986. Transactivation of the human immunodeficiency virus long terminal repeat sequence by DNA viruses. *Proc Natl Acad Sci USA* 83:9759–9763; Davis MG, Kenney SC, Kamine J, Pagano JS, Huang ES. 1987. Immediate-early gene region of human cytomegalovirus *trans*-activates the promoter of human immunodeficiency virus. *Proc Natl Acad Sci USA* 84:8642–8646; Raffanti A, Svenningsson A, Resnick L. 1991. HIV-1 infection of CD4-negative cells via

HTLV pseudovirons. *AIDS* 5:769–783; Zhu Z, Chen SSL, Huang A. 1990. Phenotypic mixing between human immunodeficiency virus and vesicular stomatitis virus or herpes simplex virus. *J AIDS* 3:215–219.

11.   Skolnik PR, Kosloff BR, Hiersch MS. 1988. Bidirectional interactions between human immunodeficiency virus type 1 and cytomegalovirus. *J Infect Dis* 508–514; Raffanti et al. 1991. *AIDS* 5:769–783.

12.   McLean AR, Nowak MA. 1992. Models of interactions between HIV and other pathogens. *J Theor Biol* 155:69–86.

13.   Perillo RP, Regenstein FG, Roodman ST. 1986. Chronic hepatitis B in asymptomatic homosexual men with antibody to the human immunodeficiency virus. *Ann Intern Med* 105:382–383; McDonald JA, Harris S, Waters JA, Thomas HC. 1987. Effect of human immunodeficiency virus (HIV) infection on chronic hepatitis B viral antigen display. *J Hepatol* 4:337–342; Krogsgaard K, Lindhardt BO, Nielsen JO, et al. 1987. The influence of HTLV-III infection on the natural history of hepatitis B virus infection in male homosexual HBsAg carriers. *Hepatology* 1:37–41.

14.   Webster A. 1991. Cytomegalovirus as a possible cofactor in HIV disease progression. *J AIDS* 4 (Suppl. 1): S47–S52; Lang DJ, Kovacs AAS, Zaia JA, et al. 1989. Seroepidemiologic studies of cytomegalovirus and Epstein-Barr virus infections in relation to human immunodeficiency virus type 1 infection in selected recipient populations. *J AIDS* 2:540–549; Webster A, Lee CA, Cook DG et al. 1989. Cytomegalovirus infection and progression towards AIDS in haemophiliacs with human immunodeficiency virus infection. *Lancet* ii:63–66.

15.   Hamilton-Dutoit SJ, Pallesen G, Franzmann MB, et al. 1991. AIDS-related lymphoma. Histopathology, immunophenotype, and association with Epstein-Barr virus as demonstrated by *in situ* nucleic acid hybridization. *Am J Pathol* 138:149–163.

16.   Wolf BC, Martin AW, Neiman RS, et al. 1990. The detection of Epstein-Barr virus in hairy cell leukemia cells by *in situ* hybridization. *Am J Pathol* 136:717–723.

17.   Resnick L, Herbst JS, Raab-Traub N. 1990. Oral hairy leukoplakia. *J Am Acad Dermatol* 22:1278–1282.

18.   Bailey JA, Brown LS Jr. 1990. Serologic investigations in a New York City cohort of parenteral drug users. *J Natl Med Assoc* 82:405–408.

19.   Frenkel LD, Gaur S, Tsolia M, Scudder R, Howell R, Kesarwala H. 1990. Cytomegalovirus infection in children with AIDS. *Rev Infect Dis* 12 Suppl 7P: S820-S826.

20. Hattori T, Koito A, Takatsuki K, et al. 1989. Frequent infection with human T-cell lymphotropic virus type I in patients with AIDS but not in carriers of human immunodeficiency virus type 1. *J AIDS* 2:272–276.

21. Lee HH, Weiss SH, Brown LS, et al. 1990. Patterns of HIV-1 and HTLV-I/II in intravenous drug abusers from the middle Atlantic and central regions of the USA. *J Infect Dis* 162:347–352.

22. Bartholemew C, Saxinger WC, Clark JW, et al. 1987. Transmission of HTLV-1 and HIV among homosexual men in Trinidad. *JAMA* 257:2604–2608; Cleghorn F, Mahabir T, Blattner W, Bartholomew C. 1988. Progression to AIDS in homosexual men infected with HIV in Trinidad: A 3 1/2 year follow-up. *West Indian Med J* 37:42–45.

23. Van Griensven GJ, de Vroome EM, de Wolf F, Goudsmit J, Roos M, Coutinho RA. 1990. Risk factors for progression of human immunodeficiency virus (HIV) infection among seroconverted and seropositive homosexual men. *Am J Epidemiol* 132:203–210.

24. European Study Group on AIDS. 1989. Risk factors for male to female transmission of HIV. *Br Med J* 298:411–415.

25. Montagnier L, Gruest J, Chameret S, et al. 1984. Adaptation of lymphadenopathy associated virus (LAV) to replication in EBV-transformed B lymphoblastoid cell lines. *Science* 225:63–66.

26. Blumberg RS, Paradis TJ, Crawford D, Byington RE, Hirsch MS, Schooley RT. 1987. Effects of human immunodeficiency virus (HIV) on the cytotoxic response to Epstein Barr virus (EBV) transformed B lymphocytes. *AIDS Res Hum Retroviruses* 3:303–315; Pagano JS, Kenney S, Markovitz D, Kamine J. 1988. Epstein-Barr virus and interactions with human retroviruses. *J Virol Methods* 21:229–239; Tozzi V, Britton S, Ehrnst A, Lenkei R, Strannegaard O. 1989. Persistent productive HIV infection in EBV-transformed B lymphocytes. *J Med Virol* 27:19–24. Davis MG, Kenney SC, Kamine J, Pagano JS, Huang ES. 1987. Immediate-early gene region of human cytomegalovirus *trans*-activates the promoter of human immunodeficiency virus. *Proc Natl Acad Sci USA* 84:8642–8646.

27. Raffanti et al. 1991. *AIDS* 5:769–783.

28. Spector DH, Wade E, Wright DA, et al. 1990. Human immunodeficiency virus pseudotypes with expanded cellular and species tropism. *J Virol* 64:2298–2308; Zhu et al. 1990. *J AIDS* 3:215–219.

29. Finkle C, Tapper MA, Knox KK, Carrigan DR. 1991. Coinfection of cells with the human immunodeficiency virus and cytomegalovirus in lung tissues of patients with AIDS. *J AIDS* 4:735–737; Webster. 1991. *J AIDS* 4 (Suppl 1): S47–S52.

30. Webster. 1991. *J AIDS* 4 (Suppl 1): S47–S52.
31. Gobert B, Amiel C, Tang J, Barbarino P, Bene MC, Faure G. 1990. CD4-like molecules in human sperm. *FEBS Lett* 261/2:339–342.
32. Bagasra O, Freund M, Weidmann J, Harley G. 1988. Interaction of human immunodeficiency virus with human sperm *in vitro. J AIDS* 1:431–435.

CHAPTER 5   AUTOIMMUNE PROCESSES AND IMMUNE SUPPRESSION IN AIDS

AIDS Progresses Independently of HIV

1. Ziegler JL, Stites DP. 1986. Hypothesis: AIDS is an autoimmune disease directed at the immune system and triggered by a lymphotropic retrovirus. *Clin Immunol Immunopathol* 41:305.
2. Quoted in Hudson R, Davidson S. 1986. *Rock Hudson: His Story.* New York: William Morrow, 187.
3. Savitz EJ. 1991. No magic cure. The war on AIDS produces few gains, except on Wall Street. *Barron's*, 16 Dec, 10–11, 22–29; Krieger JN, Coombs RW, Collier AC, et al. 1991. Recovery of human immunodeficiency virus type 1 from semen: Minimal impact of stage of infection and current antiviral chemotherapy. *J Infect Dis* 163:386–388.

The Range of Autoimmune Processes in AIDS

1. Reichert CM, O'Leary TJ, Levens DL, Simrell CR, Macher AM. 1983. Autopsy pathology in the acquired immune deficiency syndrome. *Am J Pathol* 112:357–382, case 5.
2. Conte R, Giovanardi L, Tazzari PL, De Rosa V, Rodorigo G, Re MC. 1988. Non-HLA lymphocytotoxic antibodies in HIV-seropositive and HIV-seronegative hemophiliacs. *Vox Sang* 54:47–51.
3. Pollack MS, Callaway C, Leblanc D, et al. 1983. Lymphocytotoxic antibodies to non-HLA antigens in the sera of patients with acquired immunodeficiency syndrome (AIDS). *Prog Clin Biol Res* 133:209–213; Pruzanski W, Jacobs H, Laing LP. 1983. Lymphocytotoxic antibodies against peripheral blood B and T lymphocytes in homosexuals with AIDS and ARC. *AIDS Res* 1:211–220; Wicher V, Wicher K, Esparza B. 1983. Specificity of lymphocytotoxic antibodies in AIDS and pre-AIDS patients. *AIDS Res* 1:139–148; Kloster BE, Tomar RH, Spira TJ. 1984. Lymphocytotoxic antibodies in the acquired immune deficiency syndrome. *Clin Immunol Immunopathol* 30:330–335; Williams RC, Masur H, Spira

TJ. 1984. Lymphocyte-reactive antibodies in acquired immune deficiency syndrome. *J Clin Immunol* 4:118; Kiprov DD, Bush DF, Simpson DM, et al. 1984. Antilymphocyte serum factors in patients with acquired immunodeficiency syndrome. In: Gottlieb M, Groopman J, eds. 1984. *Acquired Immune Deficiency Syndrome.* New York: Alan R Liss, 299–308; Tomar RH, John PA, Hennig AK, Kloster B. 1985. Cell targets of antilymphocyte antibodies in AIDS and LAS. *J Immunol Immunopathol* 37:37–41; McDougal JS, Hubbard M, Nicholson JKA, et al. 1985. Immune complexes in the acquired immunodeficiency syndrome (AIDS): Relationship to disease manifestation, risk group, and immunologic defect. *J Clin Immunol* 5:130; Cohen AJ, Philips TM, Kessler CM. 1986. Circulating coagulation inhibitors in the acquired immunodeficiency syndrome. *Ann Intern Med* 104:175–179; Ozturk GE, Kohler PF, Horsburgh CR, Kirkpatrick CH. 1987. The significance of antilymphocyte antibodies in patients with acquired immune deficiency syndrome (AIDS) and their sexual partners. *J Clin Immunol* 7:130–133; De La Barrera S, Fainboim L, Lugo S, Picchio R, Muchink JR, De Bracco MME. 1987. Anti–class II antibodies in AIDS patients and AIDS-risk groups. *Immunology* 62:599–603; Stricker RB, McHugh TM, Moody DJ, et al. 1987. An AIDS-related cytotoxic antibody reacts with a specific antigen on stmulated CD4 T-cells. *Nature* 327, 710–712. Stricker RB, McHugh TM, Marx PA, et al. 1987. Prevalence of an AIDS-related autoantibody against CD4+ T cells in humans and monkeys. *Blood* 70:127A; Conte et al. 1988. *Vox Sang* 54:47–51; Israël-Biet D, Venet A, Beldjord K, Andrieu JM, Even P. 1990. Autoreactive cytotoxicity in HIV-infected individuals. *Clin Exp Immunol* 81:18–24.

4. Kiprov DD, Anderson RD, Morand P, et al. 1985. Antilymphocyte antibodies and seropositivity for retroviruses in groups at high risk for AIDS. *N Engl J Med* 312:1517.

5. Pruzanski et al. 1983. *AIDS Res* 1:211–220.

6. Kloster et al. 1984. *Clin Immunol Immunopathol* 30:330–335.

7. Dorsett B, Cronin W, Chuma V, Ioachim HL. 1985. Anti-lymphocyte antibodies in patients with the acquired immune deficiency syndrome. *Am J Med* 78:621–626; Kloster et al. 1984. *Clin Immunol Immunopathol* 30:330–335; Kiprov et al. 1984. In: Gottlieb and Groopman, eds. 1984. *Acquired Immune Deficiency Syndrome,* 299–308; Zarling JM, Ledbetter JA, Sias J, et al. 1990. HIV-infected humans but not chimpanzees, have circulating cytotoxic T lymphocytes that lyse uninfected CD4+ cells. *J Immunol* 144:2992–2998.

8. Kloster et al. 1984. *Clin Immunol Immunopathol* 30:330–335; Dorsett et al. 1985. *Am J Med* 78:621–626; Ozturk et al. 1987. *J Clin Immunol* 7:130–133; Daniel V, Schimpf K, Opelz G. 1989. Lymphocyte autoantibodies and alloantibodies in HIV-positive haemophilia patients. *Clin Exp Immunol* 75:178–183.

9. Zarling et al. 1990. *J Immunol* 144:2992–2998.

10. Ameglio F, Benedetto A, Marotta P, et al. 1988. A high proportion of sera of heroin addicts possesses anti-HLA class I and class II reactivity. *Clin Immunol Immunopathol* 46:328–334.

11. Okudaira K, Goodwin JS, Williams RC Jr. 1982. Anti-Ia antibody in sera of normal subjects after *in vivo* antigenic stimulation. *J Exp Med* 156:255–267.

12. Daniel et al. 1989. *Clin Exp Immunol* 75:178–183.

13. Conte et al. 1988. *Vox Sang* 54:47–51.

14. Daniel et al. 1989. *Clin Exp Immunol* 75:178–183.

15. Darrow WW, Echenberg DF, Jaffe HW, et al. 1987. Risk factors for human immunodeficiency virus (HIV) infections in homosexual men. *AJPH* 77:479–483.

16. European Study Group on AIDS. 1989. Risk factors for male to female transmission of HIV. *Br Med J* 298:411–415; Padian NS, Shiboski SC, Jewell NP. 1990. The effect of number of exposures on the risk of heterosexual HIV transmission. *J Infect Dis* 161:883–887.

17. Naz RK, Ellauri M, Phillips TM, Hall J. 1990. Antisperm antibodies in human immunodeficiency virus infection: Effects on fertilization and embryonic development. *Biol Reprod* 42:859–868.

18. Naz et al. 1990. *Biol Reprod* 42:859–868.

19. James K, Hargreave TB. 1984. Immunosuppression by seminal plasma and its possible clinical significance. *Immunol Today* 5:357; Bagasra O, Freund M, Weidmann J, Harley G. 1988. Interaction of human immunodeficiency virus with human sperm in vitro. *J AIDS* 1:431–435; Mathur S, Goust J-M, Williamson HO, et al. 1981. Cross-reactivity of sperm and T-lymphocyte antigens. *Am J Reprod Immunol* 1:113–118; Mavligit GM, Talpaz M, Hsia FT, et al. 1984. Chronic immune stimulation by sperm alloantigens: Support for the hypothesis that spermatazoa induce immune dysregulation in homosexual males. *JAMA* 251:237–241; Rubinstein P, Walker M, Mollen N, Laubenstein LJ, Friedman-Kien AE. 1989. Immunogenetic findings in patients with epidemic Kaposi's sarcoma. In: Ma P, Armstrong D, eds. *AIDS and Infections of Homosexual Men.* 2d ed. Boston: Butterworths, 403–418; Witkin SS, Sonnabend J. 1983. Immune responses to spermatozoa in homosexual men. *Fertil Steril* 39:337–342; Adams LE, Donovan-

Brand R, Friedman-Kien A, Ramahi K, Hess EV. 1988. Sperm and seminal plasma antibodies in acquired immune deficiency (AIDS) and other associated syndromes. *Clin Immunol Immunopathol* 46:442–449.

20. Kiprov et al. 1985. *N Engl J Med* 312:1517; Adams et al. 1988. *Clin Immunol Immunopathol* 46:442.

21. Reichert et al. 1983. *Am J Pathol* 112:357–382, case 5; Welch K, Finkbeiner W, Alpers CE, et al. 1984. Autopsy finding in the acquired immune deficiency syndrome. *JAMA* 252:1152–1159; Niedt GW, Schinella RA. 1985. Acquired immunodeficiency syndrome. Clinicopathologic study of 56 autopsies. *Arch Pathol Lab Med* 109:727–734.

22. Kiprov et al. 1985. *N Engl J Med* 312:1517; Kloster et al. 1984. *Clin Immunol Immunopathol* 30:330–335; Dorsett et al. 1985. *Am J Med* 78:621–626; Tomar et al. 1985. *J Immunol Immunopathol* 37:37–41; Ozturk et al. 1987. *J Clin Immunol* 7:130–133; Ameglio et al. 1988. *Clin Immunol Immunopathol* 46:328–334; Daniel et al. 1989. *Clin Exp Immunol* 75:178–183.

23. Stricker et al. 1987. *Nature* 327:710–712; Stricker et al. 1987. *Blood* 70:127A; Chams V, Jouault T, Fenouillet E, Gluckman J-C, Kaltzmann D. 1988. Detection of anti-CD4 autoantibodies in the sera of HIV-infected patients using recombinant soluble CD4 molecules. *AIDS* 2:353–361; Thiriart C, Goudsmit J, Schellekens P et al. 1988. Antibodies to soluble CD4 in HIV-1 infected individuals. *AIDS* 2:345–351; Zarling et al. 1990. *J Immunol* 144:2992–2998.

24. Engleman EG, Benike CG, Glickman E, Evans RL. 1981. Antibodies to membrane structures that distinguish suppressor/cytotoxic and helper T lymphocyte populations block the mixed leucocyte reaction in man. *J Exp Med* 154:193–199; Pierres A, Naquet P, Van Agthoven A, et al. 1984. A rat anti-mouse T4 monoclonal antibody (H129.19) inhibits the proliferation of Ia-reactive T cell clones and delineates two phenotypically distinct T4+, Lyt-2.3-, and T4-, Lyt-2.3+ subsets among anti-Ia cytolytic T cell clones. *J Immunol* 132:2775–2781; Wofsy D, Seaman WE. 1985. Successful treatment of autoimmunity in NZB/NZW F1 mice with monoclonal antibody to L3T4. *J Exp Med* 161:378–382; Carrera AC, Sanchez-Madrid F, Lopez-Botet M, Bernabeu C, De Landazuri MO. 1987. Involvement of the CD4 molecule in a post-activation event on T cell proliferation. *Eur J Immunol* 17:179–186; Portoles P, Janeway CA Jr. 1989. Inhibition of the responses of a cloned CD4+ T cell line to different class II major histocompatibility complex ligands by anti-CD4 and by anti-receptor Fab frag-

ments. *Eur J Immunol* 19:83–87; Cosimi AB, Delmonico FL, Wright JK, et al. 1990. Prolonged survival of nonhuman primate renal allograft recipients treated only with anti-CD4 monoclonal antibody. *Surgery* 108:406–413.

25. Zarling et al. 1990. *J Immunol* 144:2992–2998; Stricker et al. 1987b. *Blood* 70:127A.

26. Morrow WJW, Isenberg DA, Sobol RE, Stricker RB, Kieber-Emmons T. 1991. AIDS virus infection and autoimmunity: A perspective of the clinical, immunological, and molecular origins of the autoallergic pathologies associated with HIV disease. *Clin Immunol Immunopathol* 58:163–180.

An Introduction to Autoimmunity

1. Westall FC, Robinson AB, Caccam J, Jackson JM, Eylar EH. 1971. Esssential chemical requirements for induction of allergic encephalomyelitis. *Nature* 229:22–24.

2. Fujinami RS, Oldstone MBA. 1985. Amino acid homology between the encephalitogenic site of myelin basic protein and virus: Mechanism for autoimmunity. *Science* 230:1043–1045; Oldstone MBA. 1987. Molecular mimickry and autoimmune disease. *Cell* 50:819–820; Jahnke U, Fischer EH, Alvord EC. 1985. Sequence homology between certain viral proteins and proteins related to encephalomyelitis and neuritis. *Science* 230:1043–1045; Srinivasappa J, Saegusa J, Prabhakar BS, et al. 1986. Molecular mimickry: Frequency of reactivity of monoclonal antiviral antibodies with normal tissues. *J Virology* 57:397–401.

3. Salk JS, Romine JS, Westall FC, Wiederholt WC. 1980. Myelin basic protein studies in experimental allergic encephalomyelitis and multiple sclerosis: A summary with theoretical considerations of multiple sclerosis etiology. In: Davison AN, Cuzner ML, eds. 1980. *Experimental Allergic Encephalomyelitis and Multiple Sclerosis*. London: Academic Press, 141–156.

4. Ellouz F, Adam A, Ciorvaru R, Lederer E. 1974. Minimal structural requirements for adjuvant activity of bacterial peptidoglycan deriviatives. *Biochem Biophys Res Commun* 59:1317–1325.

5. Nagai Y, Akiyama K, Suzuki K, et al. 1978. Structural specificity of synthetic peptide adjuvant for induction of experimental allergic encephaloymelitis. *Cell Immunol* 35:158–167.

6. Westall FC, Root-Bernstein RS. 1983. An explanation of prevention and suppression of experimental allergic encephalomyelitis. *Mol Immunol* 20:169–177; Westall FC, Root-Bernstein RS. 1986.

The cause and prevention of post-infectious and post-vaccinal encephalopathies in light of a new theory of autoimmunity. *Lancet* ii:251–252; Root-Bernstein RS. 1991. Self, nonself, and the paradoxes of autoimmunity. In: Tauber AI, ed. 1991. *Organism and the Development of Self.* Boston: Kluwer, 159–209.

7. Root-Bernstein RS, Westall FC. 1990. Serotonin binding sites. II. Muramyl dipeptide binds to serotonin binding sites on myelin basic protein, LHRH, and MSH-ACTH 4–10. *Brain Res Bull* 25:827–841.

8. Root-Bernstein. 1991. In: Tauber, ed. 1991. *Organism and Self*, 159–209.

9. Plotz PH. 1983. Autoantibodies are anti-idiotype antibodies to antiviral antibodies. *Lancet* ii:824–826; Plotz PH. 1990. What drives autoantibodies? The evidence from spontaneous human autoimmune diseases. In: De Vries RRP, Cohen IR, van Rood JJ, eds. 1990. *The Role of Micro-Organisms in Non-Infectious Diseases.* London: Springer-Verlag, 111–121.

Idiotype–Anti-idiotype Theories of Autoimmunity in AIDS

1. Ziegler JL, Stites DP. 1986. Hypothesis: AIDS is an autoimmune disease directed at the immune system and triggered by a lymphotropic retrovirus. *Clin Immunol Immunopathol* 41:305; Andrieu JM, Even P, Venet A. 1986. AIDS and related syndromes as a viral-induced autoimmune disease of the immune system: an anti-MHC II disorder. Therapeutic implications. *AIDS Res* 2:163–174; Vega MA, Guigo R, Smith TF. 1990. Autoimmune responses in AIDS. *Nature* 345:26; Young JA. 1990. HIV and HLA similarity. *Nature* 333:215; Bjork RL Jr. 1991. HIV-1: Seven facets of functional molecular mimicry. (Minireview). *Immunology Lett* 28:91–95.

2. Hoffmann GW, Cooper-Willis A, Chow M. 1986. A new symmetry: A anti-B is anti-(B anti-A), and reverse enhancement. *J Immunol* 137:61–68.

3. Hoffmann GW, Kion TA, Forsyth RB, Soga KG, Cooper-Willis A. 1988. The N-dimensional network. In: Perelson AS, ed. 1988. *Theoretical Immunology.* Part 2. Redwood City CA: Addison-Wesley, 219–319; Hoffmann GW, Grant MD. 1989. When HIV meets the immune system: alloimmunity, autoimmunity, and AIDS. *Lect Notes Biomath* 83:386–401; Hoffmann GW. 1991. A response to Duesberg with reference to an idiotypic network model of AIDS immunopathogenesis. *Res Immunol* 141:701–709; Hoff-

mann GW, Kion TA, Grant MD. 1990. An idiotypic network model for AIDS pathogenesis. *Proc Natl Acad Sci USA* 88:3060–3064.

4. Root-Bernstein RS, Hobbs SH. 1993. Similarities between CD4 protein and the proteins of infectious agents associated with AIDS: Possible inducers of lymphocytotoxic autoimmunity. Submitted.

Multiple-Antigen-Mediated Autoimmunity and AIDS Dementia

1. Westall FC, Root-Bernstein RS. 1986. Cause and prevention of postinfectious and postvaccinal neuropathies in light of a new theory of autoimmunity. *Lancet* ii:251–252.

2. Pollock TM, Morris JA. 1983. Seven year survey of disorders attributed to vaccination in northwest Thames region. *Lancet* i:753–757.

3. Nielsen SL, Petito CK, Urmacher CD, Posner JB. 1984. Subacute encephalitis in acquired immune deficiency syndrome: A postmortem study. *Am J Clin Pathol* 82:678–682.

4. Janssen RS, Saykin AJ, Kaplan JE, et al. 1988. Neurological complications of human immunodeficiency virus infection in patients with lymphadenopathy syndrome. *Ann Neurol* 23:49–55; Morton R, Graham DI, Briggs JD, Hamilton DNH. 1982. Principal neuropathological and general necropsy findings in 24 renal transplant patients. *J Clin Pathol* 34:31–39; Schneck SA. 1965. Neuropathological features of organ transplantation. *J Neuropathol Exp Neurol* 24:415–429; Root-Bernstein RS. 1991. AIDS dementia: Misplaced blame? *Sci News* 140:131, 140.

5. Navia BA, Jordan BD, Price RW. 1986. The AIDS dementia complex: I. Clinical features. *Ann Neurol* 19:517–524; Navia BA, Cho E-S, Petito CK, Price RW. 1986. The AIDS dementia complex. II. Neuropathology. *Ann Neurol* 19:525–535.

6. Petito CK, Navia BA, Cho E-S, Jordan BD, George DC, Price RW. 1985. Vacuolar myelopathy pathologically resembling subacute combined degeneration in patient with the acquired immunodeficiency syndrome. *New Engl J Med* 312:874–879.

7. Root-Bernstein RS. 1990. Multiple-antigen-mediated autoimmunity (MAMA) in AIDS: A possible model for postinfectious autoimmune complications. *Res Immunol* 141:321–339.

8. Fujinami RS, Oldstone MBA. 1985. Amino acid homology between the encephalitogenic site of myelin basic protein and virus: Mechanism for autoimmunity. *Science* 230:1043–1045; Jahnke U, Fischer EH, Alvord EC Jr. 1985. Sequence homology between

certain viral proteins and proteins related to encephalomyelitis and neuritis. *Science* 229:282–284.

9. Root-Bernstein RS, Hobbs SH. 1993. Viral antibodies precipitate myelin basic protein. Submitted.

10. Gold JWM, Armstrong D. 1989. Opportunistic infections in AIDS patients. In: Ma P, Armstrong D, eds. 1989. *AIDS and Infections of Homosexual Men.* 2d ed. Boston: Butterworths, 325–336.

11. Evans AS. 1989. Does HIV cause AIDS? An historical perspective. *J AIDS* 2:107–115, see 112.

12. Hooper C. 1991. AIDS brain research broadens in search of mechanisms. *J NIH Res* 3:17–19.

13. Janssen et al. 1988. *Ann Neurol* 23:49–55.

Multiple-Antigen-Mediated Autoimmunity and Immunosuppression

1. Root-Bernstein RS, Hobbs SH. 1993. Homologies between CD4 protein and the proteins of infectious agents associated with AIDS: Possible inducers of lymphocytotoxic autoimmunity. Submitted.

2. Jeannet M, Stalder H. 1978. Lymphocytotoxic antibodies in spontaneous cytomegalovirus infection. *Br Med J* i:509; Lamelin JP, Revillar JP, Chalopin JM, et al. 1977. Cold lymphocytotoxic antibodies in nasopharyngeal carcinoma. *Br J Canc* 35:426–432; Charlesworth JA, Quin JW, MacDonald GJ, et al. 1978. Complement, lymphotoxins, and immune complexes in infectious mononucleosis: Serial studies in uncomplicated cases. *Clin Exp Immunol* 34:241; Fong S, Vaughan JH, Tsoukas CD, et al. 1982. Selective induction of autoantibody secretion in human bone marrow by Epstein-Barr virus. *J Immunol* 129:1941; Searles RP, Davis LE, Hermanson S, Forelich CJ. 1981. Lymphocytotoxic antibodies in Guillain-Barre syndrome. *Lancet* i:273; Fujinami RS, Nelson JA, Walker L, Oldstone MBA. 1988. Sequence homology and immunologic cross-reactivity of human cytomegalovirus with HLA-DR beta chain: A means for graft rejection and immunosuppression. *J Virol* 62:100–105; Barocci S, Nocera A, Carozzi S, et al. 1987. Identification of autoreactive lymphocytotoxic antibodies in sensitised dialysis and kidney transplant patients. *Nephrol Dial Transplant* 2:266–270.

3. Root-Bernstein RS. 1993. Idiotype-antiidiotype immune complexes due to infectious agents associated with AIDS. Submitted.

4. Root-Bernstein RS. 1993. Idiotype-antiidiotype immune complexes of hepatitis B virus, Mycobacterial, and CD13(monocyte) antibodies in AIDS and Lupus. Submitted.

5. Root-Bernstein RS, Hobbs SH. 1991. Homologies between *My-coplasma* adhesion peptide, CD4, and class II MHC proteins: A possible mechanism for HIV-*Mycoplasma* synergism in AIDS. *Res Immunol* 142:519–523.

6. Root-Bernstein RS, Hobbs SH. 1993. Sperm proteins as possible inducers of lymphocytotoxic autoimmunity in AIDS. Submitted; Root-Bernstein RS, Hobbs SH. 1993. Similarities between factor VIII, factor IX, von Willebrand's factor, and HLA class II proteins as possible inducers of lymphocytotoxic autoimmunity. Submitted.

7. Gobert B, Amiel C, Tang J, Barbarino P, Bene MC, Faure G. 1990. CD4-like molecules in human sperm. *FEBS Lett* 261/2:339–342.

8. Bagasra O, Freund M, Weidmann J, Harley G. 1988. Interaction of human immunodeficiency virus with human sperm in vitro. *J AIDS* 1:431–435.

9. Black FT. 1983. *Ureaplasma urealyticum* and *Mycoplasma hominis*. In: Holmes KK, Mardh P-A, eds. 1983. *International Perspectives on Neglected Sexually Transmitted Diseases*. Washington: Hemisphere Publishing Co., 37–54.

10. Ashida ER, Schofield FL. 1987. Lymphocyte major histocompatibility complex-encoded class II structures may act as sperm receptors. *Proc Natl Acad Sci (USA)* 84:3395.

11. See the references to "The Range of Autoimmune Processes in AIDS."

CHAPTER 6   WHO IS AT RISK FOR AIDS AND WHY

AIDS Risk Groups and the Variable Latency of Disease

1. Seligmann M, Chess L, Fahey JL, et al. 1984. AIDS—An immunologic reevaluation. *New Engl J Med* 311:1286–1296.

2. Sheer R. 1987. AIDS threat to all—how serious? *LA Times*, 14 Aug, pt 1, p 1.

3. Fernandez-Cruz E, Fernandez A, Gutierrez C, et al. 1988. Progressive cellular immune impairment leading to development of AIDS: Two-year prospective study of HIV infection in drug addicts. *J Exp Immunol* 72:190–195.

4. Kaplan JE, Spira TJ, Fishbein DB, et al. 1988. A six year follow up of HIV infected homosexual men with lymphadenopathy. *JAMA* 260:2694–2697.

5. Goedert JJ, Kessler CM, Aledort LM, et al. 1989. A prospective study of human immunodeficiency virus type 1 infection and the

development of AIDS in subjects with hemophilia. *N Engl J Med* 321:1141–1148.

6. Centers for Disease Control. 1991. HIV/AIDS Surveillance Report. July, 1–18.

7. Goedert et al. 1989. *N Engl J Med* 321:1141–1148; Lee CA, Phillipas A, Elford J, et al. 1989. The natural history of human immunodeficiency virus infection in a haemophiliac cohort. *Br J Haematol* 73:228–234.

8. Schinaia N, Ghirardini A, Chiarotti F, Gringeri A, Mannucci PM, Italian Group. 1991. Progression to AIDS among Italian HIV-seropositive haemophiliacs. *J AIDS* 5:385–391; Darby S, Doll R, Thakrar B, Rizza CR, Cox DR. 1990. Time from infection with HIV to onset of AIDS in patients with haemophilia in the UK. *Stat Med* 9:681–689; Giesecke J, Scalia-Tomba G, Berglund O, Berntorp E, Schulman S, Stigendal L. 1988. Incidence of symptoms and AIDS in 146 Swedish haemophiliacs and blood transfusion recipients infected with human immunodeficiency virus. *Br Med J* 297:99–102.

9. Goedert et al. 1989. *N Engl J Med* 321:1141–1148; Lee et al. 1989. *Br J Haematol* 73:228–234; Schinaia et al. and the Italian Group. 1991. *AIDS* 5:385–391; Darby et al. 1990. *Stat Med* 9:681–689; Giesecke et al. 1988. *Br Med J* 297:99–102.

10. European Collaborative Study. 1988. Mother-to-child transmission of HIV infection. *Lancet* ii:1039–1043; Italian Multicentre Study. 1988. Epidemiology, clinical features, and prognostic factors of paediatric HIV infection. *Lancet* ii:1043–1046; Blanche S, Rouzioux C, Moscato MG, et al. 1989. A prospective study of infants born to women seropositive for human immunodeficiency virus type 1. *New Engl J Med* 320:1643–1648; Van de Perre P, Simonon A, Msellati P, et al. 1991. Postnatal transmission of human immunodeficiency virus type 1 from mother to infant. A prospective cohort study in Kigali, Rwanda. *New Engl J Med* 325:593–598; Hira SK, Kmanga J, Bhat GJ, et al. 1989. Perinatal transmission of HIV-1 in Zambia. *Br Med J* 299:1250–1252; Halsey NA, Boulos R, Holt E, et al. 1990. Transmission of HIV-1 infections from mothers to infants in Haiti. Impact on childhood mortality and malnutrition. CDS/JHU AIDS Project Team. *JAMA* 264:2088–2092; Anderson RM, May RM, Boily MC, Garnett GP, Rowley JT. 1991. The spread of HIV-1 in Africa: Sexual contact patterns and the predicted demographic impact of AIDS. *Nature* 352:581–589.

11. Seligmann et al. 1984. *New Engl J Med* 311:1286–1296.

Immunosuppressive Risks of Some Homosexual Men

1. Witchell A. 1991. Let there be light. *NY Times Mag*, 17 Nov, 32–34.
2. Waddles H. 1991. All those with AIDS deserve compassion. *Lansing State J*, 24 Nov, pt A, p 13.
3. Buimovici-Klein E, Lange M, Ong KR, Grieco MH, Cooper LZ. 1988. Virus isolation and immune studies in a cohort of homosexual men. *J Med Virol* 25:371–385.
4. Nerurkar LS, Biggar RJ, Goedert JJ, et al. 1987. Antiviral antibodies in the sera of homosexual men: Correlation with their lifestyle and drug usage. *J Med Virol* 21:123–135; Jaffe HW, Choi K, Thomas PA, et al. 1983. National case-control study of Kaposi's sarcoma and *Pneumocystis carinii* pneumonia in homosexual men: Part 1, Epidemiologic results. *Ann Intern Med* 99:145–151.
5. Jaffe et al. 1983. *Ann Intern Med* 99:145–151; Auerbach DM, Darrow WW, Jaffe HW, Curran JW. 1984. Cluster of cases of the acquired immune deficiency syndrome. *Am J Med* 76:487–492; Biggar RJ, Melbye M, Ebbesen P, et al. 1984. Low T-lymphocyte ratios in homosexual men. *JAMA* 251:1441–1446; Pifer LLW, Wang Y-F, et al. 1987. Borderline immunodeficiency in male homosexuals: Is life-style contributory? *South Med J* 80:687–697; Beral V, Bull D, Darby S, et al. 1992. Risk of Kaposi's sarcoma and sexual practices associated with faecal contact in homosexual or bisexual men with AIDS. *Lancet* 339:632–636.
6. Corey L, Holmes KK. 1980. Sexual transmission of hepatitis A in homosexual men: Incidence and mechanism. *New Engl J Med* 302:436–438.
7. Detels R, English P, Visscher BR, et al. 1989. Seroconversion, sexual activity, and condom use among 2915 seronegative men followed for up to 2 years. *J AIDS* 2:77–83.
8. Rubinstein P, Walker M, Mollen N, Laubenstein LJ, Friedman-Kien AE. 1989. Immunogenetic findings in patients with epidemic Kaposi's sarcoma. In: Ma P, Armstrong D, eds. 1989. *AIDS and Infections of Homosexual Men*. 2d ed, Boston: Butterworths, 403–418; Kiprov DD, Anderson RD, Morand P, et al. 1985. Antilymphocyte antibodies and seropositivity for retroviruses in groups at high risk for AIDS. *N Engl J Med* 312:1517; Mathur S, Goust J-M, Williamson HO, et al. 1981. Cross-reactivity of sperm and T-lymphocyte antigens. *Am J Reprod Immunol* 1:113–118; Mavligit GM, Talpaz M, Hsia FT, et al. 1984. Chronic immune stimulation by sperm alloantigens: Support for the hypothesis that spermatazoa induce immune dysregulation in homosexual males.

JAMA 251:237–241; Morrow WJW, Whalley AS, Leahy C, Rosenberg J, Stricker RB, McCutchan JA. 1989. Anti-histone H2B antibodies in HIV seropositive individuals. In V International Conference on AIDS, Montreal, Canada, 1989 [Abstract WCP 138]: 614; Naz RK, Ellauri M, Phillips TM, Hall J. 1990. Antisperm antibodies in human immunodeficiency virus infection: Effects on fertilization and embryonic development. *Biol Reprod* 42:859–868; Witkin SS, Sonnabend J. 1983. Immune responses to spermatozoa in homosexual men. *Fertil Steril* 39:337–342.

9. Alexander NJ, Anderson DJ. 1987. Immunology of semen. *Fertil Steril* 47:192–205; Turner MJ, White JO, Soutter WP. 1987. Seminal plasma and AIDS. *Immunol Today* 8:257; Turner MJ, White JO, Soutter WP. 1990. Human seminal plasma inhibits the lymphocyte response to infection with Epstein-Barr virus. *Gynecol Oncol* 37:60–65.

10. Adler MW, ed. 1988. *Diseases in the Homosexual Male*. London: Springer-Verlag; Ma, Armstrong, eds. 1989. *AIDS*.

11. Scholefield JH, Sonnex C, Talbot IC, et al. 1989. Anal and cervical intraepithelial neoplasia: Possible parallel. *Lancet* ii:765–767.

12. Yardley JH, Hendrix TR. 1980. Immunologic status in patients with giardiasis. *Gastroenterology* 78:421–422.

13. Jaffe et al. 1983. *Ann Intern Med* 99:145–151.

14. Descotes J. 1988. *Immunotoxicology of Drugs and Chemicals*. Amsterdam: Elsevier, 114–124; Ferrante A, Goh DHB. 1984. The effect of antimalarial drugs on human natural killer cells *in vitro*. *Parasite Immunol* 6:571.

15. Callen M. 1990. *Surviving AIDS*. New York: Harper Collins, 141, 164.

16. Jaffe et al. 1983. *Ann Intern Med* 99:145–151.

17. Pifer et al. 1987. *South Med J* 80:687–697.

18. Osterloh J, Olson K. 1986. Toxicities of alkyl nitrites. *Ann Intern Med* 104:727.

19. Bogart L, Bonsignore J, Carvalho A. 1986. Massive hemolysis following inhalation of volatile nitrites. *Am J Haematol* 22:327–329.

20. Bogart et al. 1986. *Am J Haematol* 22:327–329.

21. Pifer et al. 1987. *South Med J* 80:687–697.

22. Pifer et al. 1987. *South Med J* 80:687–697.

23. Jaffe et al. 1983. *Ann Intern Med* 99:145–151; Haverkos HW, Dougherty JA, eds. 1988. *Health Hazards of Nitrite Inhalants*. Washington, DC: National Institute on Drug Abuse (Monogr. 83); Lange WR, Haertzen CA, Hickey JE, et al. 1988. Nitrite inhalants: Patterns of abuse in Baltimore and Washington, DC. *Am J*

Drug Alcohol Abuse 14:29–40; Biggar et al. 1984. *JAMA* 251:1441; Auerbach et al. 1984. *Am J Med* 76:487.

24. Brambilla G. 1985. Genotoxic effects of drug-nitrite interaction products: Evidence for the need of risk assessment. *Pharmcol Res Commun* 17(A): 307–321.

25. Osterloh J et al. 1984. Butyl nitrite transformation *in vitro*, chemical nitrosation reactions, and mutagenesis. *J Anal Toxicol* 8:164–169; Haverkos HW, Pinsky PF, Drotman DP, Bregman DJ. 1985. Disease manifestation among homosexual men with acquired immunodeficiency syndrome: A possible role of nitrites in Kaposi's sarcoma. *Sex Transm Dis* 12:203–208; Root-Bernstein RS. 1990. Do we know the cause(s) of AIDS? *Persp Biol Med* 33:480–500.

26. Conkin D. 1991. Substance abuse high among gays, lesbians. *Bay Area Reporter.* October 27, 6, 30.

27. Conkin. 1991. *Bay Area Reporter.* Nov: 6, 30.

28. Jaffe et al. 1983. *Ann Intern Med* 99:145–151.

29. Pifer et al. 1987. *South Med J* 80:687–697.

30. Schechter AJ, Poiesz BJ, Brandt-Rauf PW, Papke O, Ball M. 1991. Dioxin levels in the blood of AIDS patients and controls. *Med Sci Res* 19:273–275.

31. Descotes. 1988. *Immunotoxicology*, 114–124; Kimbrough R, Jensen AA. 1989. *Halogenated Biphenyls, Terphenyls, Naphthalenes, Dibenzodioxins, and Selected Products.* Amsterdam: Elsevier; Syracuse Research Corporation. 1989. *Toxicological Profile for 2,3,7,8-Tetrachlorodibenzo-p-Dioxin.* Atlanta, GA: U.S. Public Health Service, Agency for Toxic Substances and Disease Registry.

32. Helferich W. 1992. UV-photoproducts of tryptophan can act as dioxin agents. Cancer Center Seminar Series, 4 Feb, Michigan State University Cancer Center.

33. Wallace BM, Lasker JS. 1992. Awakenings . . . UV light and HIV gene activation. *Science* 257: 1211–1212.

34. Pifer et al. 1987. *South Med J* 80:687–697.

35. De Paepe ME, Waxman M. 1989. Testicular atrophy in AIDS: A study of 57 autopsy cases. *Human Pathol* 20:210–214.

36. Dworkin BM et al. 1986. Selenium deficiency in the acquired immunodeficiency syndrome. *J Parenter Nutr* 10:405–407.

37. Chesters JK, Will M. 1981. Measurement of zinc flux through plasma in normal and endotoxin-stressed pigs and the effects of Zn supplementation during stress. *Br J Nutr* 46:119–130.

38. Spencer SH, Osis D, Kramer L, et al. 1982. Studies of zinc metabolism in normal men and in patients with neoplasia. In Pories WJ,

Strain WH, Woosley RL, eds. 1982. *Clinical Applications of Zinc Metabolism.* Springfield, IL: Charles C. Thomas, 101–102.

39. Pifer et al. 1987. *South Med J* 80:687–697.
40. Weiner RG. 1984. AIDS and zinc deficiency. *JAMA* 252:1409–1410; Apgar J. 1985. Zinc and reproduction. *Ann Rev Nutr* 5:43–58.

Immunosuppressive Risks of Intravenous Drug Abusers

1. Kristal AR. 1986. The impact of acquired immunodeficiency syndrome on patterns of premature death in New York City. *JAMA* 255:2306–2310.
2. Tierney, J. 1990. Newark's spiral of drugs and AIDS. *NY Times,* 16 Dec, Pt 1, pp. 1, 23.
3. Marmor M, Friedman-Kien AE, Laubenstein L, et al. 1982. Risk factors for Kaposi's sarcoma in homosexual men. *Lancet* i:1083–1087.
4. Donohoe RM, Falek AA. 1988. Neuroimmunomodulation by opiates and other drugs of abuse: Relationship to HIV infection and AIDS. In: Bridge TP et al., eds. 1988. *Psychological, Neuropsychiatric, and Substance Abuse Aspects of AIDS.* New York: Raven Press, 145–158; Heathcote J, Taylor KB. 1981. Immunity and nutrition in heroin addicts. *Drug Alcohol Depend* 8:245–255.
5. Rosen TS, Johnson HL. 1982. Children of methadone-maintained mothers: Follow-up to 18 months of age. *J Pediatr* 101:192–196.
6. Wilson GS, Desmond MM, Wait RB. 1981. Follow-up of methadone-treated and untreated narcotic-dependent women and their infants: Health, developmental, and social implications. *J Pediatr* 98:716–722.
7. Aylett A. 1978. Some aspects of nutritional state in "hard" drug addicts. *Br J Addict* 73:77–81; Gambara SE, Clarke JK. 1976. Comments on dietary intake of drug-dependent persons. *J Am Diet Assoc* 68:155–157.
8. Heathcote and Taylor. 1981. *Drug Alcohol Depend* 8:245–255.
9. Nakah AE, Frank O, Louria DB, Quinones MA, Baker H. 1979. A vitamin profile of heroin addiction. *Am J Public Health* 69:1058–1060.
10. Axelrod AE. 1973. Nutrition in relation to acquired immunity. In: Goodhart RS, Shils ME, eds. 1973. *Modern Nutrition in Health and Disease.* Philadelphia: Lea and Febiger.
11. Horsburgh CR Jr, Anderson JR, Boyko EJ. 1989. Increased incidence of infections in intravenous drug users. *Infect Control Hosp*

Epidemiol 10:211–215; Bailey JA, Brown LS Jr. 1990. Serologic investigations in a New York City cohort of parenteral drug users. *J Natl Med Assococ* 82:405–408; Kristal. 1986. *JAMA* 255:2308; Firooznia H, Seliger G, Abrams RM, Valensi V, Shamoun J. 1973. Disseminated extrapulmonary tuberculosis in association with heroin addiction. *Radiology* 109:291–296; Harris PD, Garret R. 1972. Susceptibility of addicts to infection and neoplasia. *New Engl J Med* 287:310; Smolar EN, Pryjma PJ, Berger S. 1975. Cytomegalovirus infection in a heroin addict. *NY State J Med* 75:406.

12. Kringsheim B, Christoffersen M. 1987. Lymph node and thymus pathology in fatal drug addiction. *Forensic Sci Int* 34:245–254.

13. Tierney. 1990. *NY Times*, 16 Dec, pt 1, pp 1, 23.

14. Tierney. 1990. *NY Times*, 16 Dec, pt 1, pp 1, 23.

15. White AG. 1973. Medical disorders in drug addicts. 200 consecutive admissions. *JAMA* 223:1469–1471; Layon J, Idris A, Warzynski M, et al. P. 1984. Altered T-lymphocyte subsets in hospitalized intravenous drug abusers. *Arch Intern Med* 144:1376–1380; Jacobson MA, Gellermann H, Chambers H. 1988. Staphylococcus aureus bacteremia and recurrent staphylococcal infection in patients with acquired immunodeficiency syndrome and AIDS-related complex. *Am J Med* 85:172–176; Morlet A, Darke S, Guinan JJ, Wolk J, Gold J. 1990. Intravenous drug users who present to the Albion Street (AIDS) Centre for diagnosis and management of human immunodeficiency virus infection. *Med J Aust* 152:78–80.

16. Schaffer SR, Schaffer SK. 1984. Use of prophylactic antibiotics by drug abusers. *JAMA* 252:1410.

17. Chandrasekar PH, Molinari JA, Kruse JA. 1990. Risk factors for human immunodeficiency virus infection among parenteral drug abusers in a low-prevalence area. *South Med J* 83:996–1001.

Multiple Immunosuppressive Risks of Transfusion Patients

1. Glaser E. 1991. *In the Absence of Angels*. New York: GP Putnams.

2. Booth W. 1988. A rebel without a cause of AIDS. *Science* 239:1485–1488, see 1488.

3. Ward JW, Bush TJ, Perkins HA, et al. 1989. The natural history of transfusion-associated infection with human immunodeficiency virus. *N Engl J Med* 321:947–952.

4. Ward et al. 1989. *N Engl J Med* 321:947–952.

5. Ward et al. 1989. *N Engl J Med* 321:947–952.

6.   Cumming PD, Wallace EL, Schorr JB, Dodd RY. 1989. Exposure of patients to human immunodeficiency virus through the transfusion of blood components that test antibody-negative. *N Engl J Med* 321:941–946.

7.   Hardy AM, Allen JR, Morgan M, Curran JW. 1985. The incidence rate of acquired immunodeficiency syndrome in selected populations. *JAMA* 253:215–220.

8.   Blumberg N, Heal JM. 1989. Tranfusion and recipient immune function. *Arch Pathol Lab Med* 113:246–253.

9.   Ward et al. 1989. *N Engl J Med* 321:947–952.

10.   Armstrong JA, Tarr GC, Youngblood LA, et al. 1976. Cytomegalovirus infection in children undergoing open-heart surgery. *Yale J Biol Med* 49:83–91.

11.   Centers for Disease Control. 1982. Inactivated hepatitis B virus vaccine. *MMWR* 31:318; Polesky HF, Hanson MR. 1989. Transfusion-associated hepatitis C virus (non-A, non-B) infection. *Arch Pathol Lab Med* 113:232–235.

12.   Nusbacher J, Chiavetta J, Naiman R, et al. 1987. Evaluation of a confidential method of excluding blood donors exposed to human immunodeficiency virus: Studies on hepatitis and cytomegalovirus. *Transfusion* 27:207–209.

13.   Reviewed in Tegtmeier GE. 1989. Posttransfusion cytomegalovirus infections. *Arch Pathol Lab Med* 113:236–245.

14.   Aach RD, Lander JJ, Sherman LA, et al. 1978. Transfusion-transmitted viruses: Interim analysis of hepatitis among transfused and non-transfused patients. In: Vyas GN, Cohen SN, Schmid R, eds. *Viral Hepatitis*. Philadelphia: Franklin Press, 383–396; Alter HJ, Purcell RH, Holland PV, Feinstone SM, Morrow AG, Moritsugu Y. 1975. Clinical and serological analysis of transfusion-associated hepatitis. *Lancet* ii:838–841; Polesky and Hanson. 1989. *Arch Pathol Lab Med* 113:232–235.

15.   Arnold R, Goldmann SF, Pflieger H. 1980. Lymphocytotoxic antibodies in patients receiving granulocyte transfusions. *Vox Sang* 38:250–258.

16.   Crome P, Moffatt B. 1971. IgM lymphocytotoxic antibodies following multiple transfusion. *Vox Sang* 21:11–20; Bucher U, de Weck A, Spengler H, Tschopp L, Kummer H. 1973. Platelet transfusions. Shortened survival of HLA-identical platelets and failure of *in vitro* detection of anti-platelet antibodies after multiple transfusions. *Vox Sang* 25:187–192; Heinrich D, Mueller-Eckhardt C, Stier W. 1973. The specificity of leukocyte and platelet alloantibodies in sera of patients with nonhemolytic transfusion reactions. *Vox Sang* 25:442–456; Heiss F, Goldmann SF, Scheinert I. 1973.

Zur transfusionsbedingten Alloimmunisierung gegenüber Leuko-
zyten-, Thrombozyten-, und Gewebeantigenen. *Münch Med Wo-
chensch* 115:1974–1978; Howard JE, Perkins HA. 1978. The
natural history of alloimmunization to platelets. *Transfusion*
18:496–503; Lalezari P, Rakel E. 1974. Neutrophil-specific antigens: Immunology and clinical significance. *Semin Hematol*
11:281–290; Lohrmann H-P, Bull MI, Decter JA, Yankee RA,
Graw RG. 1974. Platelet transfusions from HLA compatible unrelated donors to alloimmunized patients. *Ann Intern Med* 80:9-14;
Mueller-Eckhardt C. 1977. Histokompatibilität und Thrombozy-
tentransfusion. *Blut* 34:261–270; Perkins HA, Payne R, Ferguson
J, Wood M. 1966. Nonhemolytic febrile transfusion reactions.
Quantitative effects of blood components with emphasis on isoan-
tigenic incompatibility of leukocytes. *Vox Sang* 11:578–600;
Walker RH, Lin D-T, Hartrick MB. 1989. Alloimmunization following blood transfusion. *Arch Pathol Lab Med* 113:254–261;
Holland PV. 1989. Prevention of transfusion-associated graft-
versus-host disease. *Arch Pathol Lab Med* 113:285–291.

17.   Ward et al. 1989. *N Engl J Med* 321:947–952.
18.   Bove, JR, Rigney PR, Kehoe PM, Campbell J. 1987. Look-back:
Preliminary experience of AABB members. *Transfusion* 27:201–
202; Menitove JE. 1986. Status of recipients of blood from donors
subsequently found to have antibody to HIV. *New Engl J Med*
315:1095–1096; Menitove JE. The decreasing risk of transfusion-
associated AIDS. 1989. *New Engl J Med* 321:966–968.
19.   Whyte G. 1988. The transfused population of Canturbury, New
Zealand, and its mortality. *Vox Sang* 54:65–70.

The Multiple Immunosuppressive Risks of Hemophiliacs

1.   Nilsson IM. 1984. Von Willebrand's disease from 1926–1983.
Scand J Haematol Suppl 40, 33:21–43.
2.   Aronson DL. 1983. Pneumonia deaths in haemophiliacs. *Lancet*
ii:1023; Aronson, DL. 1988. Cause of death in hemophilia patients in the United States from 1968 to 1979. *Am J Haematol*
27:7-12.
3.   Massie R, Massie S. 1975. *Journey.* New York: Knopf, 81.
4.   Passaleva A, Massai G, Morfini M, Longo G, Rossi Ferrini PL,
Ricci M. 1983. Circulating immune complexes in haemophilia and
von Willebrand's disease. *Scand J Haemotol* 31:466–474; Cohen
H, Mackie IJ, Anagnostopoulos N, et al. 1989. Lupus anticoagu-
lant, anticardiolipin antibodies, and human immunodeficiency virus in haemophilia. *J Clin Pathol* 42:629–633.

5. Aronson. 1988. *Am J Haematol* 27:7-12.
6. Aronson. 1988. *Am J Haematol* 27:7-12.
7. Tabor E. 1984. Review of transmission of hepatitis by clotting factor concentrates. *Scand J Haematol* Suppl 40, 33:303-308; Rumi MG, Colombo M, Romeo R, Colucci G, Gringeri A, Mannucci PM. 1990. Serum hepatitis B virus DNA detects cryptic hepatitis B virus infections in multitransfused hemophilic patients. *Blood* 75:1654-1658.
8. Webster A, Lee CA, Cook DG, et al. 1989. Cytomegalovirus infection and progression towards AIDS in haemophiliacs with human immunodeficiency virus infection. *Lancet* ii: 63-66.
9. Sherertz RJ, Russell BA, Reuman PD. 1984. Transmission of hepatitis A by transfusion of blood products. *Arch Intern Med* 144:1579-1580.
10. Massie, Massie. 1975. *Journey*, 381.
11. Lian EC, Larcada AM, Chiu AY. 1989. Combination immunosuppressive therapy after factor VIII infusion for acquired factor VIII inhibitor. *Ann Intern Med* 110:774-778.
12. Massie, Massie. 1975. *Journey*, 385-386.
13. Fernandez-Palazzi F, Bosch NB de, Vargas AF de. 1984. Radioactive synovectomy in haemophilic haemarthrosis. Follow-up of fifty cases. *Scand J Haematol* Suppl 40, 33:291-300.
14. Hardy AM, Allen JR, Morgan WM, Curran JW. 1985. The incidence rate of acquired immune deficiency syndrome in selected populations. *JAMA* 253:215-218; Goedert JJ, Kessler CM, Aledort LM, et al. 1989. A prospective study of human immunodeficiency virus type 1 infection and the development of AIDS in subjects with hemophilia. *New Engl J Med* 321:1141-1148.
15. Goedert et al. 1989. *New Engl J Med* 321:1141-1148.
16. Darby SC, Rizza CR, Doll R, Spooner RJ, Stratton IM, Thakrar B. 1989. Incidence of AIDS and excess of mortality associated with HIV in haemophiliacs in the United Kingdom: Report on behalf of the directors of haemophilia centres in the United Kingdom. *Br Med J* 298:1064-1068.
17. Goedert et al. 1989. *New Engl J Med* 321:1141-1148.

The Immunosuppressive Risks of Infants

1. Ward R. 1988. Mainstream scientists confront unorthodox view of AIDS. *Nature* 332:574.
2. Booth W. 1988. A rebel without a cause of AIDS. *Science* 239:1485-1488.
3. Gellert GA, Durfee MJ. 1990. HIV infection and child abuse. *New*

Engl J Med 321:685; Centers for Disease Control. 1991. *HIV/ AIDS Surveillance Report* July 1991: Table 3 (Pediatric exposure category).

4. Wilson C, Painter K. 1990. Counting the AIDS losses after 9 years. *USA Today*, 19 June, pt D, pp 1–2.

5. CDC. 1991. *HIV/AIDS Surveillance* 1–18.

6. Cobrinik RW, Hood RT, Chusid E. 1959. The effect of maternal narcotic addiction on the newborn infant. *Pediatrics* 24:288–304.

7. Goodfriend MJ, Shey IA, Klein MD. 1956. The effects of maternal narcotic addiction on the newborn. *Am J Obstet Gynecol* 79:29–36.

8. Ostrea EM Jr, Chavez CJ. 1979. Perinatal problems (excluding neonatal withdrawal) in maternal drug addiction: A study of 830 cases. *J Pediatr* 94:292–295; Vargas GC, Pildes RS, Vidyasagar D, Keith LG. 1975. Effect of maternal heroin addiction on 67 live-born neonates. *Clin Pediatr* 14:751–757.

9. Ostrea and Chavez. 1979. *J Pediatr* 94:292–295; Vargas et al. 1975. *Clin Pediatr* 14:751–757.

10. Hughes WT, Sanyal SK, Price RA. 1976. Signs, symptoms, and pathophysiology of *Pneumocystis carinii* pneumonia. *Natl Cancer Inst Monogr* 43:77–84; Robbins JRB, DeVita VT Jr, Dutz W, eds. 1976. Symposium on *Pneumocystis carinii* Pneumonitis. *Natl Cancer Inst Monogr* 43: passim; Burke BA, Good RA. 1973. *Pneumocystis carinii* infection. *Medicine* 52:23–51; Burke BA. 1976. *Pneumocystis carinii* infection: Diagnosis and pathogenesis. *Natl Cancer Inst Monogr* 43:151–154.

11. Fricker HS, Segal S. 1978. Narcotic addiction, pregnancy, and the newborn. *Am J Dis Child* 132:360–366; Vargas et al. 1975. *Clin Pediatr* 14:751–757; Ostrea and Chavez. 1979. *J Pediatr* 94:292–295.

12. Miller ME, Stiehm ER. 1983. Immunology and resistance to infection. In: Remington JS, Klein JO, eds. 1983. *Infectious Diseases of the Fetus and Newborn Infant*. Philadelphia: Saunders, 27–68; Culver, KW, Ammann AJ, Partridge JC, Wong DF, Wara DW, Cowan MJ. 1987. Lymphocyte abnormalities in infants born to drug-abusing mothers. *J Pediatr* 111:230–235; Rogers MF, Ou C-Y, Rayfield M, et al. 1989. Use of the polymerase chain reaction for early detection of the proviral sequences of human immunodeficiency virus in infants born to seropositive mothers. *New Engl J Med* 320:1649–1654.

13. Ostrea and Chavez. 1979. *J Pediatr* 94:292–295.

14. Vargas et al. 1975. *Clin Pediatr* 14:751–757.

15. Uhr JW, Dancis J, Neumann CG. 1960. Delayed-type sensitivity

in premature neonatal humans. *Nature* 187:1130–1131; Naeye RL, Diener MM, Harcke HT Jr, Blanc WA. 1971. Relationship of poverty and race to birth weight and organ and cell structure in the newborn. *Pediatr Res* 5:17–22; Miller HC, Hassanein K. 1973. Fetal malnutrition in white newborn infants: Maternal factors. *Pediatrics* 52:504–512; Carr MD, Stites DP, Fudenberg HH. 1973. Dissociation of responses to phyohemagglutinin and adult allogeneic lymphocytes in human foetal lymphoid tissues. *Nature New Biol* 241:279; Ferguson AC, Lawlor GJ Jr, Neumann CG, Oh W, Stiehm ER. 1974. Decreased rosette-forming lymphocytes in malnutrition and intrauterine growth retardation. *J Pediatr* 85:717–723; Chandra RK. 1975. Fetal malnutrition and postnatal immunocompetence. *Am J Dis Child* 129:450–454; Hallberg A, Hallberg T. 1977. Evolution of three lymphocyte markers in newborn preterm infants. *Int Arch Allergy Appl Immunol* 55:102–111.

16. Scott GB, Fischl MA, Klimas N, et al. 1985. Mothers of infants with acquired immunodeficiency syndrome. *JAMA* 253:363–366; Pawha S, Kaplan M, Fikrig S, et al. 1986. Spectrum of human T-cell lymphotropic virus type III infection in children. *JAMA* 255:2299–2305; Minkoff H, Nanda D, Menez R, Fikrig S. 1987. Pregnancies resulting in infants with acquired immunodeficiency syndrome or AIDS-related complex: Follow-up of mothers, children, and subsequently born siblings. *Obstet Gynecol* 69:288–291; Shannon KM, Ammann AJ. 1985. Acquired immune deficiency syndrome in childhood. *J Pediatr* 106:332–342; Nadal D, Hunziker UA, Shupbach J, et al. 1989. Immunological evaluation in the early diagnosis of prenatal or perinatal HIV infection. *Arch Dis Child* 64:662–669.

17. Wallace J. 1989. Case presentations of AIDS in the United States. In: Ma P, Armstrong D, eds. 1989. *AIDS and Infections of Homosexual Men*. 2d ed. Boston: Butterworths, 285–295.

18. Wilson GS, Desmond MM, Wait RB. 1981. Follow-up of methadone-treated and untreated narcotic-dependent women and their infants: Health, developmental, and social implications. *J Pediatr* 98:716–722.

19. Anonymous. 1989. The no-parent child. [Editorial] *NY Times*, 24 Dec.

20. Ostrea, Chavez. 1979. *J Pediatr* 94:292–295.

21. Starr SE, Tolpin MD, Friedman HM, Paucker K, Plotkin SA. 1979. Impaired cellular immunity to cytomegalovirus in congenitally infected children and their mothers. *J Infect Dis* 140:504–505; South MA, Alvord CA Jr. 1980. The immunology of chronic

intrauterine infection. In: Stiehm ER, Fulginiti VA, eds. 1980. *Immunologic Disorders in Infants and Children*. 2d ed. Philadelphia: Saunders, 702–714.

22. Nahmias AJ, Roizman B. 1973. Infection with herpes simplex 1 and 2 (third of three parts). *New Engl J Med* 289:781–789.

23. Monnickendam MA. 1988. Chlamydial genital infections. In: Wright DJM, ed. 1988. *Immunology of Sexually Transmitted Diseases*. Drodrecht: Kluwer, 117–162.

24. Klein JO. 1983. Mycoplasma infections. In: Remington JS, Klein JO, eds. 1983. *Infectious Diseases of the Fetus and Newborn Infant*. 2d ed. Philadelphia: Saunders, 428–449.

25. Nahmias AJ, Walls KW, Stewart JA, Herrmann KL, Flyn WR. The ToRCH complex—perinatal infections associated with toxoplasma and rubella, cytomegalo- and herpes simplex viruses. *Pediatr Res* 5:405–406.

26. Lang, DJ. 1983. Cytomegalovirus. In: Holmes KK, Mardh P-A, eds. 1983. *International Perspectives on Neglected Sexually Transmitted Diseases*. Washington: Hemisphere Publishing Co, 83–92.

27. Stagno S, Reynolds DW, Pass RF, Alford CA. 1980. Breast milk and the risk of cytomegalovirus infection. *New Engl J Med* 302:1073–1076.

28. Frenkel LD, Gaur S, Tsolia M, Scudder R, Howell R, Kesarwala H. 1990. Cytomegalovirus infection in children with AIDS. *Rev Infect Dis* 12 Suppl 7P: S820-S826.

29. Haynes DC, Golub MS, Gershwin ME, et al. 1987. Long-term marginal zinc deprivation in rhesus monkeys. II. Effects on maternal health and fetal growth at midgestation. *Am J Clin Nutr* 45:1503–1513; Mulhern SA, Taylor GL, Magruder LE, Vessey AR. 1985. Deficient levels of dietary selenium suppress the antibody respone in first and second generation mice. *Nutr Res* 5:201–210; Atkinson SA, Whelan D, Whyte RK, Lonnerdal B. 1989. Abnormal zinc content in human milk. *Am J Dis Child* 143:608–611.

CHAPTER 7   IMMUNOSUPPRESSION IN THE ABSENCE OF HIV INFECTION

Theoretical Implications of Immune Suppression Among High-Risk-Group Individuals in the Absence of HIV

1. Sarngadharan MG, Popovic M, Bruch L, Schupach J, Gallo RC. 1984. Antibodies reactive with human T-lymphotropic retroviruses (HTLV-III) in the serum of patients with AIDS. *Science* 224:506; Salahudin SZ, Markham PD, Wong-Staal F, Gallo RC.

1985. The human T-cell leukemia-lymphoma virus family. *Prog Med Virol* 32:195; Clumeck N, Robert-Guroff M, Van de Perre P, et al. 1985. Seroepidemiological studies of HTLV-III antibody prevalence among selected groups of heterosexual Africans. *JAMA* 254:2599.

2.   Centers for Disease Control. 1991. *HIV/AIDS Surveillance Report.* July: 1–18.

Immune Suppression in HIV-Negative Homosexuals

1.   Groopman JE. 1984. AIDS. *Semin Oncol* 2:1.

2.   Sarngadharan MG, Veronese F, Lee S, Gallo RC. 1985. Immunological properties of HTLV-III antigens recognized by sera of patients with AIDS and AIDS related complex and of asymptomatic carriers of HTLV-III infection. *Cancer Res* (Suppl) 45:4574s.

3.   Buimovici-Klein E, Lange M, Ong, KR, Grieco MH, Cooper LZ. 1988. Virus isolation and immune studies in a cohort of homosexual men. *J Med Virol* 25:371–385.

4.   Drew WL, Mills J, Levy J, et al. 1985. Cytomegalovirus infection and abnormal T-lymphocyte subset ratios in homosexual men. *Ann Intern Med* 103:61–63.

5.   Drew et al. 1985. *Ann Intern Med* 103:61–63.

6.   Collier AC, Meyers JD, Corey L, Murphy VL, Roberts PL, Handsfield HH. 1987. Cytomegalovirus infection in homosexual men. *Am J Med* 82:593–601.

7.   Drew et al. 1985. *Ann Intern Med* 103:61–63.

8.   Pinching AJ, McManus TJ, Jeffries DJ, et al. 1983. Studies of cellular immunity in male homosexuals in London. *Lancet* ii:126–130.

9.   Weber JN, Wadsworth J, Rogers LA, et al. 1986. Three-year prospective study of HTLV-III/LAV infection in homosexual men. *Lancet* i:1179–1182.

10.   Rogers LA, Forster SM, Pinching AJ. 1989. IgD production and other lymphocyte functions in HIV infection: Immaturity and activation of B cells at different clinical stages. *Clin Exp Immunol* 75:7–11.

11.   Novick DM, Brown DJC, Lok ASF, Lloyd JC, Thomas HC. 1986. Influence of sexual preference and chronic hepatitis B virus infection on T lymphocyte subsets, natural killer activity, and suppressor cell activity. *J Hepatol* 3:363–370.

12.   Pifer LLW, Wang Y-F, Chiang TM, et al. 1987. Borderline immunodeficiency in male homosexuals: Is life-style contributory? *South Med J* 80:687–697.

13.   Murray HW, Scavuzzo DA, Kelly CD, Rubin BY, Roberts RB.

1988. T4+ cell production of interferon gamma and the clinical spectrum of patients at risk for and with acquired immunodeficiency syndrome. *Arch Intern Med* 148:1613–1616.

14.   Bartholomew C, Saxinger WC, Clark JW, et al. 1987. Transmission of HTLV-1 and HIV among homosexual men in Trinidad. *JAMA* 257:2604–2608.

Immune Suppression in HIV-Negative Drug Abusers

1.   Brown SM, Stimmel B, Taub RN, Kochwa S, Rosenfield RE. 1974. Immunologic dysfunction in heroin addicts. *Arch Intern Med* 134:1001–1006.

2.   Kreek MJ. 1973. Medical safety and side effects of methadone in tolerant individuals. *JAMA* 223:665–668; Cushman P, Grieco MH. 1973. Hyperglobulinemia associated with narcotic addiction. *Am J Med* 54:320; Spiera H, Oreskes T, Stimmel B. 1974. Rheumatoid factor activity in heroin addicts on methadone maintenance. *Ann Rheum Dis* 33:153; Cushman P, Gupta S, Grieco MH. 1977. Immunological studies in methadone maintained patients. *Int J Addict* 12:241–253.

3.   McDonough RJ, Madden JJ, Falek A, et al. 1980. Alterations of T and null lymphocyte frequency in the peripheral blood of human opiate addicts: *in vivo* evidence for opiate receptor sites on T-lymphocytes. *J Immunol* 125:2539.

4.   McDonough et al. 1980. *J Immunol* 125:2539.

5.   Donohoe RM, Nicholson JK, Madden JJ, et al. 1986. Coordinate and independent effects of heroin, cocaine, and alcohol abuse on T-cell E-rosette formation and antigenic marker expression. *Clin Immunol Immunopathol* 41:254–264; Donohoe RM, Bueso-Ramos C, Donohoe F, et al. 1987. Mechanistic implications of the findings that opiates and other drugs of abuse moderate T-cell surface receptors and antigenic markers. *Ann NY Acad Sci* 496:711–721.

6.   Heathcote J, Taylor KB. 1981. Immunity and nutrition in heroin addicts. *Drug Alcohol Depend* 8:245–255.

7.   Brugo MA, Guffanti A, Guzzeti S, Pedretti D, Stringhetti M, Confalonieri F. 1983. Differenza nel comportamento delle popolazioni T-linfocitarie nei dipendenti da eroina e da metadone. *Boll Ist Sieroter Milan* 6:517.

8.   Novick DM, Ochshorn M, Ghali V, Croxson TS, Mercer WD, Chiorazzi, Kreek MJ. 1989. Natural killer cell activity and lymphocyte subsets in parenteral heroin abusers and long-term methadone maintenance patients. *J Pharmacol Exp Therap* 250:606–610.

9. DeShazo RD, Chadha N, Morgan JE, et al. 1989. Immunologic assessment of a cluster of asymptomatic HTLV-I-infected individuals in New Orleans. *Am J Med* 86:65–70.

10. Horsburgh CR Jr, Anderson JR, Boyko EJ. 1989. Increased incidence of infections in intravenous drug users. *Infect Control Hosp Epidemiol* 10:211–215.

11. Bailey JA, Brown LS Jr. 1990. Serologic investigations in a New York City cohort of parenteral drug users. *J Natl Med Assoc* 82:405–408.

12. Woods KF, Hanna BJ. 1986. Brain stem mucormycosis in a narcotic addict with eventual recovery. *Am J Med* 80:126; Dupont B, Drouhet E. 1985. Cutaneous, ocular, and osteoarticular candidiasis in heroin addicts: New clinical and therapeutic aspects in 38 patients. *J Infect Dis* 152:577; Podzamczer D, Gudiol F. 1986. Systemic candidiasis in heroin abuser. *J Infect Dis* 153:1182.

13. Firooznia H, Seliger G, Abrams RM, et al. 1973. Disseminated extrapulmonary tuberculosis in association with heroin addiction. *Radiology* 109:291–296; Roca RP, Yoshikawa TT. 1979. Primary skeletal infections in heroin users: A clinical characterization, diagnosis, and therapy. *Clin Orthop Relat Res* 144:238; Chandrasekar PH, Narula AP. 1986. Bone and joint infections in intravenous drug abusers. *Rev Infect Dis* 8:904.

14. Buehler JW, Devine OJ, Berkelman RL, Chevarley FM. 1990. Impact of the human immunodeficiency virus epidemic on mortality trends in young men, United States. *Am J Public Health* 80:1080; Selwyn PA, Hartel D, Wasserman W, Drucker E. 1989. Impact of the AIDS epidemic on morbidity and mortality among intravenous drug users in a New York City methadone maintenance program. *Am J Public Health* 79:1358.

Immune Suppression Among Infants in the Absence of HIV

1. Rosen TS, Johnson HL. 1982. Children of methadone-maintained mothers: Follow-up to 18 months of age. *J Pediatr* 101:192–198; Wilson GS, Desmond MM, Wait RB. 1981. Follow-up of methadone-treated and untreated narcotic-dependent women and their infants: Health, developmental, and social implications. *J Pediatr* 98:716–722; Chasnoff LJ. 1988. Drug use in pregnancy: Parameters of risk. *Pediatr Clin North Am* 35:1403–1412; Selwyn PA, Schoenbaum EE, Davenny K, et al. 1989. Prospective study of human immunodeficiency virus infection and pregnancy outcomes in intravenous drug abusers. *JAMA* 261:1289–1294.

2. Culver, KW, Ammann AJ, Partridge JC, Wong DF, Wara DW,

Cowan MJ. 1987. Lymphocyte abnormalities in infants born to drug-abusing mothers. *J Pediatr* 111:230–235.

3.   Rogers LA, Forster SM, Pinching AJ. 1989. IgD production and other lymphocyte functions in HIV infection: Immaturity and activation of B cells at different clinical stages. *Clin Exp Immunol* 75:7–11; Rogers MF, Ou C-Y, Rayfield M, et al. 1989. Use of the polymerase chain reaction for early detection of the proviral sequences of human immunodeficiency virus in infants born to seropositive mothers. *New Engl J Med* 320:1649–1654.

4.   Mayers MM, Davenny K, Schoenbaum EE, Feingold AR et al. 1991. A prospective study of infants of human immunodeficiency virus seropositive and seronegative women with a history of intravenous drug use or of intravenous drug-using sex partners, in the Bronx, New York City. *Pediatrics* 88:1248–1256.

5.   Booth W. 1988. A rebel without a cause of AIDS. *Science* 239:1485–1488.

Immune Suppression in HIV-Negative Hemophiliacs

1.   Mannucci PM, Quattrone P, Matturri L. 1986. Kaposi's sarcoma without human immunodeficiency virus antibody in a hemophiliac. *Ann Intern Med* 105:466.

2.   Unzeitig JC, Church JA, Gomperts ED, Nye CA, Pasquale S, Richards W. 1984. Abnormal T-cell subsets and mitogen responses in hemophiliacs exposed to factor concentrate. *Am J Dis Child* 138:645–648; Buehrer JL, Weber DJ, Meyer AA, et al. 1990. Wound infection rates after invasive procedures in HIV-1 seropositive versus HIV-1 seronegative hemophiliacs. *Ann Surg* 211:492–498.

3.   Lang DJ, Kovacs AAS, Zaia JA, et al. 1989. Seroepidemiologic studies of cytomegalovirus and Epstein-Barr virus infections in relation to human immunodeficiency virus type 1 infection in selected recipient populations. *J AIDS* 2:540–549.

4.   Jin Z, Cleveland RP, Kaufman DB. 1989. Immunodeficiency in patients with hemophilia: An underlying deficiency and lack of correlation with factor replacement therapy or exposure to human immunodeficiency virus. *J Allergy Clin Immunol* 83:165–170.

5.   Mahir WS, Millard RE, Booth JC, Flute PT. 1988. Functional studies of cell-mediated immunity in haemophilia and other bleeding disorders. *Br J Haematol* 69:367–370.

6.   Madhok R, Gracie JA, Smith J, Low GD, Forbes CD. 1990. Capacity to produce interleukin 2 is impaired in haemophilia in the absence and presence of HIV-1 infection. *Br J Haematol* 76:70–74.

7.   Jason J, Holman RC, Evatt BL. 1990. Relationship of partially

purified factor concentrates to immune tests and AIDS. The Hemophilia-AIDS Collaborative Study Group. *Am J Hematol* 34:262–269.

8. Jason et al. 1990. *Am J Hematol* 34:262–269.

9. Hultin MB, Dattwyler RJ, Lipton RA. 1989. Controlled prospective study of factor IX concentrate therapy and immunodeficiency. *Am J Hematol* 31:71–72.

10. Telfer NR, Matthews JM, Wojnarowska F. 1989. Skin disease in haemophiliacs with and without antibodies to the human immunodeficiency virus (HIV): Further evidence of altered disease behaviour in different risk groups? *Br J Dermatol* 120:795–799.

11. Cohen J. 1992. "Mystery virus" meets the skeptics. *Science* 257:1032–1034.

Immunosuppression in HIV-Negative Heterosexuals

1. Virella G, Fudenberg HH. 1982. Secondary immunodeficiencies. In: Twomey JJ, ed. 1982. *The Pathophysiology of Human Immunologic Disorders*. Baltimore & Munich: Uban & Schwarzenberg, 91–124, see 115–116.

2. Chatterjee SN, Fiala M, Weiner J, Stewart JA, Stacey B, Warner N. 1978. Primary cytomegalovirus and opportunistic infections: Incidence in renal transplant patients. *JAMA* 240:2446–2449; Rand KH, Pollard RB, Merigan TC. 1978. Increased pulmonary superinfections in cardiac-transplant patients undergoing primary cytomegalovirus infection. *New Engl J Med* 298:951–953; Rinaldo CR Jr, Hamoudi WH, DeBaiassio RL, Rabin B, Hakala TR, Liebert M. 1983. Cellular immune response and cytomegalovirus infection in renal transplant recipients receiving cyclosporine. *Transplant Proc* 15:2775.

3. Minnefor AB, Oleske JM. 1989. AIDS in children. In: Ma P, Armstrong D, eds. 1989. *AIDS and Infections of Homosexual Men*. 2d ed. Boston: Butterworths, 296–304.

4. Asherson GL, Webster DB, Humphrey JH. 1980. *Diagnosis and Treatment of Immunodeficiency Diseases*. Oxford: Blackwell Scientific, 186–187.

5. Floersheim GL. 1978. [Immunosuppression as a concomitant effect]. *Schweiz Med Wochenschr* 108:1449–1460; Federlin K. 1985. [Diabetes mellitus and immunology—a manifold interrelation.] *Immunitat und Infecktion* 13:193–199.

6. Leslie CA, Sapico FL, Bessman AN. 1989. Infections in the diabetic host. *Compr Ther* 15:23–32.

7. Farrag SA, Morsy TA, Makarem SS, Sarwat MA. 1988. Study on

some opportunistic parasitic infections in Egyptian diabetic patients. *J Egypt Soc Parasitol* 18:197–205.

8. Brownstein MH, Shapiro L, Skolnik P. 1973. Kaposi's sarcoma in
    community practice. *Arch Dermatol* 107:137–138.

9. Jacobs JL, Libby DM, Winters RA, Gelmont DM, Fried ED, Hart-
    man BJ, Laurence J. 1991. A cluster of *Pneumocystis carinii* pneu-
    monia in adults without predisposing illnesses. *N Engl J Med*
    324:246–250.

10. McDevitt GR Jr, Brantley MJ, Cawthon MA. 1989. Rhinocerebral
    mucormycosis: A case report with magnetic resonance imaging
    findings. *Clin Imaging* 13:317–320; Connolly JP, Mitas JA II.
    1990. Torulopsis glabrata fungemia in a diabetic patient. *South
    Med J* 83:352–353.

11. Rene R, Mas A, Villabona CM, Ricart MC, Bassa A, Tolosa F.
    1990. [Otitis externa maligna and cranial neruopathy.] *Neuro-
    logia* 5:222–227.

12. Sheft DJ, Shrago G. 1970. Esophageal moniliasis. *JAMA*
    213:1859–1862; Mann NS, Caplash VK. 1975. Monilial esopha-
    gitis. *South Med J* 68:479–480; Kodsi BE, Wickremesinghe PC,
    Kozinn PJ, Iswara K, Goldberg PK. 1976. Candida esophagitis.
    A prospective study of 27 cases. *Gastroenterology* 71:715–719,
    Orringer MB, Herbert S. 1978. Monilial esophagitis: An increas-
    ingly frequent cause of esophageal stenosis? *Ann Thoracic Surg*
    26:364–374; Odds FC. 1979. *Candida and Candidosis.* Leicester:
    Leicester University Press, passim: index diabetes.

13. Heard BE, Hassan AM, Wilson SM. 1962. Pulmonary cytomega-
    lic inclusion-body disease in a diabetic. *J Clin Patholol* 15:17–20.

14. Lauzé S. 1961. Maladie à inclusions cytomegaliques chez l'adulte.
    Union médicale du Canada 90:122–130.

15. Neu I, Rodiek S. 1977. [Significance of diabetes mellitus in the
    activation of the varicella zoster virus.] *MMW* 119:543–546.

16. Shalit P, Brennan C, Murpy V, Hooton TM. 1990. Risk of cross-
    infection through shared diabetic devices. *JAMA* 263:34–35.

Non-HIV Markers of Developing Immune Suppression in AIDS

1. Biggar RJ, Anderson HK, Ebbesen P, Melbye M, Goedert JJ.
    1983. Seminal fluid excretion of cytomegalovirus related to immu-
    nosuppression in homosexual men. *Br Med J* 286:2010–2012.

2. Fiala M, Cone LA, Chang C-M, Mocarski ES. 1986. Cytomegalo-
    virus viremia increases with progressive immune deficiency in pa-
    tients infected with HTLV-III. *AIDS Res* 2:175–181.

3. Rinaldo CR Jr, Kingsley LA, Lyter DW, et al. 1986. Association

of HTLV-III with Epstein-Barr virus infection and abnormalities of T lymphocytes in homosexual men. *J Infect Dis* 154:556–561; Rahman MA, Kingsley LA, Berinig MK, et al. 1989. Reactivation of Epstein-Barr virus during early infection with human immunodeficiency virus. *J Clin Microbiol* 29:1215–1220.

4. Munoz A, Carey V, Saah AJ, et al. 1988. Predictors of decline in CD4 lymphocytes in a cohort of homosexual men infected with human immunodeficiency virus. *J AIDS* 1:396; Sumaya CV, Boswell RN, Ench Y, et al. 1985. Enhanced serological and virological findings of Epstein-Barr virus in patients with AIDS and AIDS related complex. *J Infect Dis* 154:864; Sonnabend JA. 1989. AIDS: An explanation for its occurrence among homosexual men. In: Ma P, Armstrong D, eds. *AIDS and Infections of Homosexual Men*, 2d ed. Boston: Butterworths, 454–456.

5. Zarling JM, Ledbetter JA, Sias J, et al. 1990. HIV-infected humans but not chimpanzees, have circulating cytotoxic T lymphocytes that lyse uninfected CD4+ cells. *J Immunol* 144:2992–2998; Ozturk GE, Kohler PF, Horsburgh CR, Kirkpatrick CH. 1987. The significance of antilymphocyte antibodies in patients with acquired immune deficiency syndrome (AIDS) and their sexual partners. *J Clin Immunol* 7:130–133; Stricker RB, McHugh TM, Moody DJ, et al. 1987. An AIDS-related cytotoxic antibody reacts with a specific antigen on stimulated CD4 T-cells. *Nature* 327:710–712. Stricker RB, McHugh TM, Marx PA, et al. 1987. Prevalence of an AIDS-related autoantibody against CD4+ T cells in humans and monkeys. *Blood* 70:127A.

6. McDougal JS, Hubbard M, Nicholson JKA, et al. 1985. Immune complexes in the acquired immunodeficiency syndrome (AIDS): relationship to disease manifestation, risk grup, and immunologic defect. *J Clin Immunol* 5:130; Sonnabend. 1989. In: *AIDS and Infections of Homosexual Men*, 459.

CHAPTER 8   WHY AIDS IS EPIDEMIC NOW

AIDS Is a Social Disease

1. Weber J. 1988. AIDS and the "guilty" virus. *New Scien* 5 May, 32–33.

Coming Out of the Closet

1. Sohn N, Weinstein MA, Gonchar J. 1977. Social injuries of the rectum. *Am J Surg* 134:611–612.

2. Szunyogh B. 1958. Enema injuries. *Am J Proctol* 9:303; Large PG,

Mukheiber WJ. 1956. Injury to rectum and anal canal by enema syringes. *Lancet* ii:596; Roland CG, Roger AG. 1959. Rectal perforations after enema administration. *Can Med Assoc J* 81:815; Marino AWM. 1964. Proctologic lesions observed in male homosexuals. *Dis Colon Rectum* 7:121–128; Barone JE, Sohn N, Nealson TF. 1977. Perforations and foreign bodies in the rectum: Report of 28 cases. *Ann Surg* 184:601; Marino WM, Mancini HWN. 1978. Anal eroticism. *Surg Clin N Am* 58:513; Agnew J. 1986. Hazards associated with anal erotic activity. *Arch Sex Behav* 4:307–314.

3.   Marino. 1964. *Dis Colon   Rectum* 7:121–128.

4.   Kazal HL, Sohn N, Carrasco JI, Robilotti JG Jr, Delaney WE. 1976. The gay bowel syndrome: Clinico-pathologic correlation in 260 cases. *Ann Clin Lab Sci* 6:184–192.

5.   Sohn et al. 1977. *Am J Surg* 134:611–612; Weinstein MA, Sohn N, Robbins RD. 1981. Syndrome of pelvic cellulitis following rectal sexual trauma. *Am J Gastroenter* 75:380–381.

6.   Sohn et al. 1977. *Am J Surg* 134:611–612.

7.   Sohn et al. 1977. *Am J Surg* 134:611–612.

8.   Jaffe HW, Choi K, Thomas PA, et al. 1983. National case-control study of Kaposi's sarcoma and *Pneumocystis carinii* pneumonia in homosexual men: Part 1, Epidemiologic results. *Ann Intern Med* 99:145–151; Pifer LLW, Wang Y-F, Chiang TM, et al. 1987. Borderline immunodeficiency in male homosexuals: Is life-style contributory? *South Med J* 80:687–697; Estep R, Waldorf D. 1991. Safe and unsafe sex among male street hustlers and call men. In: Schnieder BH, Huber J, eds. 1991. *Culture and Social Relations in the AIDS Crisis.* Newbury Park, CA: Sage.

9.   Kazal et al. 1976. *Ann Clin Lab Sci* 6:184–192.

10. Corey L, Holmes KK. 1980. Sexual transmission of hepatitis A in homosexual men: Incidence and mechanism. *N Engl J Med* 302:436–438; Marotta T. 1981. *The Politics of Homosexuality.* Boston: Houghton Mifflin.

11. Weeks J. 1988. Male homosexuality: Cultural perspectives. In: Adler MW, ed. 1988. *Diseases in the Male Homosexual.* London: Springer-Verlag, 1–14.

12. Marotta. 1981. *Politics of Homosexuality.*

13. Callen M. 1990. *Surviving AIDS.* New York: Harper Collins, 4.

14. Nahmias AJ, Lee FK, Beckman-Nahmias S. 1990. Sero-epidemiological and -sociological patterns of herpes simplex virus infection in the world. *Scand J Infect Dis* Suppl 69:19–36, fig 5.

15. Callen. 1990. *Surviving AIDS.*

16. Nahmias et al. 1990. *Scand J Infect Dis* Suppl 69:19–36, fig 5. See also: Melbye M, Biggar RJ. 1992. Interactions between persons at

risk for AIDS and the general population in Denmark. *Am J Epidemiol* 135:593–602.

17. Kinsey AC, Pomeroy WB, Martin CE. 1948. *Sexual Behavior in the Human Male*. Philadelphia: WB Saunders.

18. Dritz SK. 1980. Medical aspects of homosexuality. *N Engl J Med* 302:463–464; Marmor M, Laubenstein L, William DC, et al. 1982. Risk factors for Kaposi's sarcoma in homosexual men. *Lancet* i:1083.

19. Nicol CS. 1963. Venereal diseases. Moral standards and public opinion. *Br J Vener Dis* 39:168–172.

20. Jefferies FJG. 1956. Venereal disease and the homosexual. *Br J Vener Dis* 32:17–20; Schofield M. 1973. *The Sexual Behaviour of Young Adults*. London: Allen Lane.

21. Vaisrub S. 1977. Homosexuality—a risk factor in infectious disease. *JAMA* 238:1402.

22. Larson AA. 1959. The transmission of venereal disease through homosexual practices. *Can Med Assoc J* 80:22; Tarr JD, Lugar RR. 1960. Early infectious syphilis: Male homosexual relations and mode of spread. *Calif Med* 93:35.

23. Fichtner RR, Arol SO, Bount JH, et al. 1983. Syphilis in the United States: 1967–1979. *Sex Transm Dis* 10:77.

24. Centers for Disease Control. 1981. Annual Summary, 1980. *MMWR* 29: passim.

25. CDC. 1981. *MMWR* 29: passim; Centers for Disease Control. 1989. Summary of notifiable diseases, United States, 1988. *MMWR* 37: passim.

26. Fulford KWM, Dane DS, Catterall RD, et al. 1973. Australia antigen and antibody among patients attending a clinic for sexually transmitted diseases. *Lancet* i:1470–1473; Szmuness W, Much MI, Prince AM, et al. 1975. On the role of sexual behavior in the spread of hepatitis B infection. *Ann Int Med* 83:489–495; Lim KS, Taam V, Fulford KWM, et al. 1977. Role of sexual and non-sexual practices in the transmission of hepatitis B. *Br J Vener Dis* 53:190–192; Schreeder MT, Thompson SE, Hadler SC, et al. 1982. Hepatitis B in homosexual men: Prevalence of infection and factors related to transmission. *J Infect Dis* 146:7–15.

27. Corey and Holmes. 1980. *N Engl J Med* 302:436–438.

28. Becker TM, Nahmias AJ. 1985. Genital herpes, yesterday, today, and tomorrow. *Ann Rev Med* 36:185–194.

29. Nahmias et al. 1990. *Scand J Infect Dis* Suppl 69:19–36, table II.

30. Nahmias et al. 1990. *Scand J Infect Dis* Suppl 69:19–36, tables I, II.

31. Drew WL, Mintz L, Miner RC, Sands M, Ketterer B. 1981. Prevalence of cytomegalovirus infection in homosexual men. *J Infect Dis* 143:188–192.

32. Buimovici-Klein E, Lange M, Ong KR, Grieco MH, Cooper LZ. 1988. Virus isolation and immune studies in a cohort of homosexual men. *J Med Virol* 25:371–385.

33. Mildvan D, Gelb AM, William D. 1977. Venereal transmission of enteric pathogens in male homosexuals. Two case reports. *JAMA* 238:1387–1389.

34. Centers for Disease Control. 1989. Update: acquired immunodeficiency syndrome—United States, 1981–1988. *MMWR* 38:229–236.

35. Schmerin MJ, Gelston A, Jones TC. 1977. Amebiasis: An increasing problem among homosexuals in New York City. *JAMA* 238:1386–1387.

36. Vaisrub. 1977. *JAMA* 238:1402.

37. Most H. 1968. Manhattan: "A tropic isle?" *Am J Trop Med Hyg* 17:333–354; Kean BH. 1976. Venereal amebiasis. *NY State J Med* 76:930–931.

38. Drusin LM, Genvert G, Topf-Olstein B. 1976. Shigellosis: Another sexually transmitted disease. *Br J Vener Dis* 52:348–350.

39. Dritz SK, Bach AF. 1974. *Shigella enteritis* venereally transmitted. *New Engl J Med* 291:1194; Dritz SK. 1980. Medical aspects of homosexuality. *New Engl J Med* 302:463–464.

40. Schmerin et al. 1977. *JAMA* 238:1386–1387.

41. Mildvan et al. 1977. *JAMA* 238:1387–1389.

42. Kacker PP. 1973. A case of *Giardia lamblia* proctitis presenting in a VD clinic. *Br J Vener Dis* 49:318–319; Meyers JD, Kuharic JA, Holmes KK. 1977. *Giardia lamblia* infection in homosexual men. *Br J Vener Dis* 53:54–55.

43. Miller B, Stansfield SK, Zack MM, et al. 1984. The syndrome of unexplained generalized lymphadenopathy in young men in New York City. *JAMA* 251:242–246.

44. Sonnabend JA. 1984. The etiology of AIDS. *AIDS Res* 1:1–15; Sonnabend JA. 1989. AIDS; An explanation for its occurrence among homosexual men. In: Ma P, Armstrong D, eds. 1989. *AIDS and Infections of Homosexual Men*. 2d ed. Boston: Butterworths, 449–470.

45. Marmor et al. 1982. *Lancet* i:1083; Darrow WW, Echenberg DF, Jaffe HW, et al. 1987. Risk factors for human immunodeficiency virus (HIV) infections in homosexual men. *Am J Public Health* 77:479–483.

46. Root-Bernstein RS. 1990a. Do we know the cause(s) of AIDS? *Persp Biol Med* 33:480–500; Root-Bernstein RS. 1990b. Non-HIV immunosuppressive factors in AIDS: A multifactorial, synergistic theory of AIDS aetiology. *Res Immunol* 141:815–838; Son-

nabend. 1984. *AIDS Res* 1:1–15; Sonnabend. 1989. In: Ma and Armstrong, eds. 1989. *AIDS*, 449–470.

47. Hansen H, ed. 1964–1968. *The World Almanac*. New York: New York World Telegram; Delury GE, ed. 1977–1978. *The World Almanac*. New York: Newspaper Enterprise Association.

48. Johnson O, ed. 1982–1990. *Information Please Almanac*. Boston: Houghton Mifflin; Golenpaul A, ed. 1975. *Information Please Almanac*. New York: Houghton Mifflin, 732.

49. Velimirovic B. 1987. AIDS as a social phenomenon. *Social Sci Med* 25:541–552; Cohen J, Alexander P, Wofsy C. 1988. Prostitutes and AIDS: Public policy issues. *AIDS Public Pol J* 3:16–22; Waldorf D, Murphy S, Lauderback D, Reinarman C, Marotta T. 1990. Needle sharing among male prostitutes: Preliminary findings of the Prospero Project. *J Drug Issues* 20:309–334; McKusik L, Horstman W, Coates TJ. 1985. AIDS and sexual behavior reported by gay men in San Francisco. *Am J Public Health* 75:495–496.

50. Grmek M. 1990. RC Maulitz, J. Duffin, trans. *History of AIDS*. Princeton: Princeton University Press, 168.

The Explosion of Drug Abuse

1. Centers for Disease Control. 1989. Update: Acquired Immunodeficiency Syndrome—United States, 1981–1988. *MMWR* 38:229–236.

2. Duesberg PH. 1992. The role of drugs in the origin of AIDS. *Biomed Pharmacother* 46:3–15.

3. Grmek M. 1990. RC Maulitz, J Duffin, trans. *History of AIDS*. Princeton: Princeton University Press, 166–167.

4. Bewley TH, Ben-Arie O, James IP. 1968. Morbidity and mortality from heroin dependence. 1: Survey of addicts known to Home Office. *Br Med J* 1:725–732.

5. Golenpaul A, ed. 1975. *Information Please Almanac*. Boston: Houghton Mifflin, 732.

6. Golenpaul, ed. 1975. *Information Please Almanac*, 732.

7. Hansen H, ed. 1964–1979. *The World Almanac and Book of Facts*. New York: New York World Telegram; Kristal AR. 1986. The impact of the acquired immunodeficiency syndrome on patterns of premature death in New York City. *JAMA* 255:2306–2310.

8. Hansen, ed. 1964–1979. *The World Almanac*; Johnson O, ed. 1988–1990. *Information Please Almanac*. Boston: Houghton Mifflin. Dole VP. 1973. Heroin addiction—an epidemic disease. *The*

Harvey Lectures, 1971–1972. Academic Press: New York, 199–214.

9. Shannon E, Booth C, Fowler D, McBride M. 1990. A losing battle. *Time* 3 Dec, 44.

10. Bureau of Justice Statistics. 1981. US Department of Justice, Washington DC; Bureau of Justice Statistics. 1991. US Department of Justice, Washington DC.

11. Johnson, ed. 1988–1990. *Almanac.*

12. USDHHS (United States Department of Health and Human Services). 1985. *Patterns and Trends in Drug Abuse: A National and International Perspective.* Washington, D.C.: Division of Epidemiology and Statistical Analysis, National Institute on Drug Abuse.

13. USDHHS. 1985. *Drug Abuse: A National and International Perspective.*

14. USDHHS. 1988. Statistical series quarterly report: Data from the Drug Abuse Warning Network (DAWN). Annual data 1987. Series I, no. 7. Rockville, MD: National Institute on Drug Abuse.

15. Dans PE, Matricciani RM, Otter SE, Reuland DS. 1990. Intravenous drug abuse and one academic health center. *JAMA* 263:3173–3176.

16. Bewley et al. 1968. *Br Med J* 1:725–732; Louria DB, Hensle T, Rose J. 1967. The major medical complications of heroin addiction. *Ann Intern Med* 67:1–22; Cherubin CE. 1967. The medical sequelae of narcotic addiction. *Ann Intern Med* 67:23–33.

17. Hussey H, Katz S. 1950. Infections resulting from narcotic addiction. *Am J Med* 9:186; Briggs JH, McKerron CG, Souhami RL, Taylor DJE, Andrews HJ. 1966. Severe pneumonia in heroin addicts. *Lancet* ii:964; Briggs JH, McKerron CG, Souhami RL, Taylor DJE, Andrews HJ. 1967. Severe systemic infections complicating "mainline" heroin addiction. *Lancet* ii:1227–1231; Cherubin. 1967. *Ann Intern Med* 67:23–33; Cherubin CE, Brown J. 1968. Systemic infections in heroin addicts. *Lancet* i:298–299; Louria et al. 1967. *Ann Intern Med* 67:1–22.

18. Firooznia H, Seliger G, Abrams RM, Valensi V, Shamoun J. 1973. Disseminated extrapulmonary tuberculosis in association with heroin addiction. *Radiol* 109:291–296; Centers for Disease Control. 1988. Hepatitis A among drug abusers. *MMWR* 37:279–300, 305.

19. Sapira JD. 1968. The narcotic addict as a medical patient. *Am J Med* 45:555–588; Geller SA, Stimmel B. 1973. Diagnostic confusion from lymphatic lesions in heroin addicts. *Ann Intern Med* 78:703; Miller B, Stansfield SK, Zack MM, et al. 1984. The syndrome of unexplained generalized lymphadenopathy in young men in New York City. *JAMA* 251:242–246.

20. Brown SM, Stimmel B, Taub RN, Kochwa S, Rosenfield RE. 1974. Immunologic dysfunction in heroin addicts. *Arch Intern Med* 134:1001–1006.

21. Gambara SE, Clarke JK. 1976. Comments on dietary intake of drug-dependent persons. *J Am Diet Assoc* 68:155–157; Aylett A. 1978. Some aspects of nutritional state in "hard" drug addicts. *Br J Addict* 73:77–81; Nakah AE, Frank O, Louria DB, Quinones MA, Baker H. 1979. A vitamin profile of heroin addiction. *Am J Public Health* 69:1058–1060.

22. McDonough RJ, Madden JJ, Falek, A et al. 1980. Alterations of T and null lymphocyte frequency in the peripheral blood of human opiate addicts: *in vivo* evidence for opiate receptor sites on T-lymphocytes. *J Immunol* 125:2539.

23. Centers for Disease Control. 1981. Annual Summary, 1980. *MMWR* 29: passim; CDC. 1989. Summary of notifiable diseases, United States, 1988. *MMWR* 37: passsim.

24. Buehler JW, Devine OJ, Berkelman RL, Chevarley FM. 1990. Impact of the human immunodeficiency virus epidemic on mortality trends among young men, United States. *Am J Public Health* 80:1080; National Center for Health Statistics. 1989. *Monthly Vital Stat Report*. Hyattsville, MD: Department of Health and Human Services, Public Health Service, Publication No. (PMS) 89–1120; Selwyn PA, Hartel D, Wasserman W, Drucker E. 1989. Impact of the AIDS epidemic on morbidity and mortality among intravenous drug users in a New York City methadone maintenance program. *Am J Public Health* 79:1358.

25. Haverkos HW, Lange WR. 1990. Serious infections other than human immunodeficiency virus among intravenous drug abusers. *J Infect Dis* 161:894–902.

26. Fricker HS, Segal S. 1978. Narcotic addiction, pregnancy, and the newborn. *Am J Dis Child* 132:360–366.

27. Anonymous. 1989. The no-parent child. [Editorial] *NY Times*, 24 Dec.

CHAPTER 9 SOURCES OF ACQUIRED IMMUNOSUPPRESSION IN HETEROSEXUALS: A COMPARISON OF NORTH AMERICANS AND EQUATORIAL AFRICANS

AIDS Among Heterosexuals in African and Western Nations

1. Gould SJ. 1987. AIDS is natural. *NY Times Mag*, 19 Apr, 32.

2. World Health Organization. 1991. Acquired immunodeficiency syndrome. *Weekly Epidemiol Rec* 66:passim.

3. De Cock KM, Barrere B, Diaby L, et al. 1990. AIDS—the leading cause of adult death in the West African city of Abidjan, Ivory Coast. *Science* 249:793–796.

4. Noble KB. 1990. Nigeria is spared the worst of AIDS? But for how long? the experts ask. *NY Times*, 18 Nov, 15; Piot P, Tezzo R. 1990. The epidemiology of HIV and other sexually transmitted infections in the developing world. *Scand J Infect Dis* Suppl 69:89–97; Quinn C, Piot P, McCormick JB, et al. 1987. Serologic and immunologic studies in patients with AIDS in North America and Africa. *JAMA* 257:2617–2621; Mann JM, Francis H, Quinn T, et al. 1986. Surveillance for AIDS in a Central African city: Kinshasa, Zaire. JAMA 255:3255–3259.

The Unusual Immunosuppressive Risks of African Heterosexuals

1. Clumeck N, Vandeperre P, Carael M, Rouvroy D, Nzaramba D. 1985. Heterosexual promiscuity among African patients with AIDS. *New Engl J Med* 313:182; Hrdy DB. 1987. Cultural practices contributing to the transmission of human immunodeficiency virus in Africa. *Rev Infect Dis* 9:1109–1119; Noble KB. 1990. Nigeria is spared the worst of AIDS? But for how long? the experts ask. *NY Times*, 18 Nov, 15.

2. Belsey MA. 1983. Epidemiologic aspects of infertility. In: Holmes KK, Mardh P-A, eds. 1983. *International Perspectives on Neglected Sexually Transmitted Diseases*. Washington: Hemisphere Publishing Co, 269–300.

3. Hrdy. 1987. *Rev Infect Dis* 9:1109–1119.

4. Noble. 1990. *NY Times*, 18 Nov, 15.

5. Data for figure are from the references listed in the first two sections of this chapter and, in addition: Schmutzhard E, Fuchs D, Hengster P, Hausen A, Hofbauer J, Pohl P, et al. 1989. Retroviral infections (HIV-1, HIV-2, and HTLV-I) in rural northwestern Tanzania. Clinical findings, epidemiology, and association with infections common in Africa. *Am J Epidemiol* 130:309–318; Sixbey JW, Shirley P, Chesney PJ, Buntin DM, Resnick L. 1989. Detection of a second widespread strain of Epstein-Barr virus. *Lancet* ii:761–763; Delaporte E, Louwagie J, Peeters M, et al. 1991. Evidence of HTLV-II infection in Central Africa. *AIDS* 5:771–772; Lang DJ, Garruto RM, Gajdusek DC. 1977. Early acquisition of cytomegalovirus and Epstein-Barr virus antibody in several isolated Melanesian populations. *Am J Epidemiol* 105:480–487; Nzila N, Laga M, Thiam MA, et al. 1991. HIV and other sexually transmitted diseases among female prostitutes in Kinshasa. *AIDS*

5:715–721; Saxinger WC, Blattner WA, Levine PH, et al. 1984. Human T-cell leukemia virus (HTLV-1) antibodies in Africa. *Science* 225:1473–1476; Ward M, Bailey R, Lesley A, Kajbaf M, Robertson J, Mabey D. 1990. Persisting inapparent chlamydial infection in a tracoma endemic community in the Gambia. *Scand J Infect Dis* Suppl 69:137–148.

6. Osoba AO. 1981. Sexually transmitted diseases in tropical Africa: A review of the present situation. *Br J Vener Dis* 57:89–94; Donegan EA. 1985. Epidemiology of gonococcal infection. In: Brooks GF, Donegan EA, eds. *Gonococcal Infection*. London: Edward Arnold, 181–198.

7. Bentsi C, Klufio CA, Perine PL, et al. 1985. Genital infections with *Chlamydia trachomatis* and *Neisseria gonorrhoeae* in Ghanian women. *Genitourin Med* 61:48–50.

8. Monnickendam MA. 1988. Chlamydial genital infections. In: Wright DJM, ed. 1988. *Immunology of Sexually Transmitted Diseases*. Drodrecht: Kluwer, 117–162.

9. Nahmias AJ, Lee FK, Beckman-Nahmias S. 1990. Sero-epidemiological and -sociological patterns of herpes simplex virus infection in the world. *Scand J Infect Dis* Suppl 69:19–36.

10. Clumeck N, Hermans, P, De Wit S, Lee F, Van de Perre P, Nahmias A. 1987. *Herpes Type II (HSV-2): A Possible Co-factor of the HIV Infection Among Central African Heterosexual Patients*. New York: ICAAC (Abstract).

11. Wiktor KSZ, Piot P, Mann JM, et al. 1990. Human T cell lymphotropic virus type I (HTLV-I) among female prostitutes in Kinshasa, Zaire. *J Infect Dis* 161:1073–1077.

12. Tosswill JHC, Ades AE, Peckham C, Mortimer PP, Weber JN. 1990. Infection with human T cell leukemia/lymphoma virus type I in patients attending an antenatal clinic in London. *Br Med J* 301:95–96; Centers for Disease Control. 1991. *HIV/AIDS Surveillance Report* July:1–18.

13. Zuckerman AJ. 1988. Viral hepatitis. In: Wright, ed. 1988. *Sexually Transmitted Diseases*, 51–71.

14. Harries AD. 1990. Tuberculosis and human immunodeficiency virus infection in developing countries. *Lancet* 335:387–390.

15. Harries. 1990. *Lancet* 335:387–390.

16. De Cock KM, Barrere B, Diaby L, et al. 1990. AIDS—the leading cause of adult death in the West African city of Abidjan, Ivory Coast. *Science* 249:793–796.

17. Eriki PP, Okwera A, Aisu T, Morrissey AB, Ellner JJ, Daniel TM. 1991. The influence of human immunodeficiency virus infection on tuberculosis in Kampala, Uganda. *Am Rev Respir Dis*

143:185–187; Perriens JH, Colebunders Rl, Karahunga C, et al. 1991. Increased mortality and tuberculosis treatment failure rate among human immunodeficiency virus (HIV) seropositive compared with HIV seronegative patients with pulmonary tuberculosis treated with "standard" chemotherapy in Kinshasa, Zaire. *Am Rev Respir Dis* 144:750–755.

18.   Harries. 1990. *Lancet* 335:387–390.

19.   Greenberg AE, Nguen-Dinh P, Mann JM, et al. 1988. The association between malaria, blood transfusions, and HIV seropositivity in a pediatric population in Kinshasa, Zaire. *JAMA* 259:545–549.

20.   Alonso PL, Lindsay SW, Armstrong JR, et al. 1991. The effect of insecticide-treated bed nets on mortality of Gambian children. *Lancet* 337:1499–1502; Sudre P, Breman JH, Koplan JP. 1990. Delphi survey of malaria mortality and drug resistance in Africa. *Lancet* 335:722; Nevill CG. 1990. Malaria in sub-Saharan Africa. *Soc Sci Med* 31:667–669; Foster SO, Shepperd J, Davis JH, Agle AN. 1990. Working with African nations to improve the health of their children. *JAMA* 263:3303–3305.

21.   Greenberg et al. 1988. *JAMA* 259:545–549; Desowitz RS. 1991. *The Malaria Capers*. New York: Norton.

22.   Mathé G. 1992. Is the AIDS virus responsible for the disease? *Biomed Pharmachother* 446:1–2; note 3, chapter 2.5.

23.   Terry RJ, Hudson KM. 1983. Secondary immunodeficiencies in protozoan and helminth infections. In: Chandra RK, ed. 1983. *Primary and Secondary Immunodeficiency Disorders*. Edinburgh: Churchill Livingstone, 219–231; Gilbreath MJ, Pavanand K, MacDermott RP, Wells RA, Ussery MA. 1983. Characterization of cold reactive lymphocytotoxic antibodies in malaria. *Clin Exp Immunol* 51:232–238.

24.   Biggar RJ, Gigase PL, Melbye M, et al. 1985. ELISA HTLV retrovirus antibody reactivity associated with malaria and immune complexes in healthy Africans. *Lancet* ii:520–523; Volsky DJ, Wu YT, Stevenson M, et al. 1986. Antibodies to HTLV-III/LAV in Venezuelan patients with acute malarial infections. *New Engl J Med* 314:647–648; Greenberg AE, Schable CA, Sulzer AJ, et al. 1986. Evaluation of serological cross-reactivity between antibodies to *Plasmodium* and HTLV-III/LAV. *Lancet* ii:247–249.

25.   Salmeron G et al. 1983. Immunosuppressive potential of antimalarials. *Am J Med* 75:19–24.

26.   Foster et al. 1990. *JAMA* 263:3303–3305.

27.   Foster et al. 1990. *JAMA* 263:3303–3305.

28.   Oettle AG. 1962. Geographical and racial differences in the frequencey of Kaposi's sarcoma as evidence of environmental or

genetic causes. *Unio Internationalis Contra Cancrum, Acta* 18:330–363; Davies JNP, Lothe F. 1962. Kaposi's sarcoma in African children. *Unio Internationalis Contra Cancrum, Acta* 18:394–399.

29. Templeton AC. 1972. Studies in Kaposi's sarcoma. Postmortem findings and disease patterns in women. *Cancer* 30:854–867.

30. Young LS, Yao QY, Rooney CM, et al. 1987. New type B isolate of Epstein-Barr virus from Burkitt's lymphoma and normal individuals in endemic areas. *J Gen Virol* 68:2853–2862.

31. Yeager AS. 1983. Protozoan and helminth infections. In: Remington FJS, Klein JO, eds. 1983. *Infectious Diseases of the Fetus and Newborn Infant*. 2d ed, Philadelphia: Saunders, 555–569.

32. Anonymous. 1992. Drug use in developing countries. *Lancet* 339:18 Jan.; Hardon A, van der Geest S, Geerlin H, le Grand A. 1991. *The Provision and Use of Drugs in Developing Countries*. Amsterdam: Health Action International.

33. Hrdy. 1987. *Rev Infect Dis* 9:1109–1119; Carswell JW. 1983. Injection abscesses. *Ugandan Med J* 5:16–20.

34. Foster et al. 1990. *JAMA* 263:3303–3305.

35. WHO Group of Experts. 1972. *Nutritional Anemias*. Geneva: World Health Organization Technical Report Series, No. 503, 1–27; Serdula M, Seward J. 1987. Diet, malnutrition, and mortality. In: *Proceedings of Seminar on Mortality and Society in Sub-Saharan Africa, Yaounde, Cameroun, October 19–23, 1987*, of the International Union for the Scientific Study of Population, Liege, Belgium; Kasongo Project Team. 1983. Anthropometric assessment of young children's nutritional status as an indicator of subsequent risk of dying. *J Trop Pediatr* 29:69–75.

36. Isliker H, Schürch B, eds. 1981. *The Impact of Malnutrition on Immune Defense in Parasitic Infestation*. Bern: Hans Huber; Hershko C, Karsal A, Eylon L, Izak G. 1970. The effect of chronic iron deficiency on some biochemical functions of human hemopoietic tissue. *Blood* 36:321–329.

37. Sommer A, Hussaini G, Tarwotjo I, Susanto D. 1983. Increased mortality in children with mild vitamin A deficiency. *Lancet* i:585–588; Stoltzfus RJ, Jalal F, Harvey PW, Nesheim MC. 1989. Interactions between vitamin A deficiency and Plasmodium berghei infection in the rat. *J Nutr* 119:2030–2037.

38. Berkley SF, Widy-Wirski R, Okware SI, et al. 1989. Risk factors associated with HIV infection in Uganda. *J Infect Dis* 160:22–30.

39. Fallis G, Hilditch J. 1989. A comparison of seasonal variation in birthweights between rural Zaire and Ontario. *Can J Public Health* 80:205–208.

40. Ryder RW, Nsa W, Hassig SE, et al. 1989. Perinatal transmission of the human immunodeficiency virus type 1 to infants of seropositive women in Zaire. *New Engl J Med* 320:1637–1642.

41. Stearns S. 1992. Nigerians moving higher up world heroin hierarchy. *Lansing State J*, 1 Jan.

42. Hrdy. 1987. *Rev Infect Dis* 9:1109–1119.

43. McNeill WH. 1976. *Plagues and Peoples*. Garden City, NY: Anchor Press.

44. Dodge CP. 1990. Health implications of war in Uganda and Sudan. *Soc Sci Med* 31:691–698; Meheus AZ. 1984. Practical approaches in developing nations. In: Holmes KK, Mardh P-A, Sparling PF, Wiener PJ, eds. *Sexually Transmitted Diseases*. New York: McGraw-Hill, 998–1008. Toole MJ, Waldman RJ. 1990. Prevention of excess mortality in refugee and displaced populations in developing countries. *JAMA* 263:3296–3302.

45. Hunter SS. 1990. Orphans as a window on the AIDS epidemic in Sub-Saharan Africa: Initial results and implications of a study in Uganda. *Soc Sci Med* 31:681–690.

Europe and America Are Not Africa

1. Most H. 1968. Manhattan: "A tropic isle?" *Am J Trop Med Hyg* 17:333–354; Kean BH. 1976. Venereal amebiasis. *NY State J Med* 76:930–931.

2. Quinn C, Piot P, McCormick JB, et al. 1987. Serologic and immunologic studies in patients with AIDS in North America and Africa. *JAMA* 257:2617–2621.

3. Centers for Disease Control. 1991. *HIV/AIDS Surveillance Report* July: 1–18.

Heterosexual AIDS in North America and Europe

1. Scheer R. 1987. AIDS threat to all—how serious? *LA Times*, 14 Aug, pt 1, p 1.

2. Farber C. 1992. Fatal distraction. *SPIN* June:38; World Health Organization. 1991. Acquired immunodeficiency syndrome. *Weekly Epidemiol Rec* 66: passim.

3. Haney DO. 1987. The odds on AIDS. *Lansing State J*, 2 Oct, D1.

4. Booth W. 1988. Heterosexual AIDS: Setting the odds. *Science* 240:597; Hearst N, Hulley S. 1988. Preventing the heterosexual spread of AIDS: Are we giving our patients the best advice? *JAMA* 258:2428–2432.

5. Wilson C, Painter K. 1990. Counting the AIDS losses after 9 years. *USA Today*, 19 June, D1–D2.

6. Palca J. 1992. The case of the Florida dentist. *Science* 255:392–394.

AIDS Risks of Heterosexuals in North America and Europe

1. Miller S. *The Good Mother*. New York: Harper & Row, 1986, 166.
2. Hollander X. 1978. Call me madam. *Penthouse* June:46.
3. Hollander X. 1974. Call me madam. *Penthouse* October:122.
4. Bolling DR, Voeller B. 1987. AIDS and heterosexual anal intercourse. *JAMA* 258:474; Bolling DR. 1977. Prevalence, goals and complications of heterosexual anal intercourse in a gynecologic population. *J Reprod Med* 19:120–124; Melbye M, Biggar RJ. 1992. Interactions between persons at risk for AIDS and the general population in Denmark. *Am J Epidemiol* 135:593–602.
5. MacDonald NE, Wells GA, Fisher WA, et al. 1990. High-risk STD-HIV behavior among college students. *JAMA* 263:3155–3159. [Erratum appears in *JAMA* 264:1661.]
6. Rabinowitz M, Bassan I, Robinson MJ. 1988. Sexually transmitted cytomegalovirus proctitis in a woman. *Am J Gastroenterol* 83:885–887.
7. European Study Group on AIDS. 1989. Risk factors for male to female transmission of HIV. *Br Med J* 298:411–415.
8. Melbye M, Ingerslev J, Biggar RJ, et al. 1985. Anal intercourse as a possible factor in heterosexual transmission of HTLV-III to spouses of hemophiliacs. *New Engl J Med* 312:857.
9. Maddox J. 1992. Media make AIDS wishes come true. *Nature* 358:13; Hodgkinson N. 1992. AIDS causation. *Nature* 358:447.
10. Padian NS, Shiboski SC, Jewell NP. 1990. The effect of number of exposures on the risk of heterosexual HIV transmission. *J Infect Dis* 161:883–887; European Study Group on Heterosexual Transmission of HIV. 1992. HIV transmission from men to women. *Br Med J* 304:811.
11. Evans BA, McCormack SM, Bond RA, Macrae KD, Thorp RW. 1988. Human immunodeficiency virus infection, hepatitis B virus infection, and sexual behaviour of women attending a genitourinary medicine clinic. *Br Med J* 296:473–475.
12. Wallace J. 1989. Case presentations of AIDS in the United States. In: Ma P, Armstrong D, eds. 1989. *AIDS and Infections of Homosexual Men*. 2d ed. Boston: Butterworths, 285–295.
13. Naz RK, Ellauri M, Phillips TM, Hall J. 1990. Antisperm antibodies in human immunodeficiency virus infection: Effects on fertilization and embryonic development. *Biol Reprod* 42:859–868.

14. Lorian V. 1988. AIDS, anal sex, and heterosexuals. *Lancet* i:1111.

15. Bolling and Voeller. 1987. *JAMA* 258:474.

16. Centers for Disease Control. 1992. *HIV/AIDS Surveillance Report* Jan:1–18.

17. Peterman TA, Stoneburner RL, Allen JR, Jaffe HW, Curran JW. 1988. Risk of human immunodeficiency virus transmission from heterosexual adults with transfusion-associated infections. *JAMA* 259:55–58.

18. Peterman et al. 1988. *JAMA* 259:57.

19. Kreiss JK, Kitchen LW, Prince HE, Kasper CK, Essex M. 1985. Antibody to human T-lymphotropic virus type III in wives of hemophiliacs: Evidence for heterosexual transmission. *Ann Intern Med* 102:623–626.

20. Centers for Disease Control. 1992. Jan: 1–18.

21. Centers for Disease Control. 1992. Jan: 1–18.

22. Centers for Disease Control. 1992. Jan: 1–18; Lambert B. 1987. AIDS spread seen in same patterns. *NY Times*, 11 Oct, pt 1, p 1.

23. Appleman ME, Marshall DW, Brey RL, et al. 1988. Cerebrospinal fluid abnormalities in patients without AIDS who are seropositive for the human immunodeficiency virus. *J Infect Dis* 158:193–199.

24. Sumaya CV, Boswell RN, Ench Y, et al. 1986. Enhanced serological and virological findings of Epstein-Barr virus in patients with AIDS and AIDS-related complex. *J Infect Dis* 154:864–870.

25. Anonymous. 1991. Air Force captain sentenced in drug case. *NY Times*, 4 Aug, sec 1, p 15; Yost P. 1991. Court: Military has power to ban gays. *Lansing State J*, 10 Dec, A2; Egan T. 1992. Dismissed from army as lesbian, colonel will fight homosexual ban. *NY Times*, 31 May, pt 1, p 14.

26. Scheer R. 1987. AIDS threat to all—how serious? *LA Times*, 14 Aug, pt 1, p 1.

27. Bolling, Voeller. 1987. *JAMA* 258:474; Bolling DR. 1976. Heterosexual anal intercourse—an illustrative case. *J Fam Pract* 3:557–558.

28. Earl WL. 1990. Married men and same sex activity: A field of study on HIV risk among men who do not identify as gay or bisexual. *J Sex Marital Ther* 16:251–257; Waldorf D, Murphy S, Lauderback D, et al. 1990. Needle sharing among male prostitutes. *Drug Issues* 20:309–334.

29. Nemechek PM. 1991. Anabolic steroid users—another potential risk group for HIV infection. *New Engl J Med* 325:357; Sklarek HM, Manovani RP, Erens E, Heisler D, Niederman MS, Fein AM. 1984. AIDS in a bodybuilder using anabolic steroids. *New Engl J*

Med 311:1701; Tabor MBW. 1992. Settlement in suit on HIV-tainted transfusion. *NY Times*, 30 Aug, pt 1, p 19.

30. Lambert. 1987. *NY Times*, 11 Oct, pt 1, p 1.

31. Strang J, Gossop M, Griffiths P, Powis B. 1992. HIV among south London heroin users in 1991. *Lancet* 339:1060–1061.

32. Rosenblum LS, Hadler SC, Castro KG, Lieb S, Jaffe HW. 1990. Heterosexual transmission of hepatitis B virus in Belle Glade, Florida. *J Infect Dis* 161:407–411.

33. Chandra RK, ed. 1983. *Primary and Secondary Immunodeficiency Disorders*. Edinburgh: Churchill Livingstone, 187.

34. Centers for Disease Control. 1991. July: 1–18.

35. Williamson D. 1988. Spreading the word: Pediatrician Karl Hammonds crusades for the health of inner-city youth. *Princeton Alumni Weekly* 12 Oct, 27–33, see 27.

36. Williamson. 1988. *Princeton Alumni Weekly* 12 Oct:27.

CHAPTER 10   ALTERNATIVE HYPOTHESES FOR EXPLAINING AIDS

Elaborating Possibilities

1. Root-Bernstein RS. 1989. *Discovering*. Cambridge MA: Harvard University Press; Root-Bernstein RS, McEachron DL. 1982. Teaching theories: The evolution-creation controversy. *Am Biol Teacher* 44:413–420.

HIV Is Necessary and Sufficient to Cause AIDS

1. Lusso P, Ensoli B, Markham PD, et al. 1990. Productive dual infection of human CD4+ T lymphocytes by HIV-1 and HHV-6. *Nature* 337:370–373.

2. Balter M. 1991. Montagnier pursues the Mycoplasma-AIDS link. *Science* 251:271.

3. Burkett E. 1990. HIV: Not guilty? *Tropic (Miami Herald)*, 23 Dec, 12–17.

4. Associated Press. 1992. New AIDS-like illness comes to light at conference. *Lansing State J*, 22 July, 3A.

HIV Is Necessary But Requires Cofactors

1. Shiokawa Y. 1988. [Depressed cellular immunity in prostitutes and male homosexuals predisposes to HIV.] *Nippon Saikingaku Zasshi. Jap J Bacteriol* 43:559–564.

2. Pifer LL, Wang Y-F, Chiang TM, Ahokas R, Woods DR, Joyner

RE. 1987. Borderline immunodeficiency in male homosexuals: Is life-style contributory? *South Med J* 80:687–697.

3. Montagnier L, Bernman D, Guetard D, Blanchard A, et al. 1990. Inhibition de l'infectiosité de souches prototypes du VIH par des anticorps dirigés contre une séquence peptidique de mycoplasme. *Comptes rendus Acad Sci Paris* 311:425–430; Lo S-C, Hayes MM, Wang RY-H, Pierce PF, Kotani H, Shih JW-K. 1991. Newly discovered mycoplasma isolated from patients infected with HIV. *Lancet* 338:1415–1418; Chowdhury IH, Manakata T, Koyanagi Y, Kobayashi S, Arai S, Yamamoto N. 1990. Mycoplasma can enhance HIV replication in vitro: A possible cofactor responsible for the progression of AIDS. *Biochem Biophys Res Commun* 170:1365–1370; Dawson MS, Wang R, Hayes M, et al. 1991. Detection and isolation of *Mycoplasma fermentens* from urine of patients with AIDS (abstract). *91st General Meeting of the American Society for Microbiology 1991.* 5–9 May, Dallas. Washington DC: American Society for Microbiology; Black FT. 1983. *Ureaplasma urealyticum* and *Mycoplasma hominis*. In: Holmes KK, Mardh P-A, eds. 1983. *International Perspectives on Neglected Sexually Transmitted Diseases.* Washington DC: Hemisphere Publishing Co, 37–62.

4. Balter M. 1991. Montagnier pursues the Mycoplasma-AIDS link. *Science* 251:271.

5. Lo S-C, Wear DJ. 1990. Mycoplasmal agents. *Policy Review* 54:76–77.

6. Zack JA, Cann AJ, Lugo JP, Chen IS. 1988. HIV-1 production from infected peripheral blood T cells after HTLV-I induced mitogenic stimulation. *Science* 240:1026–1029; Lusso P, Lori F, Gallo RC. 1990. CD4-independent infection by human immunodeficiency virus type 1 after phenotypic mixing with human T-cell leukemia viruses. *J Virol* 64:6341–6344.

7. Holmberg SD, Steward JA, Gerber R, et al. 1988. Prior herpes simplex virus type 2 infection as a risk factor for HIV infection. *JAMA* 259:1048–1050.

8. Lusso P, Ensoli B, Markham PD, et al. 1990. Productive dual infection of human CD4+ T lymphocytes by HIV-1 and HHV-6. *Nature* 337:370–373.

9. Seto E, Yen TS, Peterlin BM, Ou JH. 1988. Trans-activation of the human immunodeficiency virus long terminal repeat by the hepatitis B virus X protein. *Proc Natl Acad Sci USA* 85:8286–8290.

10. Montagnier L, Gruest J, Chameret S, et al. 1984. Adaptation of

lymphadenopathy associated virus (LAV) to replication in EBV-transformed B lymphoblastoid cell lines. *Science* 225:63–66; Buimovici-Klein E, Lange M, Ong KR, Grieco MH, Cooper LZ. 1988. Virus isolation and immune studies in a cohort of homosexual men. *J Med Virol* 25:371–385; Scully TB, Apolloni A, Hurren L, Moss DJ, Cooper DA. 1990. Coinfection with A- and B-type Epstein-Barr virus in human immunodeficiency virus-positive subjects. *J Infect Dis* 162:643–648.

11. Biggar RJ, Anderson HK, Ebbesen P, Melbye M, Goedert JJ. 1983. Seminal fluid excretion of cytomegalovirus related to immunosuppression in homosexual men. *Br Med J* 286:2010–2012; Drew WL, Mills J, Levy J, et al. 1985. Cytomegalovirus infection and abnormal T-lymphocyte subset ratios in homosexual men. *Ann Intern Med* 103:61–63; Buimovici-Klein et al. 1988. *J Med Virol* 25:371–385.

12. Kreiss JK, Coombs R, Plummer F, et al. 1989. Isolation of human immunodeficiency virus from genital ulcers in Nairobi prostitutes. *J Infect Dis* 160:380–384.

13. Buimovici-Klein et al. 1988. *J Med Virol* 25:371–385.

14. Fernandez-Cruz E, Fernandez AM, Gutierrez C, et al. 1988. Progressive cellular immune impairment leading to development of AIDS: Two year prospective study of HIV infection in drug addicts. *Clin Exp Immunol* 72:190–195.

15. Haverkos HW. 1988. Kaposi's sarcoma and nitrite inhalants. In: Bridge TP, ed. 1988. *Psychological, Neuropsychiatric, and Substance Abuse Aspects of AIDS*. New York: Raven Press, 165–172.

16. Donohoe RM, Falek AA. 1988. Neuroimmunomodulation by opiates and other drugs of abuse: Relationship to HIV infection and AIDS. In: Bridge et al, eds. 1988. *Aspects of AIDS*, 145–158. See also Arora PK, Fride E, Petitto J, Waggie K, Skolnick P. 1990. Morphine-induced immune alterations *in vivo*. *Cell Immunol* 126:343–353.

17. Byrd R. 1990. Test-tube study: Cocaine accelerates growth of AIDS virus. *Lansing State J*, 23 Oct, 3A; Squinto SP, Mondal D, Block AL, Prakash O. 1990. Morphine-induced transactivation of HIV-1 LTR in human neurobalstoma cells. *AIDS Res Hum Retroviruses* 6:1163–1168. Des Jarlais DC, Friedman SR, Stoneburner RL. 1988. HIV infection and intravenous drug use: Critical issues in transmission dynamics, infection outcomes, and prevention. *Rev Infect Dis* 10:151–158; Gafoor M. 1990. Alcohol is a co-factor in HIV-transmission and hastens the onset of AIDS. *Nurs Times* 86:14; Peterson PK, Molitor TW, Chao CC, Sharp B.

1990. Opiates and cell-mediated immunity. In: Watson RR, ed. 1990. *Drugs of Abuse and Immune Function*. Boca Raton, FL: CRC Press, 1–18.

18. Sullivan JL, Brewster FE, Brettler DB, et al. 1986. Hemophiliac immunodeficiency: Influence of exposure to factor VIII concentrate, LAV/HTLV-III, and herpes viruses. *J Pediatr* 108:504–510.

19. Lifson AR, Rutherford GW, Jaffe HW. 1988. The natural history of human immunodeficiency virus infection. *J Infect Dis* 158:1360–1367; Peterson et al. 1990. In: Watson, ed. 1990. *Drugs and Immune Function*, 12.

20. Evans AS. 1989. Does HIV cause AIDS? An historical perspective. *J AIDS* 2:107–113; Evans AS. 1982. The clinical illness promotion factor: A third ingredient. *Yale J Biol Med* 55:193–199.

AIDS Is a Multifactorial, Synergistic Disease

1. Benitez-Bribiesca L. 1989. El SIDA. Dogmas e Incertidumbres. *Rev Med IMSS* 27:347; Benitez-Bribiesca L. 1990. La Terapéutica del SIDA. Un reto formidable para la investigatión biomédica. *Rev Med IMSS* 28:9.

2. Zhelev ZH, Raykov Z, Alexiev C. 1990. The possible role of fungal infections in AIDS. *Med Hypotheses* 32:203–206.

3. Mathé G. 1983. Le SIDA et son association avec des tumeurs et virus humains. Cause virale et/ou immnogénetique? *Biomed Pharmacother* 37:153; Mathé G. 1992. Is the AIDS virus responsible for the disease? *Biomed Pharmacother* 46:1–2.

4. Rubin H. 1988. Etiology of AIDS. *Science* 241:1389–1390.

5. Root-Bernstein RS. 1990. Do we know the cause(s) of AIDS? *Persp Biol Med* 33:480–500; Root-Bernstein RS. 1990. Non-HIV immunosuppressive factors in AIDS: A multifactorial, synergistic theory of AIDS aetiology. *Res Immunol* 141:815–838.

6. Benitez-Bribiesca L. 1991. Son en verdad los VIH los agentes causales del SIDA? *Gac Med Mex* 127:75–84.

7. Sonnabend JA. 1984. The etiology of AIDS. *AIDS Res* 1:1–15; Sonnabend JA, Saadoun S. 1984. The acquired immunodeficiency syndrome: A discussion of etiologic hypotheses. *AIDS Res* 1:107–120; Sonnabend JA, Witkin SS, Portilo DT. 1984. A multifactorial model for the development of AIDS in homosexual men. *Ann NY Acad Sci* 837:177–183; Sonnabend JA. 1989. AIDS: An explanation for its occurrence among homosexual men. In: Ma P, Armstrong D, eds. 1989. *AIDS and Infections of Homosexual Men*. 2d ed. Boston: Butterworths, 449–470.

8. Sonnabend. 1989. AIDS. In: Ma and Armstrong, eds. 1989. *AIDS*, 449–470. Also, personal communications.

9. Lee CA, Webster A, Griffiths PD, Kernoff PBA. 1990. Symptomless HIV infection after more than ten years. *Lancet* i:425–426.

10. Hoff C, Peterson RDA. 1989. Does exposure to HLA alloantigens trigger immunoregulatory mechanisms operative in both pregnancy and AIDS? *Life Sciences* 45:iii–ix; Hoff C, Peterson RDA. 1990. Lymphocyte alloantigenic challenge as a potential cofactor in HIV infection and progression to AIDS. *American J Human Biology* 2:419–427; Hoff C, James WC, Hester RB, Nolan P, Peterson RDA. 1991. Signs of cellular immunosuppression correlate with HLA-DR phenotypes in healthy HIV-negative homosexuals: Preliminary findings. *Human Biology* 63:129–135.

11. Harris RE, Langrod J, Hebert JR, Lowinson J. Zang E, Wynder EL. 1990. Changes in AIDS risk behavior among intravenous drug abusers in New York City. *NY State J Med* 90:123–126.

12. Papadopoulos-Eleopulos E. 1988. Reappraisal of AIDS. Is the oxidation induced by the risk factors the primary cause? *Med Hypotheses* 25:151–162.

13. Descotes J. 1988. *Immunotoxicology of Drugs and Chemicals*. Amsterdam: Elsevier, 237.

14. Wu J, Levy EM, Black PH. 1989. 2-Mercaptoethanol and n-acetylcysteine enhance T cell colony formation in AIDS and ARC. *Clin Exp Immunol* 77:7–10.

15. Duesberg PH. 1987. Retroviruses as carcinogens and pathogens. Expectations and reality. *Cancer Res* 47:1199–1226; Duesberg PH. 1988. HIV is not the cause of AIDS. *Science* 241:514–517; Duesberg PH. 1989. Human immunodeficiency virus and acquired immunodeficiency syndrome: Correlation but not causation. *Proc Natl Acad Sci USA* 86:755–764; Duesberg PH. 1990. AIDS: Noninfectious deficiencies acquired by drug consumption and other risk factors. *Res Immunol* 141:5–11; Duesberg PH, Ellison BJ. 1990. Is the AIDS virus a science fiction? Immunosuppressive behavior, not HIV, may be the cause of AIDS. *Policy Review* 53:40–51; Duesberg PH. 1992. The role of drugs in the origin of AIDS. *Biomed Pharmacother* 46:3–15.

16. Cohn DL, Judson FN. 1984. Absence of Kaposi's sarcoma in hemophiliacs with the acquired immunodeficiency syndrome. *Ann Intern Med* 101:401; Hardy AM, Allen JR, Morgan WM, Curran JW. 1985. The incidence rate of acquired immunodeficiency syndrome in selected populations. *JAMA* 253:215–220; Kim HC,

Nahum K, Raska K, et al. 1987. Natural history of acquired immunodeficiency syndrome in hemophilic patients. *Am J Haematol*
24:169–176; Mahir WS, Millard RE, Booth JC, Flute PT. 1988.
Functional studies of cell-mediated immunity in haemophilia and
other bleeding disorders. *Br J Haematol* 69:367–370; Telfer NR,
Matthews JM, Wojnarowska F. 1989. Skin disease in haemophli-
acs with and without antibodies to the human immunodeficiency
virus (HIV): further evidence of altered disease behaviour in different risk groups? *Br J Dermatol* 120:795–799.

AIDS Is Caused By Autoimmunity

1. Root-Bernstein RS. 1990. Multiple-antigen-mediated autoimmunity (MAMA) in AIDS: A possible model for post-infectious autoimmune complications. *Res Immunol* 141:321–340; Andrieu
   JM, Evan P, Venet A. 1986. AIDS and related syndromes as viral-
   induced autoimmune disease of the immune system: An anti-MHC
   II disorder. Therapeutic implications. *AIDS Res* 2:163–174;
   Ziegler JL, Stites DP. 1986. Hypothesis: AIDS is an autoimmune
   disease directed at the immune system and triggered by a lymphotropic retrovirus. *Clin Immunol Immunopathol* 41:305–313; Hoffmann GW. 1990. A response to Duesberg with reference to an
   idiotypic network model of AIDS immunopathogenesis. *Res Immunol* 141:701–709; Morrow WJW, Isenberg DA, Sobol RE,
   Stricker RB, Kieber-Emmons T. 1991. AIDS virus infection and
   autoimmunity: A perspective of the clinical, immunological, and
   molecular origins of the autoallergic pathologies associated with
   HIV disease. *Clin Immunol Immunopathol* 58:163–180.
2. Anonymous. 1991. HIVER's alternative hypothesis on AIDS.
   SCRIP No. 1637:26–27; Brown P. 1991. Conflict rages over alternative AIDS theories. *New Sci* 5 Oct, 9–10; Dalgleish AG, Wilson
   S, Gompels M, et al. 1992. T-cell receptor variable gene products
   and early HIV-1 infection. *Lancet* 339:824–828.
3. Editor. 1992. AIDS: How can a pussy cat kill? *Lancet* 339:839–
   840.
4. Kion TA, Hoffmann GW. 1991. Anti-HIV and anti-anti-MHC antibodies in alloimmune and autoimmune mice. *Science* 253:1138;
   Stott EJ. 1991. Anti-cell antibody in macaques. *Nature* 353:393–
   395.
5. Maddox J. 1991. AIDS research turned upside down. *Nature*
   353:297.
6. Root-Bernstein RS, Hobbs SH. 1991. Homologies between mycoplasma adhesion peptide, CD4 and class II MHC proteins: A

possible mechanism for HIV-mycoplasma synergism in AIDS. *Res Immunol* 142:519–523.

7. Root-Bernstein RS, Hobbs SH. 1993. Sequence similarities between CD4 protein and the proteins of infectious agents associated with AIDS: Possible inducers of lymphocytotoxic autoimmunity. Submitted; Root-Bernstein RS, Hobbs SH. 1993. Sequence similarity between CD4 and CD7 proteins and sperm proteins as possible initiators of lymphocytotoxic autoimmunity. Submitted; Root-Bernstein RS. 1993. Idiotype-antiidiotype immune complexes in AIDS. Submitted.

Implications and Testing of These Alternative Hypotheses

1. Griffin BE. 1989. Burden of proof. *Nature* 338:670.
2. Maddox J. 1991. AIDS research turned upside down. *Nature* 353:297.
3. Palca J. 1991. Duesberg vindicated? Not yet. *Science* 254:376–377.

CHAPTER 11 PREVENTING AND TREATING AIDS: THE FUTURE OF AIDS RESEARCH

How Could So Many Scientists Be So Wrong?

1. Kamen MD. 1985. *Radiant Science, Dark Politics*. Berkeley: University of California Press, pp. 246–247.
2. Root-Bernstein RS. 1989. *Discovering*. Boston: Harvard University Press, pp. 269–307.
3. Kuhn TS. 1970. *Structure of Scientific Revolution*. Chicago: University of Chicago Press.
4. Root-Bernstein RS. 1989. *Discovering*. Cambridge MA: Harvard University Press, 416–420.
5. Lightman A, Gingerich O. 1992. Anomalies. *Science* 256:690–693.
6. Djerassi C. 1992. *The Pill, Pygmy Chimps, and Degas' Horse*. New York: Basic Books, 46–48.
7. Kolata G. 1992. Confronting new ideas, doctors hold onto the old. *NY Times*, 10 May, 6E; Witte MH, Kerwin A, Witte CL, Scadron A. 1989. A curriculum on medical ignorance. *Med Ed* 23:24–29; Witte MH, Kerwin A, Witte CL, Tyler JB, Witte A, Powel W. 1992. *The Curriculum on Medical Ignorance. Coursebook and Resource Manual*. Tucson AZ: University of Arizona Medical School.

Treatment Implications of Multifactorial or Cofactor Theories of AIDS

1.   Smith PG, Morrow RH, Chin J. 1988. Investigating interactions between HIV infection and tropical diseases. *Int J Epidemiol* 17:705–707.

2.   Smith et al. 1988. *Int J Epidemiol* 17:705.

3.   Gershon AA. 1990. Viral vaccines of the future. *Pediatr Clin North Am* 37:689–707; Alter MJ, Hadler SC, Margolis HS, et al. 1990. The changing epidemiology of hepatitis B in the United States. Need for alternative vaccination strategies. *JAMA* 263:1218–1222; Solomon RE, Van Raden M, Kaslow RA, et al. 1990. Association of hepatitis B surface antigen and core antibody with acquisition and manifestations of human immunodeficiency virus type 1 (HIV-1) infection. *Am J Public Health* 80: 1475–1478.

4.   Kallenius G, Hoffner SE, Svenson SB. 1989. Does vaccination with bacille Calmette-Guerin protect against AIDS? *Rev Infect Dis* 11:349–351.

5.   Callen M. 1990. *Surviving AIDS*. New York: Harper Collins, 227–234; Centers for Disease Control. 1989. Guidelines for prophylaxis against *Pneumocystis carinii* pneumonia for persons infected with human immunodeficiency virus. *MMWR* 38: no. S-5; Alter et al. 1990. *JAMA* 263:1218; Heald A, Flepp M, Chave JP, Malinverni R, et al. 1991. Treatment for cerebral toxoplasmosis protects against Pneumocystis carinii pneumonia in patients with AIDS. The Swiss HIV Cohort Study. *Ann Intern Med* 115:760–763; Montaner JS, Lawson LM, Gervais A, et al. 1991. Aerosol pantamidine for secondary prophylaxis of AIDS-related Pneumocystis carinii pneumonia. A randomized, placebo-controlled study. *Ann Intern Med* 114:948–953.

6.   Palestine AG, Polis MA, De Smet MD, et al. 1991. A randomized, controlled trial of foscarnet in the treatment of cytomegalovirus retinitis in patients with AIDS. *Ann Intern Med* 115: 665–673; Anonymous. 1991. Foscarnet and extended survival in AIDS patients. *Am Fam Physician* 44:2244; Minor JR. 1991. Foscarnet versus ganciclovir in the management of cytomegalovirus disease in patients with AIDS. *Am J Hosp Pharm* 48:2478–2479.

7.   Descotes J. 1988. *Immunotoxicology of Drugs and Chemicals*. Amsterdam: Elsevier.

8.   Hamilton JR, Overall JC, Glasgow LA. 1976. Synergistic effect on

mortality in mice with murine cytomegalovirus and *Pseudomonas aeruginosa*, *Staphylococcus aureus*, or *Candida albicans* infections. *Infect Immun* 14:982–989.

9. Reviewed in: Westall FC, Root-Bernstein RS. 1983. An explanation of prevention and suppression of experimental allergic encephalomyelitis. *Mol Immunol* 20:169–177; Root-Bernstein RS, Yurochko F, Westall FC. Clinical suppression of experimental allergic encephalomyelitis by muramyl dipeptide "adjuvant." *Brain Res Bull* 17:473–476.

10. Root-Bernstein RS, Killen J, Lallouette P. Unpublished results.

11. Andrieu J-M, Even P, Venet A, et al. 1988. Effects of cyclosporin on T-cell subsets in human immunodeficiency virus disease. *Clin Immunol Immunopathol* 46:181–198.

12. Callaway CW, Whitney C. 1991. *Surviving with AIDS: A Comprehensive Program of Nutritional Co-Therapy.* Boston: Little Brown.

13. Luca' Moretti M. 1992. Specific behavioral factors among intravenous drug users have been shown to influence HIV seroconversion, and behavioral modifications among HIV seroconverted former intravenous drug users have been shown to delay the onset of AIDS, ARC, and AIDS related diseases. *J InterAm Med Health Assoc* 1:1–8.

14. Fernandez-Cruz E, Fernandez AM, Gutierrez C, et al. 1988. Progressive cellular immune impairment leading to development of AIDS: Two year prospective study of HIV infection in drug addicts. *Clin Exp Immunol* 72:190–195.

15. Weber R, Ledergerber W, Opravil M, Siegenthaler W, Lüthy R. 1990. Progression of HIV infection in misusers of injected drugs who stop injecting or follow a programme of maintenance treatment with methadone. *Br Med J* 301:1361–1365.

16. Buimovici-Klein E, Lange M, Ong KR, Grieco MH, Cooper LZ. 1988. Virus isolation and immune studies in a cohort of homosexual men. *J Med Virol* 25:371–385.

17. Imagawa D, Detels R. 1991. HIV-1 in seronegative homosexual men. *New Engl J Med* 325:1250–1251.

18. United Press International. 1992. Discoverer reworking view of AIDS. United Press International news release, 26 Apr.

19. Callen M. 1990. *Surviving AIDS.*

20. Root-Bernstein RS. 1991. Physic forgotten and true. *Sciences (NY Acad Sci)* Mar–Apr: 10–12.

21. Root-Bernstein RS. 1989. *Discovering.* Cambridge, MA: Harvard University Press.

Rethinking Public Health Policy

1. Root-Bernstein RS. 1990. Non-HIV immunosuppressive factors in AIDS: A multi-factorial, synergistic theory of AIDS etiology. *Res Immunol* 141:835.
2. Lorian V. 1988. AIDS, anal sex, and heterosexuals. *Lancet* i:1111.
3. Buimovici-Klein et al. 1988. *J Med Virol* 25:371; Sonnabend JA. 1989. AIDS: An explanation for its occurrence among homosexual men. In: Ma P, Armstrong D, eds. 1989. *AIDS and Infections of Homosexual Men.* 2d ed. Boston: Butterworths, 463–464.
4. Luca' Moretti M. 1992. Toward an updated AIDS prevention program. *J InterAm Med Health Assoc* 1:10–21.
5. Rubin H. 1988. Etiology of AIDS. *Science* 241:1389–1390.
6. Kane MA, Alter JH, Hadler SC, Margolis HS. 1989. Hepatitis B infection in the United States. Recent trends and future strategies for control. *Am J Med* 87:11S–13S. Gingold B. 1989. Gay bowel syndrome: An overview. In: Ma and Armstrong, eds. 1989. *AIDS*, 49–58.
7. Van Griensven GJ, de Vroome EM, Goudsmit J, Coutinho RA. 1989. Changes in sexual behaviour and the fall in incidence of HIV infection among homosexual men. *Br Med J* 298:218–221; Judson FN. 1990. The relationship of HIV infections to infections with pathogenic *Neisseria* in homosexual men. *Med Clin North Am* 74:1353–1366; Winkelstein W Jr. Wiley JA, Padian NS, Samuel M, Shiboski S, Ascher MS, Levy JA. 1988. The San Francisco Men's Health Study: Continued decline in HIV seroconversion rates among homosexual/bisexual men. *Am J Public Health* 78:1472–1474; Evans BA, McLean KA, Dawson SG, et al. 1989. *Br Med J* 298:215–218.
8. Van Griensven et al. 1989. *Br Med J* 298:218.
9. Evans et al. *Br Med J* 298:215.
10. Winkelstein et al. 1988. *Am J Public Health* 78:1472.
11. Judson et al. 1990. *J Clin North Am* 74:1353.
12. Morse DL, Truman BI, Hanrahan JP, et al. 1990. AIDS behind bars. Epidemiology of New York State prison inmate cases, 1980–1988. *NY State Med J* 90:133–138.
13. Centers for Disease Control. 1990. Surveillance for AIDS and HIV infection among Black and Hispanic children and women of childbearing age, 1981–1989. *MMWR* 39: No. SS-3, 23–30.

New Directions in Biomedical Research

1. Root-Bernstein RS. 1990. Do we know the cause(s) of AIDS? *Persp Biol Med* 33:480–500; Root-Bernstein. 1991. Non-HIV immuno-

suppressive agents in AIDS: A multifactorial synergistic theory of AIDS etiology. *Res Immunol* 141:835.

2.   Akker S, Van Den, G, Goedbloed E. 1960. Pneumonia caused by *Pneumocystis carinii* in a dog. *Trop Geogr Med* 12:54–58; Burke RA, Good RA. 1973. *Pneumocystis carinii* infection. *Medicine* 52:23–51; Medearis DN, Jr. 1964. Mouse cytomegalovirus infection. II. Observations during prolonged infections. *Am J Hyg* 80:103–112; Sheldon WH. 1959. Experimental pulmonary *Pneumocystis carinii* infection in rabbits. *J Exp Med* 110:147; Choisser RM, Ramsey EM. 1940. Etiology of Kaposi's disease. *South Med J* 33:392–396.

3.   Root-Bernstein. 1990. *Persp Biol Med* 33:480; Root-Bernstein. 1991. *Res Immunol* 141:835; Root-Bernstein RS, Hobbs SH. 1993. Sequence similarities between CD4 protein and infectious agents associated with AIDS. Submitted.

What If the HIV-Only Theory Turns Out to Be Right?

1.   Kaufmann WA. 1961. *The Faith of a Heretic* Doubleday.
2.   Root-Bernstein RS. 1989. *Discovering*. Cambridge, MA: Harvard University Press.

GLOSSARY

AIDS-RELATED COMPLEX (ARC). A variety of symptoms, such as chronic swollen glands, repeated fevers and diarrhea, unintentional weight loss, anorexia, and candidiasis, that often precede AIDS but that are not severe enough to meet the definition of AIDS.

ALLOGENEIC (ALLOANTIGENS). Antigens derived from cells or tissues of genetically similar but not identical individuals (e.g., human blood, semen, antibodies, etc.).

ANALINGUS. Licking or sucking the anus of a sexual partner.

ANOREXIA. Lack of appetite and inability to eat.

ANTIBODY. A protein secreted by some of the white blood cells (B cells) in response to antigens. Antibodies protect against infection, and are retained after infection. ELISA and Western blot tests look for specific antibodies.

ANTIGEN. Any foreign material (viruses, bacteria, cells, etc.) which, when introduced into the body, elicits an immune response.

ANTIGENEMIA. Presence of antigens from an infectious agent in the blood.

AUTOIMMUNITY. A broad class of reactions in which the immune system, which is supposed to protect against disease, itself begins to attack the body, thereby causing disease.

B LYMPHOCYTE (B CELL). The type of white blood cell that matures in the bone marrow and that secretes antibodies. These cells are part of the humoral immune system.

479

CANDIDIASIS. A chronic infection with *Candida albicans,* a normally harmless yeast, or some of its unusual relatives.

CELL-MEDIATED IMMUNITY. Immunity in which direct interaction between one type of immune system cells (T-cells) and the foreign agent must occur.

CLONE. A set of cells all of which are genetically identical, and all of which have been derived from a single original precursor. The immune system responds to infection by cloning cells that produce specific antibodies or T cell proteins appropriate to the infection.

COCULTIVATION. A test for virus infection in which cells possibly harboring a virus are grown with easily infected cells that can express large quantities of virus.

COFACTOR. Any agent that can promote the activity of a virus (such as HIV) or other infectious agent during a disease process.

COINFECTION. Infection with more than one microbe simultaneously.

COMPLEMENT. A set of chemicals in the blood that interact with antibodies to destroy antigens.

COMPLEMENTARITY. A relationship between two objects such that one fits into another exactly like hand-and-glove or mold-and-cast.

CROSS-REACTIVE. Antibodies are generally specific for a particular antigen, but sometimes an antibody will react to both a virus and a human cell. In this instance the antibody is said to be cross-reactive.

DEMENTIA. Loss of brain functions that can be manifested as loss of memory, learning ability, and motor control, among other things. Dementia is common among AIDS patients.

DEMYELINIZATION. A process in which the myelin sheath, which acts like an insulator on some nerves, is destroyed, resulting in faulty nerve and brain function.

DISSEMINATED. Said of a disease that has spread out of its normal area or organ of infection to many regions of the body.

ELISA. An acronym for Enzyme-Linked Immunosorbent Assay, a test used to detect antibodies produced by the body against HIV or other microbes.

ENDEMIC. Said of a disease that has been prevalent in a region for a long time.

ENTERIC. Describing infections of the intestines.

EPIDEMIC. A rapidly spreading disease that affects many people.

EPIPHENOMENON. A secondary or additional symptom or complication arising during the course of a sickness, treatment, or experiment. A result, not a cause of an effect.

ETIOLOGY. The study of the origins and causes of disease.

FISTING. Sticking the fingers or hand into the anus of a sexual partner.

GRAFT-VERSUS-HOST DISEASE (GVHD). A disease that results when white blood cells from a donor attack the new host following an organ transplant or some types of transfusions.

HIV-NEGATIVE. Describes a person who has no antibody to Human Immunodeficiency Virus.

HIV-POSITIVE. Describes a person who does have antibody to Human Immunodeficiency Virus, and who is therefore presumed to be infected.

HLA. Human Leukocyte Antigens. Proteins found on the cell surfaces of white blood cells that distinguish one type of immune system cell from another.

HOMOLOGOUS. Having a region of identity. In the case of proteins, homologies refer to regions with identical amino acid sequences. (Amino acids are the subunits that make up proteins).

HUMORAL IMMUNITY. The part of the immune system that produces circulating antibodies, which act at a distance from the cells (B cells) that produce them.

IMMUNOSUPPRESSION. The impairment of the normal or proper functioning of the immune system and its protective activities against disease.

INTRAVENOUS. Delivered or injected into the veins.

KAPOSI'S SARCOMA (KS). The oldest known opportunistic disease associated with AIDS. A cancer of the skin and lymph that is characterized by purplish or reddish nodules, often spread all over the body.

LATENT INFECTION. A quiet, clinically inapparent infection in which viruses insert their genes into the genes of the host cells, and then remain inactive until some further stimulus turns on virus production.

LYMPHADENOPATHY. Persistent, generalized swollen glands lasting more than three months.

LYMPHOCYTES. The white bloods cells in blood that are responsible for providing protection against infection and disease.

LYMPHOCYTOTOXIC. Deadly to white blood cells.

LYMPHOCYTOTOXIC ANTIBODIES (LCTA). Antibodies that kill lymphocytes.

MACROPHAGE. A type of monocyte cell that scavenges foreign particles and cells, digests them, and presents the resulting antigens to T cells to initiate an immune response. Macrophages, like T helper cells, can be infected with HIV and other viruses associated with AIDS.

MHC. Major Histocompatibility Complex. A set of proteins found on some lymphocytes that are responsible for determining what is antigenic (foreign) and what is "self."

MITOGENS. Agents that cause lymphocytes to undergo mitosis (cell division).

MORBIDITY. The proportion of disease due to a specific cause.

MORTALITY. The proportion of death due to a specific cause.

MULTIFACTORIAL. Resulting from the interaction of several agents (factors) simultaneously.

NATURAL KILLER (NK) LYMPHOCYTES. White blood cells that directly attack and destroy foreign cells and tissues.

OPPORTUNISTIC DISEASES. Diseases that occur only in people whose immune systems are immunosuppressed.

PANDEMIC. A worldwide epidemic.

PATHOGEN. Infectious agent that can cause disease.

PATHOGENESIS. Production and development of disease.

PNEUMOCYSTIS CARINII PNEUMONIA (PCP). An opportunistic protozoal infection of the lungs causing pneumonia often associated with AIDS.

POLYMERASE CHAIN REACTION (PCR) TEST. A test for the presence of DNA specific to a particular infectious agent that involves the massive replication of a gene sequence specific to that agent.

PSEUDOVIRONS. Virus particles that contain elements of more than one type of virus. They may result from transactivation and coinfection.

RETROVIRUS. The class of viruses that use RNA instead of DNA for their genetic material. HIV and HTLVs are retroviruses.

SERONEGATIVE. No antibodies to an infectious agent (such as HIV) have been formed or are present in the blood of a person. Seronegativity indicates that the person has not been exposed to this disease agent.

SEROPOSITIVE. Antibodies to a particular organism (such as HIV) have been formed and are present in the blood of a person. Seropositivity indicates prior infection with the disease agent.

SYNDROME. A collection or pattern of disease symptoms that often occur together.

SYNERGISM. The interaction of two agents to produce an effect that is more than the additive effects of the individual agents.

T-HELPER (T-4) LYMPHOCYTE. A specific type of T cell that activates other lymphocytes in the presence of antigen. These are the cells that are infected by HIV (and other viruses) during AIDS. Their destruction cripples the immune system allowing opportunistic infections to take hold.

T-HELPER/T-SUPPRESSOR (T4/T8) RATIO. The proportion of T-helper and T-suppressor cells present in blood. A normal value is about 2 (meaning that an individual usually has twice as many T helper cells as T suppressors), but most AIDS patients have ratios less than 1 (meaning that more than half of their T-helper cells have been destroyed).

T-LYMPHOCYTE (T CELL). A white blood cell that matures in the thymus gland and which is involved in cell-mediated immunity.

T-SUPPRESSOR LYMPHOCYTES. A type of T cell that is responsible for controlling the activity of T helper cells and other immune reactions.

TRANSACTIVATION. A process in which turning on one latent virus infection also turns on others in the same cells.

WESTERN BLOT TECHNIQUE. A test identifying specific antibodies. It is thought to be more specific than ELISA tests.

GROUPS SUPPORTING REEVALUATION OF THE HIV-AIDS HYPOTHESIS

The Group for the Scientific Reappraisal of the HIV/AIDS Hypothesis. C/o Charles A. Thomas, Jr., Ph.D. (619) 272-3884. Publishes *Rethinking AIDS,* a quarterly newsletter, available from 2040 Polk Street, Suite 321, San Francisco, CA 94109. The Group has an international membership comprised mainly of scientists and physicians who question the role of HIV in AIDS.

HEAL (Health Education AIDS Liaison). C/o Michael Ellner, 16 East 16th Street, New York, NY 10003. Phone (212) 674-HOPE. Acts as "HIV debriefers," supplying information packets and cassette tapes on living with AIDS, alternatives to AZT treatments, etc.

The HIV Connection? C/o Ed Vargas, 1072 Folsom Street, Suite 321, San Francisco, CA 94102. (415) 552-9160. An organization of businessmen, community leaders, physicians, and people with AIDS applying pressure to government to investigate alternatives to the HIV-only theory of AIDS.

Project AIDS International. C/o Mark Alampi or Jeremy Selvey, 8033 Sunset Blvd #2640, Los Angeles, CA 90046. (213)467-3352. An international collaborative group that collects, validates, and disseminates information about AIDS, emerging treatments, and alternative research.

Cure Now. C/o Jerry Tarranova, P.O. Box 29386, Los Angeles, CA 90029. (213) 660-7563. Publishes a quarterly newsletter focussing on alternative treatments of AIDS.

The Foundation for Alternative AIDS Research (Stichting Alternatief Aidsonderzoek, S.A.A.O.). Jan van der Tooren, President. P. O. Box 1447, NL 1200 BK Hilversum, The Netherlands. Telephone 31-35 24 30 84. The most important of the European collaboratives currently questioning the HIV dogma.

INDEX

Acer, David, 46–47, 314–15
Ackerman, James, 10–11
Acrodermatitis enteropathica, 139
Acyclovir, 133, 356
Addictive drugs. *See* Drug abuse; Intravenous drug abusers
Adenovirus pneumonia, 107
Adenoviruses, 27, 107, 157, 160, 172, 212
Adjuvant peptide, 195
Adjuvants, 194–95, 347
Adulterants in drugs, 124–25
Africa: AIDS in, 222, 300–301, 305–306; antiparasitic imadazole drugs used in, 133–34; comparison of immunosuppressive risks in, with Western countries, 310–12; deaths from AIDS in, 300; drug abuse in, 308; HIV infection in, 70–71, 300, 311; HIV-negative AIDS patients in, 29; immunosuppressive risks of African heterosexuals, 301–12; infectious diseases in, 303–306; latency for AIDS in, 222; malaria in, 156, 157; malnutrition in, 307–308; medical procedures and health care in, 306–307, 309; origination of HIV in, 70–71, 311; rate of AIDS transmission from mother to infant in, 222; sexually transmitted diseases in, 301–303; sickle cell anemia in,

145; social and political revolutions in, 308–309
African Americans. *See* Blacks
Age: immunosuppression associated with, 146–47; as risk for AIDS, 320–21
AIDS: in Africa, 300–301, 305–306; alternative hypotheses for, 327–49; antigenic overload theory of, 148–50; autoimmune theory of, 184–219, 344–47; calculations of spread of, and changing definitions, 67; clinical picture of, 7–10, 13; cofactor theory of, 25–28, 87–88, 92–94, 97, 114, 149–50, 173, 197, 327, 330–31, 332–38, 354–63; definition of, 13, 17, 58–68, 77, 84, 101, 111–13, 114; in developing countries, 354–55; disease progression of, 7–8, 184–85, 279; diseases associated with, 150–60; drowned-man analogy of, 65–66, 112, 260–61; as epidemic, 67, 281–98; etiological criteria for, 100–104; first reports of, in medical community, 7–8; HIV as cause of, 22, 24, 29, 56, 61, 68–77, 84–85, 101, 111, 328–32, 348, 371–73; HIV transmission and, 30–38; HIV-negative AIDS patients, 28–30, 261; HIV-related markers for